AF614757

METHODS IN MOLECULAR BIOLOGY™

For further volumes:
http://www.springer.com/series/7651

Liver Proteomics

Methods and Protocols

Edited by

Djuro Josic

Warren Alpert Medical School, Brown University, Providence, RI, USA

Douglas C. Hixson

Division of Hematology and Oncology, Rhode Island Hospital, Providence, RI, USA

Editors
Djuro Josic
Warren Alpert Medical School
Brown University
Providence, RI, USA

Douglas C. Hixson
Division of Hematology and Oncology
Rhode Island Hospital
Providence, RI, USA

ISSN 1064-3745 ISSN 1940-6029 (electronic)
ISBN 978-1-61779-958-7 ISBN 978-1-61779-959-4 (eBook)
DOI 10.1007/978-1-61779-959-4
Springer New York Heidelberg Dordrecht London

Library of Congress Control Number: 2012942932

Printed on acid-free paper

Humana Press is a brand of Springer
Springer is part of Springer Science+Business Media (www.springer.com)

Preface

The liver is responsible for a wide range of critical functions essential to life. This large blood-filled organ is composed of several different cell types, including hepatocytes, the major epithelial cell type, endothelial cells lining the blood-filled liver sinusoids, cholangiocytes forming the bile ducts, liver-specific macrophages known as Kupffer cells, and Stellate cells, a source of several growth factors and a mediator of liver fibrosis. Because of the important roles performed by the liver in health and disease, characterization of the liver proteome is an essential step in fully elucidating its function. The methods described in the chapters of this book represent the latest strategies being used to characterize the liver proteome at the global, cellular, subcellular, posttranslational, and functional level.

In the first three chapters of this book (Bausch-Fluck et al., Zhang et al., and Lee et al.), some of the current approaches being used for global proteomic analysis of the liver are reviewed. Many of these have been applied to the Human Liver Proteome Project (HLPP), the first initiative organized around a single organ. The scientific objectives of HLPP are to characterize the proteome of the healthy human liver expression profiling (cellular localization, posttranslational modifications, and protein–protein interactions) and to develop tools for the study of the liver proteome (e.g., sample banking and/or antibodies). In early 2010, the human liver proteome (HLP) data set contained 6,788 identified proteins, including 3,721 proteins newly identified in the liver, and it is constantly growing. The list of these proteins is now publicly available. The human liver transcriptome (HLT) data set consists of 11,205 expressed genes, and these data indicate that, in order to characterize the complete HLP, still a significant amount of work is necessary. Currently, the HLP is the largest proteome data set for a human organ and the first with a corresponding transcriptome derived from the same sample.

As a highly metabolic cell, the hepatocyte is packed with an array of organelles including mitochondria (over 1,000/cell), endoplasmic reticulum (15% of the cell volume), lysosomes, peroxisomes, some 50 Golgi complexes, and an extensive cytoskeletal framework. This high density of organelles reflects the many different functions carried out by hepatocytes, functions ranging from the synthesis and secretion of plasma proteins, iron storage, detoxification of xenobiotics, scavenging of plasma glycoproteins and other biologically important molecules, secretion of bile, removal of ammonia produced by protein breakdown, and metabolism of carbohydrates. Understanding how disease, hepatotoxins, or injury affects these organelles will be facilitated by knowledge of their protein expression profiles under normal conditions in the liver. Methods for acquiring such information are described in two chapters by Timperio et al. and Petrushkova and Lisitsa. Approaches for looking at specific subsets within peroxisomes for total liver, specific cell types, or specific organelles are described in four chapters by Islinger et al., He et al., Cao and Liang, and Chen et al. Methods for profiling proteins with posttranslational modifications (phosphoproteome, glycoproteome, and acetylome) are described in the next four chapters by Zhang et al., Demirkan et al., Santamaria et al., and Chang and Yu-Lee.

The next several chapters are devoted to strategies being used to identify biomarkers for liver disease and injury. At present, there are over 100 forms of liver disease caused by a variety of different agents. Although some types of liver pathology are still not well understood, the majority stems from well-known factors that include acute or chronic hepatitis induced by hepatitis viruses B, C, and D (chapters by Cunha and Coelho and Wu and Song), fatty liver disease (Rodrigez-Suarez et al.), cirrhosis and chronic inflammation resulting from alcohol abuse (Newton et al.), and acute or chronic damage caused by exposure to xenobiotics, which include environmental toxins, many of which are potential carcinogens, and drugs such as acetaminophen. Since xenobiotics are primarily metabolized and detoxified by hepatocytes, the liver becomes vulnerable to acute or chronic damage leading to impaired liver function or complete liver failure. Indeed, many liver carcinogens are activated by the metabolic efforts of hepatocytes to convert them to a less toxic, more soluble, and thus more easily excreted form. The growth of data sets generated from global and targeted proteomic and metabolomic profiling (chapter by D'Alessandro et al.) has increased the feasibility of identifying biomarkers characteristic of human liver injury and disease, an undertaking justified by the need for biomarkers that can become the basis for more specific and sensitive methods for diagnosis, prognosis, and treatment of acute and chronic liver disease. Promising results from comparative proteomic analysis of rat and mouse models of liver disease give credence to ongoing efforts aimed at identifying biomarkers for human liver disease.

In addition to its role in xenobiotic metabolism, the liver synthesizes most of the proteins found in plasma. Albumin and fibrin account for 64% of the plasma proteins, with another 35% contributed by other globular proteins used for the transport of ions, lipids, and hormones. The remaining 1% is made up of regulatory proteins, enzymes, proenzymes, enzyme inhibitors, hormones, and growth factors. Since the concentration and structure of many plasma proteins synthesized in the liver change in response to disease, plasma offers a readily accessible source of biomarkers for liver disease. The major limitation at present is the difficulty in detecting disease-associated changes in liver-derived proteins present at levels that are several orders of magnitude lower than major plasma proteins such as albumin, transferrin, and inter-alpha trypsin inhibitor. In the three concluding chapters, current strategies are described for identifying potential liver biomarkers in plasma (Wong and Luk and D'Amici et al.) and other biological fluids such as urine (Conde-Vancells and Falcon-Perez).

Providence, RI, USA *Djuro Josic*

Providence, RI, USA *Douglas C. Hixson*

Acknowledgement

We thank Professor Jasminka Giacometti, Department of Biotechnology, University of Rijeka, Croatia, for the assistance in editing this book.

Contents

Contributors

AFSANEH ABDOLZADE-BAVIL • *Beckton & Dickinson GmbH, Heidelberg, Germany*

PHILIP C. ANDREWS • *Department of Biological Chemistry, The University of Michigan, Ann Arbor, MI, USA*

DAMARIS BAUSCH-FLUCK • *Institute of Molecular Systems Biology, Swiss Federal Institute of Technology (ETH) Zurich, Zurich, Switzerland; National Center of Competence in Research (NCCR) Neural Plasticity and Repair Center for Proteomics, UZH/ETH Zurich, Zurich, Switzerland*

RUI CAO • *College of Life Sciences, Hunan Normal University, Changsha, China*

XUEQUN CHEN • *Department of Physiology, Wayne State University, Detroit, MI, USA*

CAROL CHUANG • *Department of Molecular and Cellular Biology, Baylor College of Medicine, Houston, TX, USA*

ANA V. COELHO • *Unidade de Biologia Molecular, Centro de Malária e outras Doenças Tropicais, Instituto de Higiene e Medicina Tropical, Universidade Nova de Lisboa, Lisbon, Portugal; Laboratório de Espectrometria de Massa, Instituto de Tecnologia Química e Biológica, Universidade Nova de Lisboa, Lisbon, Portugal; Departamento de Biologia, Universidade de Évora, Évora, Portugal*

JAVIER CONDE-VANCELLS • *Metabolomics Unit, CIC bioGUNE, CIBERehd, Bizkaia Technological Park, Derio, Spain*

FERNANDO J. CORRALES • *Division of Hepatology and Gene Therapy, Proteomics Unit, Centre for Applied Medical Research (CIMA), University of Navarra, Pamplona, Spain*

CELSO CUNHA • *Unidade de Biologia Molecular, Centro de Malária e outras Doenças Tropicais, Instituto de Higiene e Medicina Tropical, Universidade Nova de Lisboa, Lisbon, Portugal; Laboratório de Espectrometria de Massa, Instituto de Tecnologia Química e Biológica, Universidade Nova de Lisboa, Lisbon, Portugal; Departamento de Biologia, Universidade de Évora, Évora, Portugal*

ANGELO D'ALESSANDRO • *Tuscia University, Largo dell'Università snc, Viterbo, Italy*

GIAN MARIA D'AMICI • *Department of Environmental Sciences, University of Tuscia, Largo dell'Università snc, Viterbo, Italy*

GOKHAN DEMIRKAN • *Department of Pediatrics, Brown University and Rhode Island Hospital, Providence, RI, USA; Department of Molecular Biology, Cell Biology, and Biochemistry, Brown University, Providence, RI, USA*

FELIX ELORTZA • *Proteomics Platform, CIC bioGUNE, CIBERehd, ProteoRed, Bizkaia Technology Park, Derio, Spain*

JUAN M. FALCON-PEREZ • *IKERBASQUE, Basque Foundation for Science, Bilbao, Spain*

JOAQUÍN FERNÁNDEZ-IRIGOYEN • *Proteomics Unit, Biomedical Research Center, Navarra Health Service, Pamplona, Spain*

MARTINA SRAJER GAJDOSIK • *Department of Chemistry, J.J. Strossmayer University, Osijek, CroatiaProteomics Core, COBRE Center for Cancer Research Development, Rhode Island Hospital and Brown University, Providence, RI, USA*

FEDERICA GEVI • *Tuscia University, Largo dell'Università snc, Viterbo, Italy*

PHILIP A. GRUPPUSO • *Department of Pediatrics, Brown University and Rhode Island Hospital, Providence, RI, USA; Department of Molecular Biology, Cell Biology, and Biochemistry, Brown University, Providence, RI, USA*

JINTANG HE • *State Key Laboratory of Protein and Plant Gene Research, College of Life Sciences, Peking University, Beijing, People's Republic of China*

ANDREAS HOFMANN • *Institute of Molecular Systems Biology, Swiss Federal Institute of Technology (ETH) Zurich, Zurich, Switzerland; National Center of Competence in Research (NCCR) Neural Plasticity and Repair Center for Proteomics, UZH/ETH Zurich, Zurich, SwitzerlandNovartis Pharma AG, Basel, Switzerland*

MARKUS ISLINGER • *Department of Anatomy and Cell Biology, Medical Cell Biology, University of Heidelberg, Heidelberg, Germany; Centre for Cell Biology and Department of Biology, University of Aveiro, Aveiro, Portugal*

ARUL JAYARAMAN • *Artie McFerrin Department of Chemical Engineering, Texas A&M University, College Station, TX, USA; Department of Biomedical Engineering, Texas A&M University, College Station, TX, USA*

JIANGUO JI • *State Key Laboratory of Protein and Plant Gene Research, College of Life Sciences, Peking University, Beijing, People's Republic of China*

DJURO JOSIC • *Proteomics Core, COBRE Center for Cancer Research Development, Rhode Island Hospital and Brown University, Providence, RI, USA; Department of Biotechnology, University of Rijeka, Rijeka, Croatia*

YU-CHEN LEE • *Department of Molecular Pathology, University of Texas, M.D. Anderson Cancer Center, Houston, TX, USA*

SONGPING LIANG • *College of Life Sciences, Hunan Normal University, Changsha, China*

SVEN LIEBLER • *Department of Anatomy and Cell Biology, University of Heidelberg, Heidelberg, Germany*

SUE-HWA LIN • *Department of Molecular Pathology, University of Texas, M.D. Anderson Cancer Center, Houston, TX, USA*

ANDREY V. LISITSA • *Institute of Biomedical Chemistry, Russian Academy of Medical Sciences, Moscow, Russia*

SIQI LIU • *Center of Proteomic Analysis, Beijing Institute of Genomics, Chinese Academy of Sciences, Beijing, China*

YASHU LIU • *State Key Laboratory of Protein and Plant Gene Research, College of Life Sciences, Peking University, Beijing, People's Republic of China*

XIAOMIN LOU • *Center of Proteomic Analysis, Beijing Institute of Genomics, Chinese Academy of Sciences, Beijing, China*

JOHN M. LUK • *Cancer Science Institute of Singapore, National University of Singapore, Singapore, Singapore; Department of Pharmacology, National University of Singapore, Singapore, Singapore; Department of Surgery, National University of Singapore, Singapore, Singapore; Department of Surgery, The University of Hong Kong, Hong Kong, China*

JOSE M. MATO • *Metabolomics Unit, CIBERehd, Bizkaia Technology Park Building, Derio, Spain*

GARY L. NELSESTUEN • *Department of Biochemistry, Molecular Biology and Biophysics, University of Minnesota, Minneapolis, MN, USA*

BILLY W. NEWTON • *Artie McFerrin Department of Chemical Engineering, Texas A&M University, College Station, TX, USA; Department of Biomedical Engineering, Texas A&M University, College Station, TX, USA*

NATALIA A. PETUSHKOVA • *Institute of Biomedical Chemistry, Russian Academy of Medical Sciences, Moscow, Russia*

SHASHI K. RAMAIAH • *Pfizer Global Research and Development, Drug Safety Research & Development, Chesterfield, MO, USA*

SARA RINALDUCCI • *Department of Environmental Sciences, University of Tuscia, Largo dell'Università snc, Viterbo, Italy*

EVA RODRIGUEZ-SUAREZ • *Proteomics Platform, CIC bioGUNE, CIBERehd, ProteoRed, Bizkaia Technology Park, Derio, Spain*
PETER ROEPSTORFF • *Department of Biochemistry and Molecular Biology, University of Southern Denmark, Odense, Denmark*
DAVID H. RUSSELL • *Department of Chemistry, Texas A&M University, College Station, TX, USA*
WILLIAM K. RUSSELL • *Department of Chemistry, Texas A&M University, College Station, TX, USA*
ARTHUR R. SALOMON • *Department of Molecular Biology, Cell Biology, and Biochemistry, Brown University, Providence, RI, USA*
VIRGINIA SÁNCHEZ-QUILES • *Division of Hepatology and Gene Therapy, Proteomics Unit, Centre for Applied Medical Research (CIMA), University of Navarra, Pamplona, Spain*
ENRIQUE SANTAMARÍA • *Proteomics Unit, Biomedical Research Center, Navarra Health Service, Pamplona, Spain*
ALAN R. SINAIKO • *Department of Pediatrics, University of Minnesota, Minneapolis, MN, USA*
GUANG SONG • *Beijing Institute of Genomics, Chinese Academy of Sciences, Beijing, People's Republic of China*
HAIDAN SUN • *Center of Proteomic Analysis, Beijing Institute of Genomics, Chinese Academy of Sciences, Beijing, China*
ANNA MARIA TIMPERIO • *Department of Environmental Sciences, University of Tuscia, Largo dell'Università snc, Viterbo, Italy*
ALFRED VÖLKL • *Department of Anatomy and Cell Biology, University of Heidelberg, Heidelberg, Germany*
QINGSONG WANG • *State Key Laboratory of Protein and Plant Gene Research, College of Life Sciences, Peking University, Beijing, People's Republic of China*
QUANHUI WANG • *Center of Proteomic Analysis, Beijing Institute of Genomics, Chinese Academy of Sciences, Beijing, China*
YANZHUANG WANG • *Department of Molecular, Cellular and Developmental Biology, The University of Michigan, Ann Arbor, MI, USA*
GERHARDT WEBER • *FFE Service GmbH, Margeritenweg, Kirchheim, Germany*
BERND WOLLSCHEID • *Institute of Molecular Systems Biology, Swiss Federal Institute of Technology (ETH) Zurich, Zurich, Switzerland; National Center of Competence in Research (NCCR) Neural Plasticity and Repair Center for Proteomics, UZH/ETH Zurich, Zurich, Switzerland*
KWONG-FAI WONG • *Cancer Science Institute of Singapore, National University of Singapore, Singapore*
LIN WU • *Beijing Institute of Genomics, Chinese Academy of Sciences, Beijing, People's Republic of China*
LI-YUAN YU-LEE • *Department of Medicine, Baylor College of Medicine, Houston, TX, USA*
XUMIN ZHANG • *Department of Biochemistry and Molecular Biology, University of Southern Denmark, Odense, Denmark*
YAN ZHANG • *Department of Biochemistry, Molecular Biology and Biophysics, University of Minnesota, Minneapolis, MN, USA; Department of Laboratory Medicine, Harvard Medical School, Children's Hospital Boston, Boston, MA, USA*
LELLO ZOLLA • *Department of Environmental Sciences, University of Tuscia, Largo dell'Università snc, Viterbo, Italy*

Chapter 1

Cell Surface Capturing Technologies for the Surfaceome Discovery of Hepatocytes

Damaris Bausch-Fluck, Andreas Hofmann, and Bernd Wollscheid

Abstract

Proteins expressed at the cell surface define how cells can functionally interact with their microenvironment in time and space. The cell surface subproteome, or surfaceome, represents a cellular information gateway not only enabling the processing of environmental molecular cues but also limiting cellular interaction capacities. Therefore, the array of antibody-detectable cell surface proteins is widely used to phenotype and categorize cells. Quantitative differences in surfaceome markers can not only indicate different developmental cellular stages but also serve as markers of disease. In fact, cell surface proteins are promising biomarker candidates, since they are often, apart from their plasma membrane expression, secreted, shed, or released otherwise from the tissue into the bloodstream. From minute amounts of blood these informative proteins can be detected and quantified by ELISA or highly sensitive state-of-the art targeted mass spectrometric techniques. However, the identification of the complete surfaceome and its constituents is hampered by a lack of suitable technologies to detect these proteins at the cell surface location. Antibodies for the detection of cell surface proteins are only available for a subset of the potentially expressed surfaceome members. The mass spectrometry-based cell surface capturing (CSC) technology and recently developed variants overcome these limitations by selectively enriching and identifying cell surface proteins that are either *N*-glycosylated (Glyco-CSC, Cys-Glyco-CSC), or have an extracellularly exposed and conformationally available lysine (Lys-CSC). Here, we outline the CSC technology and its variants in a detailed step-by-step protocol for soluble and adherent cells. Representative results from the application of the CSC technologies to the hepatocyte cell line Hepa1-6 illustrate the complementary nature of the CSC technologies, which enables a systems biology view of the surfaceome.

Key words: Phenotyping cells, Cell surface capturing, Lys-CSC, Cys-Glyco-CSC, Liver proteomics, Surfaceome, Cell surface glycoproteins, Mass spectrometry

1. Introduction

Cell surface proteins are of clinical interest as therapeutic targets due to their accessibility at the cell surface for affinity binders, such as antibodies or small molecules (1). Around 70% of target

Djuro Josic and Douglas C. Hixson (eds.), *Liver Proteomics: Methods and Protocols*, Methods in Molecular Biology, vol. 909, DOI 10.1007/978-1-61779-959-4_1,

proteins of currently approved drugs by the Food and Drug Administration (FDA) are membrane proteins, or proteins located in the extracellular space (2). For decades now, immunophenotyping of cells mainly relied on cell surface proteins as classification antigens. The expression pattern of antibody-detectable proteins has allowed for the classification of immune cells in particular, and the construction of the hematopoietic (3) differentiation tree based mainly on CD antigens (4). Despite these advances, the biochemical detection and parallel analysis of cell surface proteins have been proven difficult due to their hydrophobicity and relatively low abundance. Antibody-based protein detection workflows, such as flow cytometry and immunofluorescence, have been the methods of choice for their individual detection, but a comprehensive view of the surfaceome of cells is still lacking.

Mass spectrometry-based technologies offer the advantage of multiplexed and unbiased detection of proteins independent of existing antibody collections. The original cell surface capturing (CSC) technology takes advantage of the fact that a majority of cell surface proteins are glycosylated (5, 6). Through the mild oxidation of cell surface-exposed glycan structures with meta-sodium periodate, *cis*-diols in the glycan structure are reacted into aldehydes, a new chemical moiety now available at the cell surface for further coupling reactions. These newly generated aldehydes can be specifically reacted with hydrazide-containing bifunctional linker molecules, such as Biocytin Hydrazide, in order to covalently attach a biotin label to these cell surface-exposed glycoproteins, as a tag for subsequent affinity enrichment. Cells are then harvested and lysed by Dounce homogenization in a hypotonic lysis buffer and microsomal membranes are collected by ultracentrifugation, after having separated the nucleus from the cellular lysate. Upon solubilization and digestion of the membrane proteins, biotinylated glycopeptides are enriched with streptavidin bead affinity capturing. Unspecific, non-biotinylated peptides and molecules are removed by extensive washing protocols and formerly *N*-glycosylated peptides are specifically released with peptide *N*-glycosidase F (PNGase F), which introduces a mass shift of 0.9846 Da at the initially glycosylated asparagine. This significant mass shift can be detected in the mass spectrometer for the specific identification of the N-glycosylation site (7).

Based on this initial CSC protocol, we developed recently two variants of the CSC technology in order to further increase surfaceome coverage. The Cys-Glyco-CSC technology takes advantage of the fact that glycosylated peptides often contain cysteine residues within disulfide bridges (8). These "piggyback"peptides of cysteine containing glycopeptides can therefore be captured as well, providing additional confidence in the specific protein identification, as well as structural information. The Cys-Glyco-CSC differs from the original CSC in the time point of reduction and alkylation of

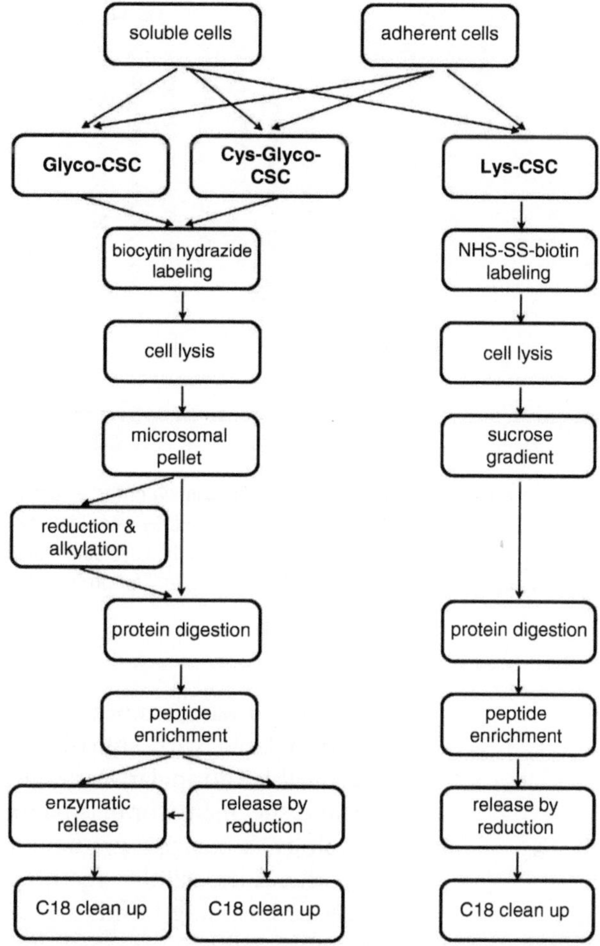

Fig. 1. Overview of Glyco-CSC, Cys-Glyco-CSC, and Lys-CSC workflow. The Glyco-CSC involves biocytin hydrazide labeling, cell lysis, collection of the microsomal pellet, reduction, alkylation and digestion of the proteins, solid phase enrichment of glycopeptides, and enzymatic release. The Cys-Glyco-CSC skips the protein reduction and alkylation step and instead releases cysteine peptides by chemical reduction from the solid phase. The Lys-CSC applies a sulfo-NHS-SS-biotin labeling, cell lysis, collection of plasma membranes in a sucrose gradient, protein digestion, solid phase enrichment of lysine-containing peptides, and release by chemical reduction. All peptides are C18 cleaned up prior to mass spectrometric analysis. All three protocols are applicable to soluble or adherent cells.

the protein mixture. In the former case, disulfide bonds are preserved until the solid phase enrichment step, in which peptides bound via disulfide bonds to glycosylated peptides—the piggyback peptides—are indirectly enriched together with the glycopeptides. Cysteine-containing non-glycosylated peptides are released by chemical reduction prior to PNGase F treatment (Fig. 1).

The Lys-CSC technology is a second CSC technology variant enabling the direct mass spectrometric identification of lysine containing cell surface-exposed proteins (8). In the Lys-CSC technology

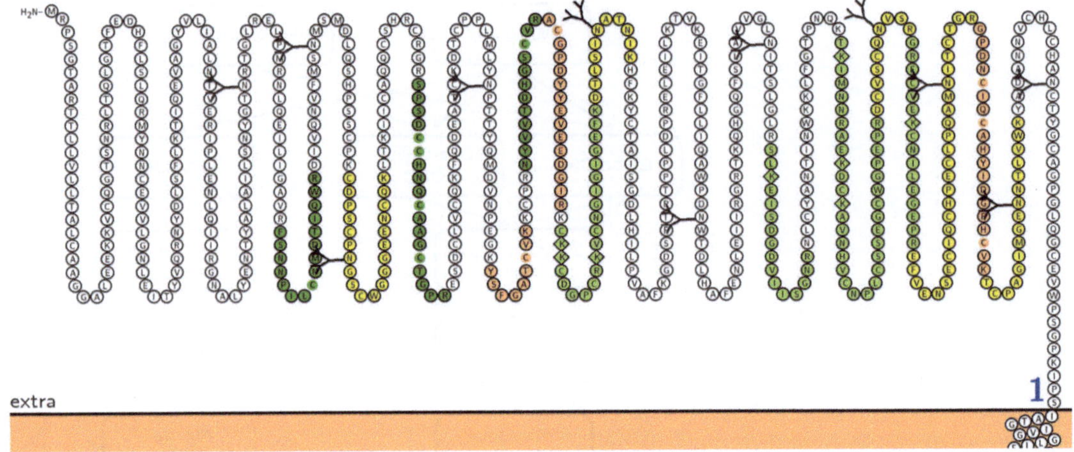

Fig. 2. Schematic representation of the extracellular domain of the epidermal growth factor receptor. Seventeen distinct peptides are colored according to the method with which they were identified (*yellow* Glyco-CSC, *orange* Cys-Glyco-CSC, *light green* Lys-CSC, *dark green* Lys-CSC piggyback). Modified cysteine residues are marked with a *white border*, modified asparagines and lysines with a *blue circle*. Software "Protter" for protein visualization courtesy of U. Omasits (manuscript in preparation).

workflow, primary amines (mostly ε-amines of lysine side chains) are biotinylated with sulfo-NHS-SS-biotin, a mostly membrane-impermeable biotinylation reagent. In order to increase the sensitivity and specificity for plasma membrane proteins in the Lys-CSC technology, a sucrose gradient to enrich for plasma membranes over other intracellular membranes is integrated into the CSC workflow. Disulfide bonds are also preserved within this protocol to keep the sulfo-NHS-SS-biotin tag intact and to enrich for piggyback peptides, bound via a disulfide bond to a biotinylated lysine peptide. Biotinylated peptides are coupled to streptavidin beads and unspecific peptides are removed by thorough washing. The release of specific peptide is performed by chemical reduction, which leaves a 3-(carbamidomethylthio)propanoyl tag on formerly biotinylated lysines (Fig. 1). In the same step disulfide-bridged cysteine-containing peptides are released as well and subsequently alkylated.

The advantage of the original Glyco-CSC technology is its high degree of specificity for *N*-glycosylated peptides and the focused, nearly exclusive identification of bona fide cell surface proteins (7, 3). However, the exclusive enrichment of *N*-glycosites precludes the identification of cell surface proteins that are not, or only, *O*-glycosylated, and for glycosites that are not within a mass spectrometrically identifiable peptide. Therefore, the CSC technology variants Cys-Glyco-CSC and Lys-CSC are offering complementary strategies for increasing the sequence coverage of identified proteins and for the identification of proteins that would escape the Glyco-CSC strategy (Fig. 2).

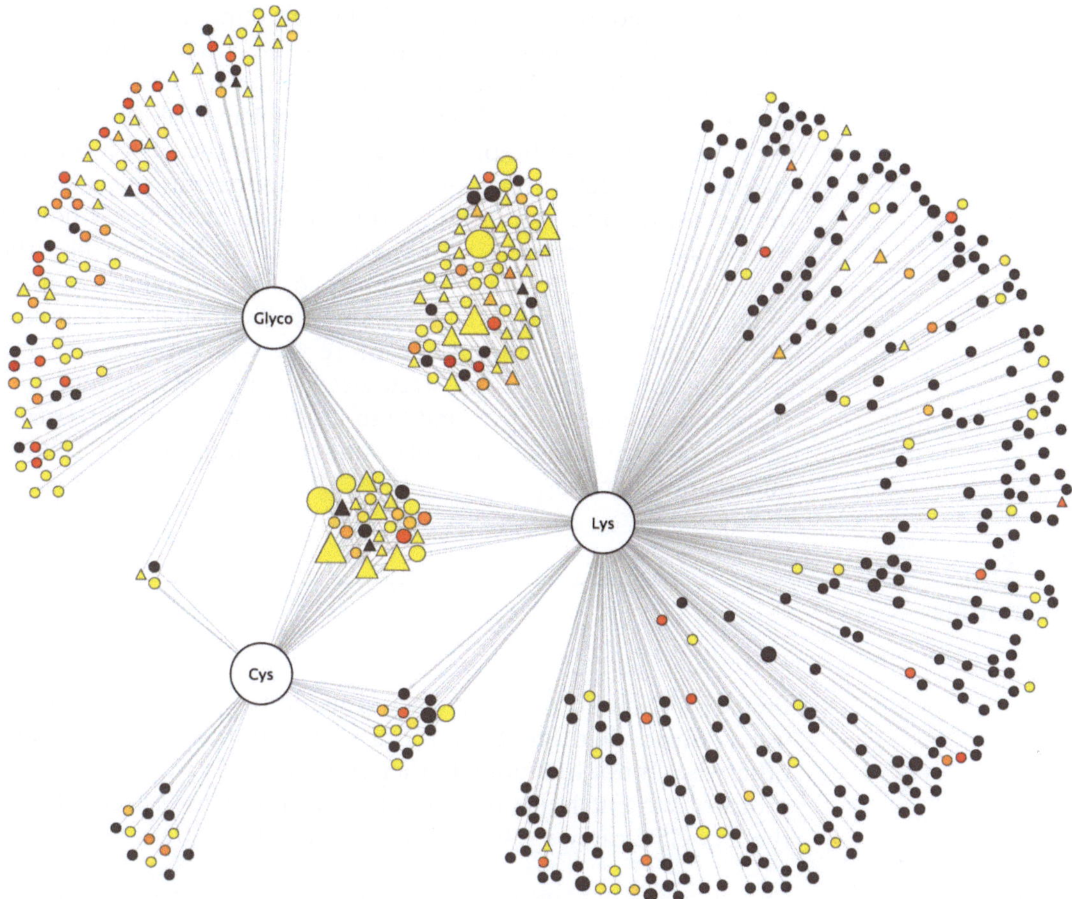

Fig. 3. Cytoscape representation of detected surfaceome of Hepa1-6 cells. Identified proteins are depicted in *circles*, those with a CD nomenclature are shown as *triangles*. Commonly or exclusively identified proteins between the three different CSC variants are represented by connection to the respective method. The *color* indicates the number of predicted transmembrane domains by Phobius, whereas *grey* means no TM domain and the number of TM domains increases from *yellow* to *red*, ranging from 1 to 14 TM domains. The size of the protein *circles* and *triangles* is shown dependent on the number of peptides identified, ranging from 1 to 77. For proteins identified with different methods, the individual peptides are summed up.

Here we describe in detail the application of CSC technology and its variants to the hepatoma cell line Hepa1-6. The Cys-Glyco-CSC analysis revealed in the glyco fraction 211 proteins with at least one peptide containing the consensus sequence N[115]XS/T and in the cys fraction 61 proteins with at least one cysteine-containing peptide. Twenty-seven proteins were exclusively identified in the cysteine fraction and would have escaped identification by only applying the standard Glyco-CSC. In the Lys-CSC we identified 402 proteins with 1,383 peptides, of which 745 have a modified lysine and 98.5% contain either a modified lysine or a cysteine (Fig. 3, Supplementary Tables 1–5).

The added benefit of specifically targeting the surfaceome using the CSC technology is further illustrated in a comparison to a recently published dataset from experiments geared towards the full proteome coverage of Hepa1-6 cells by Pan et al. (9), consisting of more than 4,000 proteins. By targeting the surfaceome we identified 199 cell surface proteins not reported in this state-of-the-art study to be expressed on this particular cell line, amongst which are such essential proteins for liver metabolism like the hepatocyte growth factor receptor, or the insulin receptor (Supplementary Tables 1 and 2). Due to the inability of current MS technologies to efficiently capture the whole proteome of a cell, subproteome targeting strategies like the CSC technologies enable rapid, cost-effective, and high-resolution insights into surfaceome biology and a molecular understanding of this cellular information gateway.

2. Materials

Prepare all solutions with analytical grade reagents and ultrapure water. Store all reagents at room temperature, unless indicated otherwise. If not mentioned otherwise, materials are used in either protocol variant. Items only required for the (Cys-)Glyco-CSC or the Lys-CSC carry a respective label in brackets.

2.1. Glycoprotein Biotinylation (Glyco-CSC/Cys-Glyco-CSC)

1. Glyco labeling buffer: PBS pH 6.5, 0.1% (v/v) FCS. Add a tablet for 500 ml PBS (Invitrogen, Carlsbad, CA, USA) to 450 ml water, adjust the pH to 6.5 with 85% (w/v) H_3PO_4 and add 0.5 ml fetal bovine serum. Mix and make up to 500 ml with water. Store at 4°C.
2. $NaIO_4$ stock solution: 1.6 mM $NaIO_4$ in labeling buffer. Resuspend 21 mg $NaIO_4$ in 0.8 ml labeling buffer and make up to 1 ml with labeling buffer. Keep $NaIO_4$ in the dark, since it is light sensitive (see Note 1).
3. MACSmix tube rotator (Miltenyi Biotec, Bergisch Gladbach, Germany).
4. Biocytin hydrazide solution: 6.5 mM biocytin hydrazide (Biotium, Hayward, CA, USA) in labeling buffer. Add 30 mg biocytin hydrazide to 10 ml labeling buffer, mix, and make up to 12 ml with labeling buffer (see Note 1).

2.2. Lysine Biotinylation (Lys-CSC)

1. Iodoacetamide stock solution: See Item 7 under Subheading 2.4.
2. Lysine labeling buffer: 2 mM sulfo-NHS-SS-biotin ((sulfosuccinimidyl-2-(biotinamido)-ethyl-1, 3′-dithiopropionate),

Thermo Fisher Scientific, Perbio Science, Rockford, IL, USA) and 4 mM iodoacetamide in PBS. Add 5.55 mg sulfo-NHS-SS-biotin and 40 μl of iodoacetamide stock solution to 4 ml PBS, mix, and make up to 5 ml with PBS (see Note 1).

2.3. Membrane Preparation

1. Hypotonic lysis buffer (Glyco-CSC): 10 mM Tris pH 7.5, 0.5 mM $MgCl_2$. Add 1.21 g Tris base and 47.61 mg $MgCl_2$ to 900 ml water and adjust pH to 7.5 with HCl. Mix and make up to 1 l with water. Store at 4°C.
2. Optional: Hypotonic lysis buffer for Cys-Glyco-CSC: 10 mM Tris pH 7.5, 0.5 mM $MgCl_2$, 10 mM iodoacetamide. Prepare hypotonic lysis buffer as described in item 1, add 20 μl of iodoacetamide stock solution (item 7, subheading 2.4), mix, and make up to 1 l with water. Store in the dark at 4°C.
3. Hypotonic lysis buffer (Lys-CSC): 10 mM Tris pH 7.5, 0.5 mM $MgCl_2$, 10 mM iodoacetamide. Add 1.21 g Tris base and 47.61 mg $MgCl_2$ to 900 ml water and adjust pH to 7.5 with HCl. Mix and make up to 1 l with water. Store at 4°C.
4. 500 mM MES stock solution pH 6: Add 48.81 g MES hydrate to 400 ml water. Adjust pH with NaOH and make up to 500 ml with water.
5. Membrane preparation buffer: 50 mM MES pH 6, 10 mM $MgCl_2$, 450 mM NaCl, 280 mM sucrose. Add 100 ml of the 500 mM MES stock solution pH 6, 0.95 g $MgCl_2$, 26.30 g NaCl, and 95.84 g sucrose to 700 ml water. Mix and make up to 1 l with water. Aliquot in 50 ml tubes and store at −20°C (see Note 2).
6. Membrane wash buffer: 25 mM Na_2CO_3. Add 2.65 g Na_2CO_3 to 900 ml water and make up to 1 l with water.
7. Sucrose buffer (Lys-CSC): 35% sucrose in membrane preparation buffer and water. Add 6.03 g to 8 ml water, mix, and fill up to 10 ml with water. Add 10 ml of membrane preparation buffer. Store at −20°C (see Note 2).

2.4. Protein Digestion

1. 100 mM NH_4HCO_3 buffer: 100 mM NH_4HCO_3. Add 7.91 g NH_4HCO_3 to 900 ml water and make up to 1 l with water (see Note 3).
2. 2,2-Thiodiethanol stock solution (BHES) (Lys-CSC): 100 mM BHES. Add 12.22 mg BHES to 900 μl water. Mix and make up to 1 ml with water. Store at −20°C.
3. RapiGest stock solution: 1% (w/v) RapiGest surfactant (Waters, Milford, MA, USA). Add 10 mg RapiGest to 900 μl water. Mix, make up to 1 ml with water, and store at 4°C.
4. VialTweeter (Hielscher, Teltow, Germany) for intense sonication of small volumes.

5. BCA Protein Assay Kit (Pierce, Rockford, IL, USA).
6. Tris(2-carboxyethyl) phosphine (TCEP) stock solution: 100 mM TCEP. Add 28.66 mg TCEP to 700 μl water, mix, and make up to 1 ml with water. Aliquot and store at –20°C.
7. Iodoacetamide stock solution: 500 mM iodoacetamide. Add 92.48 mg iodoacetamide to 700 μl water, mix, and make up to 1 ml with water. Keep iodoacetamide in the dark, since it is light sensitive. Aliquot and store at –20°C.
8. Digestion buffer (Glyco-CSC, Cys-Glyco-CSC): 100 mM NH_4HCO_3, 1 mM iodoacetamide, 1 mM 2,2′-thiodiethanol. Add 12.2 mg 2,2′-thiodiethanol to 90 ml 100 mM NH_4HCO_3 buffer, mix, and add 200 μl iodoacetamide stock solution. Make up to 100 ml with 100 mM NH_4HCO_3.
9. Digestion buffer (Lys-CSC): 100 mM NH_4HCO_3, 1.25 mM iodoacetamide, 1.25 mM BHES. Add 1 μl iodoacetamide stock solution, 5 μl BHES stock solution to 320 ml 100 mM NH_4HCO_3 buffer, mix, and add 80 ml RapiGest stock solution.
10. Sequence grade modified trypsin (Promega, Madison, WI, USA). Store at –80°C.
11. 500,000 U/ml glycerol free PNGase F (New England BioLabs, Ipswich, MA, USA). Store at 4°C.

2.5. Affinity Chromatography

1. Streptavidin Plus UltraLink Resin (Pierce, Rockford, IL, USA). Store at 4°C.
2. 1 ml Mobicol columns with 35 μm pore size filter (MoBiTec, Goettingen, Germany).
3. Vac-Man laboratory vacuum manifold (Promega, Madison, WI, USA) and vacuum pump.
4. NaCl washing buffer: 5 M NaCl. Add 292.20 g NaCl to 600 ml water, mix, and make up to 1 l with water.
5. EDTA stock solution: 500 mM ethylenediaminetetraacetic acid (EDTA). Add 14.61 g EDTA to 80 ml water, adjust pH to 8 with NaOH, and make up to 100 ml with water.
6. Detergent buffer: 137 mM NaCl, 50 mM Tris pH 7.8, 10% glycerol, 0.5 mM EDTA pH 8, 0.1% Triton X-100. Add 6.06 g Tris base to 700 ml water and adjust pH to 7.8 with HCl. Then, add 8.01 g NaCl, 100 ml glycerol, 1 ml EDTA stock solution, and 10 ml Triton X-100 and make up to 1 l with water.
7. $NaHCO_3$ washing buffer: 100 mM $NaHCO_3$ pH 11. Add 8.40 g $NaHCO_3$ to 900 ml water. Adjust pH with NaOH and make up to 1 l with water (see Note 3).
8. Isopropanol solution: 80% (v/v) 2-propanol. Mix 400 ml 2-propanol and 100 ml water.

9. Dithiothreitol (DTT) stock solution: 100 mM DTT. Add 15.4 mg DTT to 900 μl water. Mix and make up to 1 ml with water. Store at –20°C.
10. Elution buffer: 100 mM NH_4HCO_3, 10 mM TCEP, 1 mM DTT. Add 15.4 mg DTT to 90 ml 100 mM NH_4HCO_3 buffer and mix. Add 2 ml TCEP stock solution and make up to 100 ml with 100 mM NH_4HCO_3 buffer.
11. 500,000 U/ml glycerol free PNGase F (New England BioLabs, Ipswich, MA, USA). Store at 4°C.

2.6. Desalting of Peptide Solution

1. 10% formic acid: 10% (v/v) formic acid. Add 1 ml formic acid to 9 ml water (see Note 5).
2. C18 UltraMicroSpin columns (The Nest Group, Southborough, MA, USA) with 0.03–30 μg capacity.
3. 80% acetonitrile: LC-MS grade water, 80% (v/v) acetonitrile, 0.1% (v/v) formic acid. Add 80 ml acetonitrile and 0.1 ml formic acid to 15 ml LC-MS grade water. Mix and make up to 100 ml with LC-MS grade water (see Note 5).
4. Sample buffer: LC-MS grade water, 5% (v/v) acetonitrile, 0.1% (v/v) formic acid. Add 5 ml acetonitrile and 0.1 ml formic acid to 90 ml LC-MS grade water. Mix and make up to 100 ml with LC-MS grade water (see Note 5).
5. 50% acetonitrile: LC-MS grade water, 50% (v/v) acetonitrile, 0.1% (v/v) formic acid. Add 50 ml acetonitrile and 0.1 ml formic acid to 45 ml LC-MS grade water. Mix and make up to 100 ml with LC-MS grade water (see Note 5).

3. Methods

3.1. Biotinylation of Cell Surface Proteins

The biotinylation of cell surface-exposed glycans or lysins can either be done with cells in solution or directly on the cell culture dish when working with adherent cells. The following considerations should be taken into account for performing either procedure. Detaching cells with trypsin might harm epitopes exposed at the cell surface. Better suitable are shearing forces (without harming the cell integrity) or EDTA to detach cells from culture dishes. Loosely attached cells could detach during the labeling procedure and in some cases it is therefore beneficial to collect the cells in a falcon tube before starting with the labeling procedure. Extracellular matrices containing glycan structures interfere with the labeling procedure. In this case the cells should be removed from the plate and labeled in solution or replated without extracellular support prior to labeling. Steps only necessary to perform in case of one protocol variant are labeled respectively.

3.1.1. Soluble Cells

1. Resuspend a cell pellet of approximately 10^8 cells in a 50 ml tube with 40 ml labeling buffer and afterwards pellet cells by centrifugation. Repeat this washing procedure twice (see Notes 6 and 7).
2. Glyco-CSC/Cys-Glyco-CSC: Resuspend the cells with 49.5 ml labeling buffer and add 500 μl of the $NaIO_4$ stock solution to the cell suspension. The final concentration of the solution is 1.6 mM $NaIO_4$. Incubate the cells at 4°C for 15 min on an MACSmix tube rotator.
3. Lys-CSC: Resuspend with 5 ml of the lysine labeling buffer. Incubate the cells for 30 min at room temperature on an MACSmix tube rotator. Transfer the cell suspension in a 50 ml tube and make up to 40 ml with PBS.
4. Glyco-CSC/Cys-Glyco-CSC: Pellet the cells by centrifugation and discard the supernatant. Afterwards, wash the cells twice by resuspending the cells in 40 ml labeling buffer and subsequent centrifugation.
5. Glyco-CSC/Cys-Glyco-CSC: Resuspend the cell pellet with 12 ml biocytin hydrazide solution and transfer the cell suspension in a 15 ml tube. Incubate the cells at 4°C for 60 min on an MACSmix tube rotator.
6. Transfer the cell suspension in a 50 ml tube and make up to 40 ml with labeling buffer (see Note 9). Pellet the cells by centrifugation and discard the supernatant. Afterwards, wash the cells twice with 40 ml labeling buffer.
7. Resuspend the cell pellet with 10 ml hypotonic lysis buffer and incubate the cells on ice for 10 min (see Note 10).
8. Optional for Cys-Glyco-CSC: Perform step 6 with hypotonic lysis buffer containing iodoacetamide (see Item 6 under Subheading 1.2).

3.1.2. Adherent Cells

1. Aspirate culture media of five cell culture dishes (Ø 14 cm), with approximately 80–90% confluence. Wash cells twice with labeling buffer (see Notes 6 and 7).
2. Glyco-CSC/Cys-Glyco-CSC: Add 500 μl of the $NaIO_4$ stock solution to 49.5 ml labeling buffer to a final concentration of 1.6 mM $NaIO_4$. Add 10 ml to each plate and incubate the cells at 4°C in the dark.
3. Lys-CSC: Add 2 ml of lysine labeling buffer to each plate. Incubate the cells for 30 min at room temperature on a slow, linearly tilting shaker (see Note 8).
4. Aspirate buffer. Wash twice with 10 ml labeling buffer per plate.
5. Glyco-CSC/Cys-Glyco-CSC: Add 2.4 ml biocytin hydrazide solution to each plate. Incubate the cells for 60 min on a slow, linearly tilting shaker (see Note 8).
6. Glyco-CSC/Cys-Glyco-CSC: Aspirate the biocytin hydrazide solution. Wash the cells twice with labeling buffer.

7. Glyco-CSC/Cys-Glyco-CSC: Add 2 ml hypotonic lysis buffer to each plate. Incubate at 4°C for 10 min. Afterwards, scrape the cells off the plate and collect them in a 15 ml falcon tube. Add 2 ml hypotonic lysis buffer to collect the remaining cells of all plates. The final volume of the cell suspension should not exceed 12 ml (see Note 10).
8. Optional for Cys-Glyco-CSC: Perform step 6 with hypotonic lysis buffer containing iodoacetamide (see Item 2 under Subheading 2.3).
9. Lys-CSC: Add 4 ml hypotonic lysis buffer (Lys-CSC) to the first plate. Scrape cells off the plate and transfer the cell suspension to the next plate. Continue with scraping and transferring the cell suspension to the next plate until all plates are scraped. Use 2 ml hypotonic lysis buffer (Lys-CSC) to wash all plates again and collect the remaining cells. The final volume of cell suspension should not exceed 8 ml. Incubate the cell suspension at 4°C for 10 min (see Note 10).

3.2. Cell Lysis and Membrane Preparation

1. Transfer the cells in a Dounce homogenizer and homogenize the cells on ice with 40 strokes (see Note 11).
2. Transfer the cell homogenate in a 15 ml tube and centrifuge at 2,000 × *g* for 10 min to remove cell debris and nuclei (see Note 12).
3. Glyco-CSC/Cys-Glyco-CSC: Distribute the supernatant equally in two ultracentrifuge tubes and mix in a 1:1 ratio with membrane preparation buffer. Fill up the ultracentrifuge tubes with a 1:1 mix of hypotonic lysis buffer and membrane preparation buffer. Then, centrifuge at 100,000 × *g* for 60 min in order to pellet the cell membranes (see Note 13).
4. Glyco-CSC/Cys-Glyco-CSC: Discard the supernatants, combine the membrane pellets in one ultracentrifuge tube, and incubate the membrane pellets on ice in 400 μl membrane wash buffer for 30 min.
5. Glyco-CSC/Cys-Glyco-CSC: Fill up the ultracentrifuge tube with hypotonic lysis buffer and centrifuge at 100,000 × *g* for 60 min.
6. Lys-CSC: Collect supernatant from step 2 and mix in a 1:1 ratio with membrane preparation buffer.
7. Lys-CSC: Add 7 ml sucrose buffer each to two ultracentrifuge tubes. Add very carefully 4 ml of the mixture from step 9 on top of the sucrose cushion. The sucrose cushion and the sample should not mix and should stay in two separate layers (see Note 13).
8. Lys-CSC: Centrifuge at 150,000 × *g* for 60 min in order to accumulate plasma membrane patches at the phase border and pellet other cellular membranes.

9. Lys-CSC: Collect carefully with a pipette 1 ml from the phase border containing the plasma membranes and transfer it into another ultracentrifuge tube. Fill the tube with hypotonic lysis buffer (Lys-CSC) and centrifuge at 100,000 × *g* for 20 h (see Note 25).

3.3. Protein Digestion

1. Discard the supernatant and transfer the membrane pellet in a 2 ml tube.
2. Glyco-CSC/Cys-Glyco-CSC: Add 340 μl of the 100 mM NH_4HCO_3 buffer and 40 μl of the 1% RapiGest stock solution (see Note 14).
3. Optional for Cys-Glyco-CSC: Use digestion buffer instead of pure NH_4HCO_3.
4. Lys-CSC: Add 320 μl of digestion buffer and 80 μl of the 1% RapiGest stock solution (see Note 14).
5. Indirectly sonicate the tube in the VialTweeter at 4°C until the membrane pellet is completely dissolved (see Note 15).
6. Determine the protein concentration of the solution with the BCA Protein Assay Kit.
7. Optional for Cys-Glyco-CSC: Skip step 8 and continue with step 9.
8. Glyco-CSC/Cys-Glyco-CSC: Add 8 μl of TCEP stock solution to the protein solution and incubate at room temperature for 30 min in order to reduce protein disulfide bonds. Then, add 12 μl of iodoacetamide stock solution to the protein solution in order to alkylate free thiol groups and incubate at room temperature in the dark for 30 min.
9. Lys-CSC: Add 1 μl PNGase F to solubilized membranes and deglycosylate proteins for 2 h at 37°C.
10. Add trypsin in a 1:50 protein:protease ratio to the protein solution and incubate at 37°C overnight on the MACSmix tube rotator.

3.4. Affinity Enrichment of Biotinylated Glycopeptides

1. Inactivate trypsin by heating the digestion solution at 95°C for 10 min (see Notes 16 and 17).
2. Add 350 μl of the streptavidin beads into a Mobicol column and place the Mobicol column on the vacuum manifold (see Note 18). Wash the streptavidin beads with 5 ml of the 100 mM NH_4HCO_3 buffer by cycles of resuspending streptavidin beads with 400 μl NH_4HCO_3 buffer and subsequently aspirating the liquid.
3. Add the digest solution to the streptavidin beads and rotate the Mobicol column on an MACSmix tube rotator for 1 h at 4°C. Transfer beads to new Mobicol column (see Notes 19 and 21).

4. Wash the streptavidin beads with 10 ml of the 100 mM NH_4HCO_3 buffer by cycles of resuspending the streptavidin beads with 400 μl NH_4HCO_3 buffer and subsequently aspirating the liquid on the vacuum manifold. Wash in the same manner the beads consecutively with each of the following washing buffers: 10 ml of NaCl washing buffer, 10 ml of detergent buffer (Glyco-CSC/Cys-Glyco-CSC), 10 ml of 100 mM NH_4HCO_3 buffer, 10 ml of $NaHCO_3$ washing buffer, 2 ml of isopropanol solution (Glyco-CSC/Cys-Glyco-CSC), and 10 ml of prewarmed (60°C) 100 mM NH_4HCO_3 buffer (see Note 20). Finally, transfer the streptavidin beads in a new Mobicol column and wash the streptavidin beads with 10 ml of the 100 mM NH_4HCO_3 buffer (see Note 21).
5. Optional for Cys-Glyco-CSC: Resuspend streptavidin beads in 400 μl elution buffer and incubate for 60 min at 37°C on an MACSmix tube rotator. Collect released cysteine peptides in a 2 ml tube. Add 1 μl iodoacetamide stock solution to the eluate and incubate at room temperature in the dark for 30 min.
6. Optional for Cys-Glyco-CSC: Resuspend streptavidin beads in 400 μl NH_4HCO_3, add 1 μl of iodoacetamide stock solution, and incubate at room temperature in the dark for 30 min. Wash beads with 10 ml of NH_4HCO_3.
7. Glyco-CSC/Cys-Glyco-CSC: Resuspend the streptavidin beads with 400 μl NH_4HCO_3 and add 1 μl of the glycerol-free PNGase F. Incubate the suspension at 37°C overnight on an MACSmix tube rotator (see Note 22).
8. Lys-CSC: Resuspend the streptavidin beads with 445 μl NH_4HCO_3 and add 5 μl DTT stock solution and 50 μl of TCEP stock solution. Incubate the suspension for 60 min at 37°C on an MACSmix tube rotator.
9. Place the Mobicol column in a 2 ml tube and collect released peptides by centrifugation. Afterwards, resuspend the streptavidin beads with 400 μl NH_4HCO_3 and collect the flow through in a 2 ml tube. Combine the two eluates.

3.5. Desalting of the Peptide Sample

1. Acidify the peptide eluates with 10% formic acid to a pH of 2–3 (see Note 23).
2. Place a C18 column in a 2 ml tube and condition the column by adding 100 μl of 80% acetonitrile into the C18 column. Centrifuge at 100 × *g* for 1 min and afterwards discard the flow through. Repeat this procedure once with 80% acetonitrile and then flush the column four times with 100 μl sample buffer (see Note 24).
3. Load the glycopeptides on the C18 column and afterwards flush the column three times with 100 μl sample buffer to remove salts.

4. Place the C18 column in a 2 ml tube and elute the glycopeptides by adding 100 μl 50% acetonitrile and centrifuging at 100×*g* for 1 min. Repeat this elution step once and dry the combined eluates in a SpeedVac concentrator.
5. Resuspend the dried peptides with 20 μl sample buffer and store the peptide sample frozen until analysis by LC-MS.

3.6. LC-MS Analysis and Data Analysis

1. We commonly separate glycopeptides by C18 reversed-phase liquid chromatography, ionize the peptides by nanospray-ESI, and acquire MS and MS/MS spectra on a sensitive high-mass-accuracy mass spectrometer.
2. Raw data files from the MS instruments are usually converted to mzXML files and searched with Sorcerer-Sequest against the appropriate protein database. Database search criteria include the variable modification of 0.984020 Da for asparagines (representing formerly *N*-glycosylated asparagines after deamidation through the PNGase F treatment), 145.019749 Da for lysines (representing 3-(carbamidomethylthio)propanoyl-tagged lysines) (Lys-CSC), and static modification of 57.021464 Da for cysteines (representing carbamidomethyl-containing cysteines after alkylation with iodoacetamide). Peptide and protein identifications are statistically validated with the *Trans*-Proteomic Pipeline TPP (10), containing PeptideProphet (11) and ProteinProphet (12).

4. Notes

1. Prepare buffers immediately before use.
2. Store the membrane preparation buffer and the sucrose buffer at –20°C, because working under unsterile conditions can lead to contamination of the buffer due to the high sucrose concentration.
3. Prepare bicarbonate buffers freshly because CO_2 escapes over time and the pH value will increase.
4. BHES is used to prevent overalkylation, which could occur at N-terminal amines, at histidines or at lysine side chains (13).
5. Always use glassware to transfer formic acid, since formic acid leaches out plasticizer from plasticware.
6. The amount of cells used for the experiment is dependent on the size of the cell type. Small cells (e.g., stem cells) have a smaller cell surface and have therefore presumably fewer cell surface proteins per cell. In this case it is advised to use more cells per experiment. Using 200 g of cellular biomass should in general produce a good yield.

7. Carry out all centrifugation steps of the protocol at 4°C and put tubes on ice during waiting times. The washing steps help to remove residual cell culture medium and cell debris. When working with adherent cells, put the plates on ice or in the fridge as often as possible.
8. Make sure that the biocytin hydrazide solution (Cys-Glyco-CSC) or sulfo-NHS-SS-biotin solution (Lys-CSC) wets the whole plate during shaking. Alternatively increase the volume of the biocytin hydrazide solution, by keeping the biocytin hydrazide concentration constant.
9. When cells are transferred into a different tube, always try to keep the loss of cells as small as possible. For example, after the transfer of the cell suspension in the 50 ml tube, remaining cells in the 15 ml tube can be resuspended with labeling buffer and then added to the cells in the 50 ml tube until the volume of 40 ml is reached.
10. Slowly move the piston of the Dounce homogenizer completely up and down. Reduce the speed if foam is generated or if the homogenizer warms up. The cell lysis can also be carried out by other means, for example by sonication with the VialTweeter with 20 impulses at half maximum amplitude and half maximum cycle time.
11. Optionally the cells can be solubilized in only 2 ml of hypotonic lysis buffer and stored at –20°C for up to 6 months.
12. The supernatant, containing cell membranes, should be a bit cloudy after the centrifugation step. If the supernatant is completely clear, the homogenization step should be repeated.
13. We usually use a Beckman SW41 swinging bucket rotor. Make sure to fill the ultracentrifuge tubes completely with liquid; otherwise tubes could collapse in the buckets during centrifugation.
14. Loosen the membrane pellet from the tube bottom with the pipette tip at each side and try to transfer the membrane pellet as a whole.
15. Depending on the size of the membrane pellet and the cell line it may require long, intense sonication to completely solubilize the membrane pellet. In order to avoid thermal sample degradation, it is recommended to apply cycles of sonication and cooling on ice.
16. Make sure to thoroughly inactivate trypsin. Heat inactivation of trypsin worked best in our hands, but protease inhibitors or acidification of the sample may also be an option to inactivate trypsin activity.
17. The digest could be stored at –20°C for up to 6 months.

18. Use filter with 35 μm pore size. Smaller pore size filters tend to clog easily.
19. If you are interested in non-biotinylated peptides of the digestion solution, you can collect these peptides by centrifugation of the Mobicol column in a 2 ml tube.
20. Optionally, warm up buffers to 60°C, which may in some cases increase the efficiency to remove unspecific peptides.
21. Streptavidin beads can be easily transferred into a new Mobicol column by resuspending the streptavidin beads with 400 μl of the 100 mM NH_4HCO_3 buffer. The transfer of the streptavidin beads into a new Mobicol column removes peptides, which are unspecifically bound to the Mobicol column.
22. Optionally, incubate the streptavidin beads at 37°C overnight in 50 mM sodium phosphate buffer supplemented with 1% NP-40, which is the manufacturer-recommended buffer. However, PNGase F also efficiently releases glycopeptides from streptavidin beads in 100 mM NH_4HCO_3 buffer.
23. The acidification of the peptide solution is necessary for the subsequent desalting step by C18 reversed-phase chromatography. Carefully add a few microliters of the 10% formic acid and let CO_2 escape by gentle mixing with a pipette tip. Test the pH value of the glycopeptides solution by transferring 1 μl of the solution on a pH indicator paper.
24. Flushing the C18 column with sample buffer can also be carried out on the vacuum manifold by continuously aspirating the liquid.
25. The duration of the centrifugation should be at least overnight (16 h) to pellet as many membrane patches as possible. Longer centrifugation durations might be beneficial.

References

1. Schiess R, Wollscheid B, Aebersold R (2009) Mol Oncol 3:33–44
2. Yildirim MA, Goh KI, Cusick ME, Barabasi AL, Vidal M (2007) Nat Biotechnol 25:1119–1126
3. Gundry RL, Boheler KR, Van Eyk JE, Wollscheid B (2008) Proteomics Clin Appl 2:892–903
4. Zola H (2006) Mol Med 12:312–316
5. Schiess R, Mueller LN, Schmidt A, Mueller M, Wollscheid B, Aebersold R (2009) Mol Cell Proteomics 8:624–638
6. Apweiler R, Hermjakob H, Sharon N (1999) Biochim Biophys Acta 1473:4–8
7. Wollscheid B, Bausch-Fluck D, Henderson C, O'Brien R, Bibel M, Schiess R, Aebersold R, Watts JD (2009) Nat Biotechnol 27:378–386
8. Hofmann A, Gerrits B, Schmidt A, Bock T, Bausch-Fluck D, Aebersold R, Wollscheid B. Blood 116:e26–e34
9. Pan C, Kumar C, Bohl S, Klingmueller U, Mann M (2009) Mol Cell Proteomics 8:443–450
10. Keller A, Eng J, Zhang N, Li XJ, Aebersold R (2005) Mol Syst Biol 1:2005.0017
11. Keller A, Nesvizhskii AI, Kolker E, Aebersold R (2002) Anal Chem 74:5383–5392
12. Nesvizhskii AI, Keller A, Kolker E, Aebersold R (2003) Anal Chem 75:4646–4658
13. Boja ES, Fales HM (2001) Anal Chem 73:3576–3582

Chapter 2

An On-Target Desalting and Concentration Sample Preparation Protocol for MALDI-MS and MS/MS Analysis

Xumin Zhang, Quanhui Wang, Xiaomin Lou, Haidan Sun, Peter Roepstorff, and Siqi Liu

Abstract

2DE coupled with MALDI-MS is one of the most widely used and powerful analytic technologies in proteomics study. The MALDI sample preparation method has been developed and optimized towards the combination of simplicity, sample-cleaning, and sample concentration since its introduction. Here we present a protocol of the so-called *S*ample loading, *M*atrix loading, and on-target *W*ash (SMW) method which fulfills the three criteria by taking advantage of the AnchorChip™ targets. Our method is extremely simple and no pre-desalting or concentration is needed when dealing with samples prepared from 2DE. The protocol is amendable for automation and would pave the road for high-throughput MALDI-MS or MS/MS-based proteomics studies with guaranteed sensitivity and high identification rate. The method has been successfully applied to mouse liver proteome study and so far has been employed in other proteome studies by world-wide researchers.

Key words: AnchorChip™, Sample loading, Matrix loading and on-target wash method, MALDI sample preparation of mouse liver proteins

1. Introduction

MALDI together with ESI have been so far the most widely used in mass spectrometry as ionization methods in modern proteomics study. Today, many researchers are working with low amount of proteins, e.g., low-femtomole or even attomole level. For this sake, a good sample preparation method with high sensitivity and salt tolerance should be highly desired.

Although ESI analysis is much more sensitive to contaminants than MALDI analysis, the additional sample clean is not always

Djuro Josic and Douglas C. Hixson (eds.), *Liver Proteomics: Methods and Protocols*, Methods in Molecular Biology, vol. 909, DOI 10.1007/978-1-61779-959-4_2,

Fig. 1. Structure of AnchorChip™ targets plate. The sample support is covered by a teflon-like film (in), among which there are some interrupts (in) used as the anchors to concentrate samples. There are 384 sample spots and 96 calibrant spots, and every 4 sample anchors are centered by a calibrant anchor.

necessary since in most cases ESI-MS analysis are coupled online with RP-HPLC, which desalt, concentrate, and separate sample simultaneously.

Numerous MALDI sample preparation methods have been developed since its introduction. These include dried-droplet, crushed-crystal, thin-layer, and sandwich methods (1–4). Despite the fact that MALDI is tolerant to impurity to some extent, the pre-desalting and concentration is always necessary especially for low amount of sample.

A millstone development is the introduction of the prestructured MALDI-MS sample support, the so-called AnchorChip™ targets plate (Fig. 1). The plate is coated with a thin layer of teflon-like material with small circular interruptions at which the underlying stainless steel surface gets accessible. The design allows sample concentration by limiting matrix deposit to the small interrupts. Some sample preparation methods have been developed based on AnchorChip™ technology towards high sensitivity and automation (5, 6). We present a protocol of the so-called SMW method which is composed of three steps: *S*ample loading, *M*atrix loading, and on-targets *W*ash (7). Taking advantage of the sample concentration on AnchorChip™ targets and on-target desalting effect on CHCA prepared sample, this method proves to be simple, efficient, and sensitive for both MALDI-MS and MS/MS analyses. This method was successfully applied into our study of mouse liver proteomics study and demonstrated better sensitivity and simpler and easier sample handling than other available AnchorChip-based sample preparation samples (7). Moreover, it has been verified and adopted in variety of proteomics projects by world-wide researchers (8–18).

2. Material

2.1. In-Gel Digestion

1. Purified water (Millipore, Bedford, MA, USA).
2. Gel washing solvent: acetonitrile (ACN)/H_2O (30:70, v/v).
3. Gel drying solvent: 100% ACN.
4. Reduction solution: 10 mM DTT in 50 mM NH_4HCO_3.
5. Alkylation solution: 10 mg/ml iodoacetamide (IAM) in 50 mM NH_4HCO_3.
6. Trypsin solution: 10 ng/μl Trypsin (Promega, Madison, Wisconsin, USA) in 25 mM NH_4HCO_3.
7. Digestion buffer: 25 mM NH_4HCO_3.
8. Termination buffer: 5% TFA (v/v).

2.2. MALDI Sample Preparation

1. MTP AnchorChip™ targets plate, 600/384 (600 μm diameter, 384 anchors) (Bruker, Bremen, Germany) (see Note 1).
2. Safe-Lock™ tubes (Eppendorf, Hamburg, Germany) (see Note 2).
3. Purified water (Millipore, Bedford, MA, USA).
4. Plate wash solution: 90% ethanol (EtOH) (v/v).
5. EtOH.
6. Alpha-Cyano-4-hydroxycinnamic acid (CHCA) (Sigma-Aldrich, Steinheim, Germany).
7. Matrix stock solution: 10 μg/μl CHCA in 70% ACN and 0.1% TFA (v/v) (see Note 3).
8. Matrix dilution solution: 90% ACN + 0.1% TFA (v/v).
9. Matrix working solution: 0.5 μg/μl CHCA in 70% ACN and 0.1% TFA (v/v) (see Note 4).
10. Desalting solution: 0.5% TFA (v/v) for individual anchors or 0.02% TFA (v/v) for whole-plate desalting.
11. Peptide calibration standard II (Bruker, Bremen, Germany). It contains nine peptides: Bradykinin(1–7), Angiotensin_II, Angiotensin_I, Substance_P, Bombesin, Renin_Substrate, ACTH_clip(1–17), ACTH_clip(18–39), and Somatostatin 28, with a mass range of 700–4,000 Da (Table 1).

Table 1
The mass distribution of the peptide calibration standard II

Calibrants	[M+H]+
Bradykinin(1–7)	757.399
Angiotensin_II	1,046.541
Angiotensin_I	1,296.684
Substance_P	1,347.735
Bombesin	1,619.822
Renin_Substrate	1,758.932
ACTH_clip(1–17)	2,093.086
ACTH_clip(18–39)	2,465.198
Somatostatin(28)	3,147.471

12. Calibration stock solution: 1 pM/μl peptide calibration standard II in 0.5% TFA (v/v).
13. Calibration working solution: 20 fM/μg peptide calibration standard II in matrix working solution.
14. Recrystallization solution: 0.1 μg/μl CHCA in 70% ACN and 0.1% TFA (v/v).

3. Methods

The mouse liver proteins were extracted from the sacrificed male mice with 2DE compatible lysis buffer [10 mM Tris–HCl, pH 7.4, 8 M urea, 4% CHAPS (w/v), 10 mM DTT, 1 mM PMSF, 2 mM EDTA]. The first dimensional separation was accomplished on 18-cm IPG strips (100 μg/strip) using an IPGphor (GE-Health, Uppsala, Sweden). The focused protein strips were transferred onto 12% SDS-PAGE gels using Ettan DALT II system (GE-Health, Uppsala, Sweden) for the second dimensional separation. All the procedures were carried out according to manufacturer's manual. An adapted silver staining method was used to visualize the protein spots (7).

3.1. In-Gel Digestion

1. Wash the stained gels with purified water for three times.
2. Excise the 2DE spots (1 mm diameter) from the gel and transfer to Eppendorf tubes or 96-well plate when working with a number of gel spots.

3. Wash the gel spots with 50 μl water and then 30% ACN (v/v) with vortex in Eppendorf Mixer (Eppendorf, Hamburg, Germany). After 10 min vortex, the solution is removed and discarded.
4. Add 50 μl gel drying solution to dehydrate the gel spots with vortex. After 10 min vortex, the solution is removed and discarded. The gel would turn white and opaque, otherwise, repeat this step.
5. Put the tubes in Speed-Vacuum (Heto-holten, Copenhagen, Denmark) for 5 min to remove the remaining trace ACN.
6. Add 20 μl reduction solution to the each tube and spin down to make sure the gel spot is completely rehydrated.
7. Incubate for 45 min at 56°C to accomplish the reduction.
8. Remove the solution and wait until the tube cool down to room temperature.
9. Add 20 μl alkylation solution to each tube and immediately put the tube in dark for 1 h at room temperature.
10. Remove the solution and wash the gel spots with 50 μl water and then gel washing solvent with vortex in Eppendorf Mixer. Repeat the wash for another two times.
11. Remove the solution and add 50 μl gel drying solution to dehydrate the gel spots with vortex. The gel would turn white and opaque, otherwise, repeat this step.
12. Put the tubes in Speed-Vacuum for 5 min to remove the remaining trace ACN.
13. Add 1 μl trypsin solution to the dry gel spots, and incubate it for 20 min on ice to rehydrate (see Note 5).
14. Add 9 μl digestion solution to cover the gel spot and transfer to 37°C incubator for overnight digestion.
15. Add 1 μl termination solution to terminate the digestion (see Note 6).

3.2. Plate Cleaning (*See* Note 7)

1. Rinse the AnchorChip™ targets plate with water and ethanol for three times.
2. Clean the AnchorChip™ targets plate in water incubation and then in plate wash solution incubation, 5 min sonication each.
3. Rinse the AnchorChip™ targets plate with ethanol and wipe away any visible droplet remaining on the plate with Kimwipes tissue.
4. Keep at room temperature until it is completely dry.

3.3. MALDI Matrix Solution Preparation

1. Weigh CHCA and dissolve it with solvent containing 70% ACN and 0.1% TFA to reach a final concentration of 10 μg/μl.

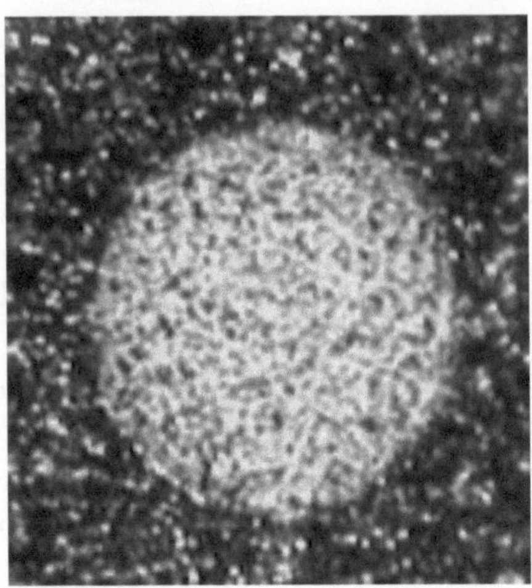

Fig. 2. The matrix crystal layer on 600 μm AnchorChip™ target resulting from 0.8 μl matrix loading solution. The photo was taken with the camera equipped to Bruker MS instrument.

2. Keep the solution in Eppendorf Mixer with vortex for 30 min and the resulting solution is the matrix stock solution (see Note 3).
3. Mix the matrix stock solution and the matrix dilution solution with the ratio of 1:19.
4. Vortex for a few seconds. The resulting solution is the matrix working solution (see Note 4).

3.4. Calibration Solution Preparation (See Note 8)

1. Dissolve the peptide calibration standard II with 0.5% TFA to reach a concentration of 1 pM/μl. The resulting solution is the calibration stock solution.
2. Dilute the calibration stock solution with 49 times the volume of matrix working solution to reach a final concentration of 20 fM/μl. The resulting solution is the calibration working solution.

3.5. Recrystallization Solution Preparation (See Note 9)

1. Mix the matrix working solution and the matrix dilution solution with the ratio of 1:4.
2. Vortex for a few seconds. The resulting solution is the recrystallization solution.

3.6. MALDI Sample Preparation

It is necessary to test the matrix working solution prior to sample analysis. For this sake, 0.8 μl matrix working solution is directly loaded onto an anchor and let it dry. The resulting matrix deposit should be similar to what we observe in Fig. 2. The matrix should deposit into small homogeneous crystals and evenly distribute on the hydrophilic anchor.

3.6.1. Preparation of Analyte

1. 1 μl digest is loaded onto the AnchorChip™ target, and let it completely dry in air and concentrate on the anchor (see Notes 10 and 11).
2. 0.8 μl matrix working solution is loaded onto the AnchorChip™ target containing dried sample. Let it dry completely (see Note 12).
3. Load 2 μl 0.5% TFA solution onto the matrix deposit and remove it with pipette after 30 s (see Notes 13 and 14).

3.6.2. Preparation of Calibrant Standard

1. 0.8 μl calibration working solution is loaded onto the calibrant anchor. Let it dry completely.
2. Load 2 μl 0.5% TFA solution onto the matrix deposit. Remove it with pipette after 30 s (see Note 14).

3.6.3. Recrystallization (See Note 9)

1. After wash with TFA solution, 0.8 μl recrystallization solution is loaded onto the matrix deposit. Let it dry completely.
2. Load 2 μl 0.5% TFA solution onto the matrix deposit. Remove it with pipette after 30 s.

The concentration and desalting capability of the SMW method are demonstrated with the microscopic observations at different preparation steps (Fig. 3) and the resulting MS spectra (Fig. 4).

In order to distinguish the matrix crystal from the salt crystal, Tris–HCl as the test solution was chosen since its needle-like crystal is completely different from the dotted crystal of CHCA. The dried concentrated sample (Fig. 3a) can be completely redissolved with matrix working solution (Fig. 3b). With the evaporation of ACN, the matrix deposits before (Fig. 3c) the salt (Fig. 3d). Subsequently, the salt crystal on the matrix layer is easily redissolved in washing solution (Fig. 3e) and removed by pipette. Due to its hydrophobic feature, the peptide sample embedded into matrix crystal survives the wash procedure.

The desalting effect is also proved with the MS spectra acquired from the sample at different steps (Fig. 4). Before desalting, the spectrum is dominated by the signals from matrix salt products which are characterized with 0 as decimal in mass range of 800–900 (Fig. 4a), whereas, after desalting, it is dominated by peptide signals characterized with 4 or 5 as decimal in mass range of 800–900 (Fig. 4b) (19), and more peptide signals are observed with significantly improved intensity and resolution. Moreover, an additional wash step helps reducing the salt signals to further lower level (Fig. 4c).

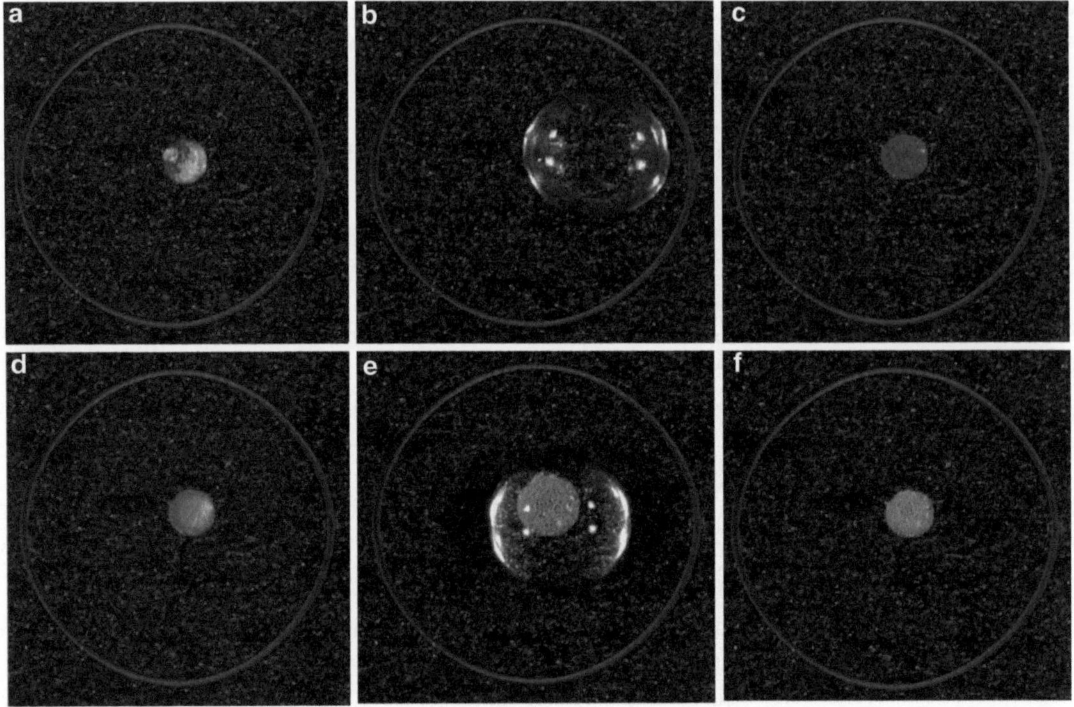

Fig. 3. SMW sample preparation. 1 μl 25 mM Tris–HCl was used as the analyte since the needle-like crystals of Tris–HCl salt can be distinguished from the dot-crystals of CHCA. (**a**) Sample dried on target; (**b**) 0.8 μl matrix solution was applied on the target; (**c**) matrix crystals deposited at 80 s after matrix loading; (**d**) salt deposited upon the matrix crystals at 160 s after matrix loading; (**e**) 1 μl 0.5% TFA is loaded on the target; (**f**) matrix crystals after removing salt with the washing solution.

3.7. Data Acquisition

3.7.1. MS Analysis

1. Load the AnchorChip™ plate into the Bruker MALDI-TOF/TOF MS (Bruker).
2. Choose and open an appropriate Reflector MS method.
3. The geometry is automatically set as "MTP AnchorChip 600-384" when the plate is recognized by the MS.
4. Teach the plate with three default Auto-Teach spots to calibrate the anchor location (see Note 15).
5. Switch to calibration mode by setting the chip number as "1".
6. Start the data acquisition when the instrument is ready.
7. After spectrum accumulation of 50 shots, the summed spectrum is automatically calibrated with the default calibrant $[M+H]^+$ values, and the calibration is accepted when the mass shifts are all less than 20 ppm.
8. Switch to sample analysis mode by setting the chip number as "0".
9. Start the data acquisition on the sample spots surrounding the analyzed calibrant anchor.

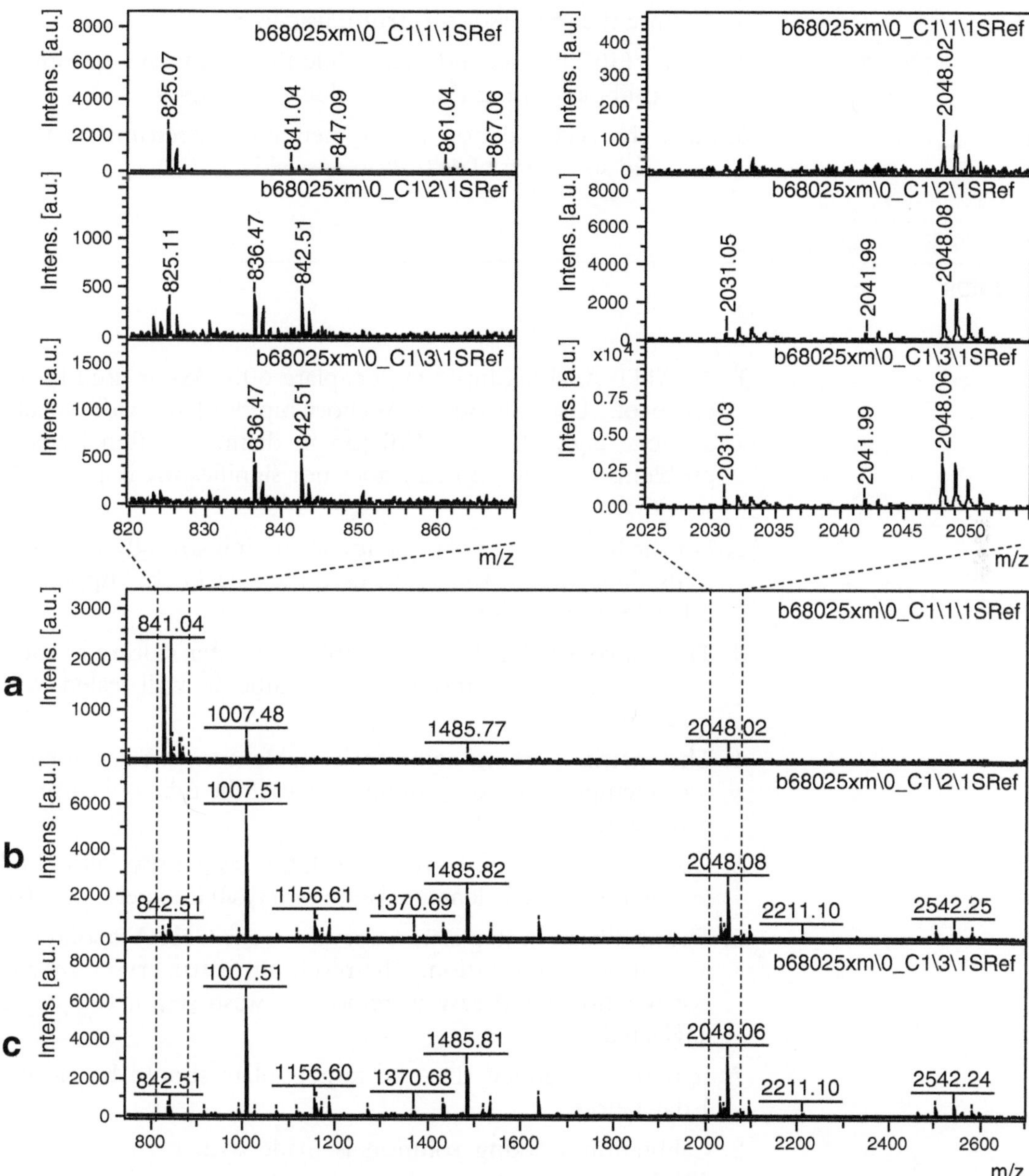

Fig. 4. The desalting effect of SMW method. From top to bottom, the spectrum is acquired from the sample before desalting (**a**), after single desalting (**b**) and after double desalting (**c**). The two zoomed spectra demonstrate the wash effect in removal of salt signals and enhancement of peptide signals.

10. Save the summed spectrum after spectrum accumulation of 100 shots.

3.7.2. MS/MS Analysis

1. Choose and open an appropriate reflector MS/MS method.
2. Input the *m/z* value of precursor in which you are interested. The *m/z* value is from the corresponding MS spectrum obtained above.

3. Acquire the precursor spectrum, usually 50 shots.
4. Click "Lift" button and meanwhile the spectrum is automatically calibrated using the m/z value of precursor.
5. Start the accumulation of fragmentation spectrum, and the summed spectrum of 400 shots is saved.

4. Notes

1. An MTP AnchorChip™ targets plate 600/384 is used in this protocol. Use of other AnchorChip™ plate with smaller anchors, e.g., 200 and 400 μm in diameter, often leads to problems in operation and does not significantly improve the sensitivity.
2. For the high content of organic solvent, it is strongly suggested to use high-quality tubes. We used the Safe-Lock™ Eppendorf tubes for this purpose.
3. The matrix stock solution is stable even after storing at room temperature for 1 month if the tube is well-sealed with parafilm.
4. The matrix working solution is stable even after storing at room temperature for a couple of days if the tube is well-sealed with parafilm.
5. Use of same volume of trypsin solution as gel spot allows full rehydration and efficient inhalation of trypsin into gel spot (20).
6. The acidification step is necessary since the CHCA is soluble in neutral or basic solution. The resulting matrix crystal may be not functional and easy to remove in wash step if this step is neglected.
7. It is recommended that the targets plate should be cleaned 1 day before.
8. Calibration working solution is made with high content of ACN to avoid oxidation during long time drying in air.
9. The recrystallization step is not necessary for in-gel digestion sample using NH_4HCO_3 as digestion buffer. It is required and powerful when sample contains metal ion containing buffer, e.g., Na^+ (21).
10. The drying process usually takes 10–20 min depending on the room temperature and humidity.
11. It is observed that the drying in air takes much longer time and in some cases is very difficult when more than 2 μl sample was loaded. Therefore, an alternative method is suggested when large volume of sample is needed. The sample is first removed

to a new tube and dried in speed-vacuum; secondly, 0.8 μl matrix working solution is used to resuspend the dried sample and then loaded onto the AnchorChip™ target. Another advantage of this method is that the oxidation on methionine and tryptophan is significantly reduced since drying is employed in vacuum.

12. The drying process of matrix working solution usually takes 2–3 min.
13. The wash step needs to repeat to reduce the contaminant to a further lower level when very low amount of sample is applied.
14. When dealing with hundreds of samples on the one plate, the wash step can be replaced by dipping the plate into 0.02% TFA (v/v) for 20 s followed by drying.
15. The teach step allows the instrument to recognize the anchor position, and facilitates the matrix spots search during automation.

References

1. Karas M, Bachmann D, Bahr U, Hillenkamp F (1987) Matrix-assisted ultraviolet-laser desorption of nonvolatile compounds. Intl J Mass Spectrom Ion Process 78:53–68
2. Xiang F, Beavis RC (1994) A method to increase contaminant tolerance in protein matrix-assisted laser-desorption ionization by the fabrication of thin protein-doped polycrystalline films. Rapid Commun Mass Spectrom 8:199–204
3. Vorm O, Roepstorff P, Mann M (1994) Improved resolution and very high-sensitivity in MALDI TOF of matrix surfaces made by fast evaporation. Anal Chem 66:3281–3287
4. Li L, Golding RE, Whittal RM (1996) Analysis of single mammalian cell lysates by mass spectrometry. J Am Chem Soc 118:11662–11663
5. Gobom J, Schuerenberg M, Mueller M, Theiss D, Lehrach H, Nordhoff E (2001) Alpha-cyano-4-hydroxycinnamic acid affinity sample preparation. A protocol for MALDI-MS peptide analysis in proteomics. Anal Chem 73:434–438
6. Thomas H, Havlis J, Peychl J, Shevchenko A (2004) Dried-droplet probe preparation on AnchorChip (TM) targets for navigating the acquisition of matrix-assisted laser desorption/ionization time-of-flight spectra by fluorescence of matrix/analyte crystals. Rapid Commun Mass Spectrom 18:923–930
7. Zhang X, Shi L, Shu S, Wang Y, Zhao K, Xu N, Liu S, Roepstorff P (2007) An improved method of sample preparation on AnchorChip targets for MALDI-MS and MS/MS and its application in the liver proteome project. Proteomics 7:2340–2349
8. Zhang K, Wrzesinski K, Fey SJ, Mose Larsen P, Zhang X, Roepstorff P (2008) Assessing CMT cell line stability by two dimensional polyacrylamide gel electrophoresis and mass spectrometry based proteome analysis. J Proteomics 71:160–167
9. Zhang K, Wrzesinski K, Stephen JF, Larsen PM, Zhang X, Roepstorff P (2008) Comparative proteome analysis of three mouse lung adenocarcinoma CMT cell lines with different metastatic potential by two-dimensional gel electrophoresis and mass spectrometry. Proteomics 8:4932–4945
10. Wang Q, Han H, Xue Y, Qian Z, Meng B, Peng F, Wang Z, Tong W, Zhou C, Guo Y, Li G, Liu S, Ma Y (2009) Exploring membrane and cytoplasm proteomic responses of Alkalimonas amylolytica N10 to different external pHs with combination strategy of de novo peptide sequencing. Proteomics 9:1254–1273
11. Zhang X, Hojrup P (2010) Cyclization of the N-terminal X-Asn-Gly motif during sample preparation for bottom-up proteomics. Anal Chem 82:8680–8685
12. Shetty NP, Jensen JD, Knudsen A, Finnie C, Geshi N, Blennow A, Collinge DB, Jorgensen HJ (2009) Effects of beta-1,3-glucan from

Septoria tritici on structural defence responses in wheat. J Exp Bot 60:4287–4300

13. Pedersen M, Ligowska M, Hammer K (2010) Characterization of the CI repressor protein encoded by the temperate lactococcal phage TP901-1. J Bacteriol 192:2102–2110
14. Taiyoji M, Shitomi Y, Taniguchi M, Saitoh E, Ohtsubo S (2009) Identification of proteinaceous inhibitors of a cysteine proteinase (an Arg-specific gingipain) from *Porphyromonas gingivalis* in rice grain, using targeted-proteomics approaches. J Proteome Res 8: 5165–5174
15. Laino P, Shelton D, Finnie C, De Leonardis AM, Mastrangelo AM, Svensson B, Lafiandra D, Masci S (2010) Comparative proteome analysis of metabolic proteins from seeds of durum wheat (cv. Svevo) subjected to heat stress. Proteomics 10:2359–2368
16. Folmes CD, Sawicki G, Cadete VJ, Masson G, Barr AJ, Lopaschuk GD (2010) Novel O-palmitolylated beta-E1 subunit of pyruvate dehydrogenase is phosphorylated during ischemia/reperfusion injury. Proteome Sci 8:38
17. Gustafsson JO, Oehler MK, McColl SR, Hoffmann P (2010) Citric acid antigen retrieval (CAAR) for tryptic peptide imaging directly on archived formalin-fixed paraffin-embedded tissue. J Proteome Res 9:4315–4328
18. Balamurugan S (2010) Growth temperature associated protein expression and membrane fatty acid composition profiles of *Salmonella enterica* Serovar Typhimurium. J Basic Microbiol 50:507–518
19. Wool A, Smilansky Z (2002) Precalibration of matrix-assisted laser desorption/ionization-time of flight spectra for peptide mass fingerprinting. Proteomics 2:1365–1373
20. Katayama H, Nagasu T, Oda Y (2001) Improvement of in-gel digestion protocol for peptide mass fingerprinting by matrix-assisted laser desorption/ionization time-of-flight mass spectrometry. Rapid Commun Mass Spectrom 15:1416–1421
21. Zhang X, Rogowska-Wrzesinska A, Roepstorff P (2008) On-target sample preparation of 4-sulfophenyl isothiocyanate-derivatized peptides using AnchorChip targets. J Mass Spectrom 43:346–359

Chapter 3

Plasma Membrane Isolation Using Immobilized Concanavalin A Magnetic Beads

Yu-Chen Lee, Martina Srajer Gajdosik, Djuro Josic, and Sue-Hwa Lin

Abstract

Isolation of highly purified plasma membranes is the key step in constructing the plasma membrane proteome. Traditional plasma membrane isolation method takes advantage of the differential density of organelles. While differential centrifugation methods are sufficient to enrich for plasma membranes, the procedure is lengthy and results in low recovery of the membrane fraction. Importantly, there is significant contamination of the plasma membranes with other organelles. The traditional agarose affinity matrix is suitable for isolating proteins but has limitation in separating organelles due to the density of agarose. Immobilization of affinity ligands to magnetic beads allows separation of affinity matrix from organelles through magnets and could be developed for the isolation of organelles. We have developed a simple method for isolating plasma membranes using lectin concanavalin A (ConA) magnetic beads. ConA is immobilized onto magnetic beads by binding biotinylated ConA to streptavidin magnetic beads. The ConA magnetic beads are used to bind glycosylated proteins present in the membranes. The bound membranes are solubilized from the magnetic beads with a detergent containing the competing sugar alpha methyl mannoside. In this study, we describe the procedure of isolating rat liver plasma membranes using sucrose density gradient centrifugation as described by Neville. We then further purify the membrane fraction by using ConA magnetic beads. After this purification step, main liver plasma membrane proteins, especially the highly glycosylated ones and proteins containing transmembrane domains could be identified by LC-ESI-MS/MS. While not described here, the magnetic bead method can also be used to isolate plasma membranes from cell lysates. This membrane purification method should expedite the cataloging of plasma membrane proteome.

Key words: Liver plasma membrane, Magnetic beads, Concanavalin A

1. Introduction

The first step in determining the plasma membrane proteome is to isolate plasma membrane from cells or tissues. Density gradient centrifugation is one of the most commonly used procedures to separate plasma membranes from other organelles in cell homogenates.

Djuro Josic and Douglas C. Hixson (eds.), *Liver Proteomics: Methods and Protocols*, Methods in Molecular Biology, vol. 909, DOI 10.1007/978-1-61779-959-4_3, © Springer Science+Business Media New York 2012

This procedure is lengthy yet only recovers a small percentage of the plasma membranes (1). Often, the plasma membrane preparations are also contaminated with other organelles. Affinity chromatography based on unique features of this organelle should be able to improve the speed, recovery, and purity of plasma membranes. However, the agarose-based affinity matrix that use gravity to separate the affinity matrix from the contaminants are not suitable for isolating organelles, as agarose comes down with the organelle fraction. Recent application of magnetic beads to immobilize ligands for affinity purification provides a possibility to use affinity chromatography to isolate plasma membranes. Use of magnetic beads with immobilized monoclonal antibodies for isolation of highly pure plasma membrane has also been reported (2). However, this method requires a large amount of purified monoclonal antibodies and can be applied only to those cells that express the specific membrane proteins.

One of the unique features of plasma membrane proteins is that their extracellular domains are frequently glycosylated. This feature has led to the application of lectin-affinity chromatography as one of the steps in the purification of many membrane proteins. The lectins are a group of carbohydrate-binding proteins that have different sugar-binding specificities. The most commonly used lectin for binding glycosylated membrane proteins is concanavalin A (ConA), which binds the α-D-glucose and α-D-mannose present in high-mannose glycopeptides (3). We sought to combine the glycosylation of membrane proteins and the ease of separating magnetic beads from solutions by a magnet for the isolation of plasma membranes. This procedure is simpler than the traditional approach, does not require expensive centrifugation equipment, and results in good recovery and high purity of plasma membranes. In this study, we first describe the procedure of isolating rat liver plasma membranes using sucrose density gradient centrifugation as described by Neville (4). We then describe the procedure to obtain highly purified liver plasma membranes from Neville membrane using ConA magnetic beads. In addition, we also describe the procedure for the removal of peripheral proteins while the membranes are immobilized on the ConA magnetic beads. This leads to further enrichment of transmembrane proteins and heavily glycosylated integral membrane proteins, and enhances their identification by subsequent LC-ESI-MS/MS analysis. These methods should expedite the cataloging of the plasma membrane proteome.

2. Materials

1. Biotinylated ConA was purchased from Vector Laboratories (Burlinhame, CA, USA).
2. Streptavidin magnetic beads were from BioClone, Inc. (San Diego, CA, USA).

3. Frozen rat livers were purchased from Pel Freeze (Rogers, AR, USA).
4. CHAPS was from Pierce (Rockford, IL, USA).
5. Methyl a-d-mannopyranoside was from Sigma-Aldrich (St. Louis, MO, USA).
6. Tris-buffered saline (TBS) (10×) was from BioRad (Hercules, CA, USA).
7. SDS-PAGE and Western blot analyses.
   (a) NuPAGE® Novex 4–12% Bis-Tris Gel 1.0 mm, 10-well, were from Invitrogen (Carlsbad, CA, USA).
   (b) Nupage MOPS SDS running buffer (20×) was also from Invitrogen.
   (c) Nupage transfer (20×) is from Invitrogen. Dilute 20× containing 20% (v/v) methanol as 1× transfer buffer.
   (d) Nitrocellulose membrane from Millipore, Bedford, MA, and 3MM chromatography paper from Whatman (Maidstone, UK).
   (e) TBS with Tween-20 (TBS-T): TBS prepare from 10× TBS stock and add 0.5% (v/v) Tween-20 (Sigma-Aldrich).
   (f) Blocking buffer: 5% (w/v) nonfat dry milk in TBS-T.
   (g) Primary antibody dilution buffer: TBS-T supplemented with 5% (w/v) nonfat dry milk.
   (h) CEACAM1 antibody was generated previously (5).
   (i) Secondary antibody: goat anti-mouse IgG or goat anti-rabbit IgG conjugated to horseradish peroxidase (Pierce, Rockford, IL, USA).
8. Enhanced chemiluminescent (ECL) was purchased from Thermo Scientific (Rockford, IL, USA) (SuperSignal West Pico Stable peroxidase Solution and Prod # 1856136) (SuperSignal West Pico Luminol/Enhancer Solution).
9. Gel staining solution, Gelcode was from Thermo Scientific.
10. Methanol, acetic acid were purchased from Fisher Scientific (Middletown, VA, USA).
11. The intensities of the respective bands were quantified by using the Image J program.
12. Coomassie Blue Plus Protein assay reagent is from Thermo Scientific.
13. PNGaseF is from New England BioLabs (Ipswich, MA, USA). The kit contains 10× denaturation solution, 10× G7, 10× NP40, and 15,000 U (500,000 U/ml) PNGaseF.

14. Ammonium bicarbonate (NH_4HCO_3; Sigma-Aldrich) is dissolved in water at 0.1 M. Store at room temperature. Dilute as needed.
15. RedyPrep 2-D Cleanup kit (BioRad). Use according to the manufacturer's instruction.
16. Triethylammonium bicarbonate (Sigma-Aldrich) is dissolved in water at 1.0 M. Store at room temperature. Dilute if needed.
17. Acetonitrile (ACN; Sigma-Aldrich) is stored at room temperature. Highly flammable and volatile. Use under chemical hood with proper protective gear.
18. Trypsin: 50 mM NH_4HCO_3, 5 mM $CaCl_2$, and 12.5 mg/ml trypsin (porcine, sequence grade, Promega, Madison, WI, USA).
19. Dithiothreitol (DTT; Sigma-Aldrich) is dissolved in water at 0.1 M. Store at –20°C. Dilute as necessary.
20. Iodoacetamide (Sigma-Aldrich) is stored at room temperature. Highly toxic. Use under chemical hood with proper protective gear.
21. Formic acid (Sigma-Aldrich) is stored at room temperature. Highly flammable and corrosive. Use under chemical hood with proper protective gear.
22. Trifluoroacetic acid (Sigma-Aldrich) is stored at room temperature. Highly corrosive and toxic. Use under chemical hood with proper protective gear.
23. Strong cation-exchange TipTop™ cartridge (polyLC Inc., Columbia, MD, USA).
24. Ammonium formate (Sigma-Aldrich) is dissolved in water at 0.5 M. Dilute as needed.
25. Acetic acid (Sigma-Aldrich). Highly corrosive.

3. Methods

3.1. Preparation of Crude Plasma Membranes from Rat Liver

1. Two rat livers (20 g) were minced and homogenized in 500 ml of ice-cold solution containing 1 mM $CaCl_2$, protease inhibitors (Roche, Indianapolis, IN, USA), and 50 mM HEPES, pH 7.4.
2. The homogenate was filtered through two layers of cheesecloth to remove the connective tissues and the filtered homogenate was spun in a centrifuge at 1,500×*g* for 10 min. The supernatant was removed and the loose pellets at the bottom collected.
3. The pellets were resuspended by homogenizing them in buffer (50 mM HEPES, pH 7.4) in a Dounce homogenizer. The

volume of the homogenate measured. This volume was used to calculate the amount of 69% sucrose (w/w) to be added to the homogenate. The solution was brought to 44% sucrose (w/w) by adding 1.1 volumes of 69% (w/w) sucrose to the homogenate. The sucrose concentration of the homogenate, i.e., 44%, was confirmed by a refractometer.

4. The homogenates in 44% sucrose were transferred to SW28 polyallomer tubes (Beckman Coulter, Brea, CA, USA) and overlaid carefully with 42.3% (w/w) sucrose until the material came to 4 mm below the rim of the tube.
5. The samples were spun at (113,000 × *g*) at 4°C for 2 h in a Beckman ultracentrifuge. The plasma membranes that appeared on the top of the tube were collected.

Note: This membrane isolation method was developed by Neville (4) and is commonly used to isolate crude plasma membranes from rat liver. This membrane isolation procedure gave us a high yield of plasma membranes: about 30–50 mg of crude plasma membranes was usually obtained from one rat liver.

3.2. Preparation of ConA Magnetic Beads

1. 1 ml of streptavidin magnetic beads in suspension (contains 500 μl settled beads) was placed into a 1.5 ml Eppendorf test tube. The magnetic beads were separated from the solution by holding the tube near a magnet (Dynal MPC-L, Invitrogen), and the solution was discarded by pipetting. The beads were then washed twice with 1 ml of water followed by 1 ml of TBS containing 150 mM NaCl and 50 mM Tris–HCl, pH 7.4.
2. The washed beads were resuspended in 210 μl of TBS, and 70 μl of biotinylated ConA (2.5 mg/ml) was added. This suspension was mixed on a rocker for 1 h to allow binding of biotinylated ConA to streptavidin.
3. The ConA magnetic beads were then washed sequentially with 1 ml of TBS, 1 ml of TBS with 1% (v/v) Triton X-100 (Sigma-Aldrich), and three times with 1 ml of TBS. They were then resuspended in 450 μl of TBS and stored at 4°C until use.

Note: Biotinylated ConA contains an average of one biotin per ConA molecule, and the streptavidin magnetic beads have a binding capacity of approximately 2,000 pmol of biotin/ml of the magnetic beads. Because the molecular mass of ConA is 104 kDa, each milliliter of magnetic beads can bind approximately 208 μg of biotinylated ConA. After binding of biotinylated ConA onto streptavidin magnetic beads (1–2 μm size), the efficiency of the immobilization was determined by comparing the amount of protein in the flow-through fraction with that in the original solution before binding. Of the 180 μg of biotinylated ConA in the binding solution, 42 ± 3 μg were found in the flow-through, yielding a coupling efficiency of 76%.

3.3. Purification of Neville Membrane Using ConA Magnetic Beads

1. A volume of 500 μl (about 5–10 mg protein) of Neville rat liver plasma membranes, prepared as described above, was added to the prepared ConA magnetic beads.
2. The membranes were allowed to bind with ConA magnetic beads on the rocker for 1–2 h.
3. After binding, the magnetic beads were separated from the solution by holding the tube near a magnet; the unbound membranes (i.e., the flow-through fraction) were removed.
4. The beads were washed five times with 1 ml of TBS.
5. To elute the membrane proteins from the ConA magnetic beads, the ConA magnetic beads were incubated with 500 μl of 0.25 M methyl α-d-mannoside with 0.5% (w/v) CHAPS in TBS for 10 min. To increase the recovery, the elution step was repeated twice.

3.4. Alkaline Wash of Membranes on ConA Magnetic Beads to Remove Peripheral Proteins

1. If removal of peripheral proteins are preferred before eluting the membrane proteins from the ConA magnetic beads, the plasma membranes on ConA magnetic beads (from Subheading 3.3, step 4) was further washed two times with 0.1 M Na_2CO_3 (pH ~ 11).
2. The pH of the membranes was brought back to pH 7.4 by washing the ConA magnetic beads with TBS two times.
3. The membrane proteins were eluted with TBS buffer containing 0.25 M α-methylmannoside and 0.5% CHAPS.

3.5. Protein Deglycosylation by PNGaseF

1. 56 μl of 10× denaturation solution was added to 200 μg (in 400 μl) of membrane proteins eluted from ConA magnetic beads. The mixture was heated to 100°C for 10 min to denature the proteins.
2. After the sample was cooled down to room temperature, 56 μl of 10× G7 buffer, 56 μl of 10× NP40, and 3 μl of PNGaseF were added.
3. The reaction mixture was incubated for 2 h at 37°C.

Note: *N*-Glycosidase F, also known as PNGase F, is an amidase that cleaves between the innermost GlcNAc and asparagine residues of high mannose, hybrid, and complex oligosaccharides from N-linked glycoproteins.

3.6. 5′-Nucleotidase Assay

1. Proteins (10–40 μg) were incubated in glycine buffer (75 mM, pH 9.0) containing 5 mM $MgSO_4$, with or without AMP (final concentration, 2.5 mM), in a total volume of 100 μl for 20 min at 37°C.
2. At the end of the incubation, 100 μl of phosphate reaction solution containing 10% ascorbic acid and 0.42% (w/v) ammonium molybdate in 1 N H_2SO_4 was added, and the

color-reaction products were measured at 820 nm in a microtiter plate reader (Synergy HT, Bio-Tek, Winooski, VT, USA).

3. The 5′-nucleotidase activity was determined by subtracting the values obtained in the absence of AMP from those obtained in the presence of AMP and comparing them with a phosphate standard curve.

Note: 5′-Nucleotidase activity was determined by measuring the inorganic phosphate released from AMP. The assay can be performed in a 96-well microtiter plate.

3.7. Protein Concentration Determination

1. In a 96-well microtiter plate, pipet 10 μl of each sample into the wells.
2. 10 μl of various concentrations of bovine serum albumin (0–1 mg/ml) were used to establish a protein standard curve.
3. Add 200 μl of Coomassie Blue Plus Protein assay reagent (Pierce) to each sample or protein standard. The color was read at 595 nm.

3.8. SDS-PAGE and Western Blot Analyses

Equal amounts of protein from the total cell lysates, flow-through, washes, and eluted fractions were run on 4–12% SDS-PAGE. The gel was either fixed with 50% methanol and 7% acetic acid and stained with Gelcode (Pierce, Rockford, IL) or transferred to nitrocellulose membranes for Western blotting. The membranes were blotted with antibody against CEACAM1 (5). The intensities of the respective bands were quantified by using the Image J program.

3.9. In-Solution Tryptic Digestion

1. Aliquots from plasma membrane samples containing about 50 μg protein were precipitated with the RedyPrep 2-D Cleanup kit (BioRad) according to the manufacturer's instruction (6).
2. Denatured protein pellet was redissolved with 20 μl of 0.5 M triethylammonium bicarbonate, pH 8.5.
3. Add 10 mM DTT and 55 mM iodoacetamide according to the manufacturer's protocol (Applied Biosystems, Foster City, CA, USA).
4. After removing the alkylating solution, add trypsin and digest for 24 h at 37°C.
5. Dry digested peptides in a vacuum centrifuge (Vacufuge, Eppendorf, Hamburg, Germany).
6. The material is then twice resuspended in water and vacuum dried, and subsequently twice redissolved in a solution containing 0.5% (v/v) formic acid and 20% (v/v) ACN, and thereafter vacuum dried.
7. After resuspending in the same solvent (see step 6) and controlling the pH value, the peptides were isolated using a strong

cation-exchange TipTop™ cartridge (polyLC Inc.), according to the manufacturer's instructions, and eluted with ammonium formate.

8. The ammonium formate eluates were vacuum dried and redissolved in formic acid:water:ACN:TFA mixture (0.1:95:5.0:0.01 v/v).

3.10. Protein Identification by LC-MS/MS

1. Tryptic peptides were separated on a 12 cm × 75 μm I.D. C_{18} reversed-phase column (Column Engineering, Ontaria, CA, USA) containing an integrated ~4 μm ESI tip.
2. Peptides were eluted using a linear gradient starting with 100% solvent A (0.1 M acetic acid in water) to 70% solvent B (acetonitrile) over 30 min (nano-HPLC system, Agilent Technologies, Paolo Alto, CA, USA).
3. Peak parking during the time when peptides were expected was accomplished by reducing the flow rate ten times (from 200 to ~20 nl/min).
4. Eluting peptides were introduced into an LTQ linear ion trap mass spectrometer (Thermo Electron Corporation, San Jose, CA, USA) with a 1.9 kV electrospray voltage.
5. Full MS scans in the m/z range of 400–1,800 were followed by data independent acquisition of spectra for the five most abundant ions, using 30-s dynamic exclusion time.
6. Protein identification was performed in two independent experiments.
7. Database searching was performed using the peak lists in the SEQEST program (7).
8. The precursor-ion tolerance was 2.0 Da and the fragment ion tolerance was 0.8 Da.
9. Enzymatic digestion was specified as trypsin, with up to two missed cleavages allowed.
10. The search contains sequences identified as rat in NCBI's nr database (November 2006) which was created using FASTA filtering tools found in BioWorks (Thermo Electron Corporation).
11. A list of reversed sequences was created from these entries and appended to them for database searching so that false positive rates could be estimated (8).

4. Results

The scheme of using ConA magnetic beads to purify plasma membranes is shown in Fig. 1. When these ConA magnetic beads were used to enrich plasma membranes from a crude

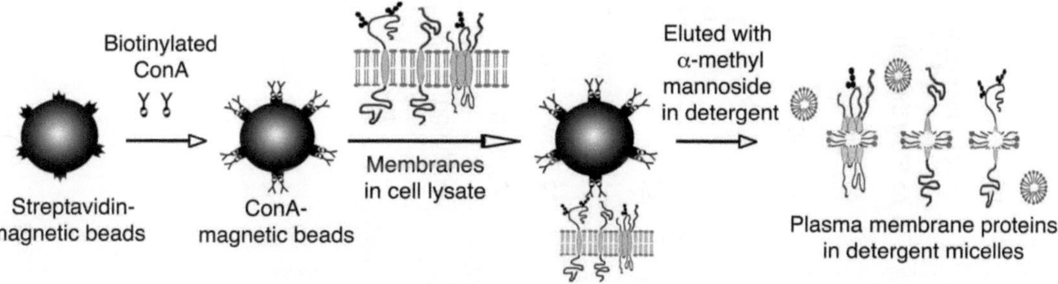

Fig. 1. Scheme of using concanavalin A (ConA) magnetic beads to purify plasma membranes. Biotinylated ConA was immobilized onto streptavidin magnetic beads. The ConA magnetic beads were used to bind plasma membranes through the binding of glycosylated proteins. The plasma membrane proteins were eluted from ConA magnetic beads by using detergents and α-methylmannoside, which is a competitive sugar for binding to ConA. (Reproduced from Lee et al. (9) with permission from Elsevier).

Table 1
Enrichment of plasma membrane from crude rat liver membranes. (Reproduced from Lee et al. (9) with permission from Elsevier)

Protein fraction	Total protein (mg)	Total 5′-nucleotidase activity (nmol AMP/min)	Specific activity (nmol AMP/mg protein/min)	Fold purification	Yield (%)
Crude membranes	3.7	80.0	21.6	1.0	100.0
Flow-through	0.5	5.0	10.0	0.5	6.3
Wash	0.8	4.0	5.0	0.2	5.0
Elution	0.7	56.0	80.0	3.7	70.0

membrane preparation prepared by the method described by Neville (4), this procedure resulted in fourfold enrichment of plasma membrane marker 5′-nucleotidase activity with 70% recovery of the activity in the crude membrane fraction of rat liver (Table 1). In agreement with the results of 5′-nucleotidase activity, immunoblotting with antibodies specific for a rat liver plasma membrane protein, CEACAM1, indicated that CEACAM1 was enriched about threefold relative to that of the crude membranes (Fig. 2). To enrich the transmembrane proteins, the membranes bound onto ConA magnetic beads were washed with an alkaline solution containing 0.1 M sodium carbonate (pH 11). The CEACAM1 protein, which is a transmembrane protein, was further enriched about four- to five folds after alkaline wash (Fig. 3). CEACAM1 is also heavily glycosylated.

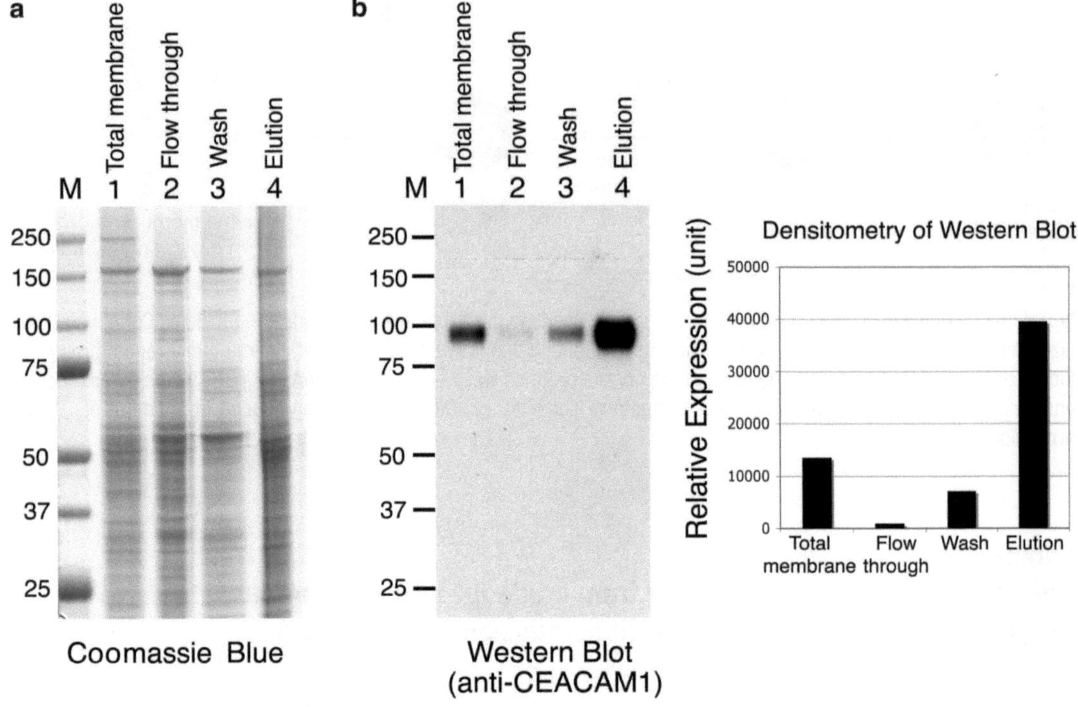

Fig. 2. Enrichment of a plasma membrane protein CEACAM1 after purifying membranes through ConA magnetic beads. Rat liver plasma membrane, prepared as described by the method developed by Neville (4), was purified with the concanavalin A (ConA) magnetic beads method as described. Equal amounts of protein (4 μg) from the original total membrane, flow-through, wash, and elution fractions were loaded onto SDS-PAGE gels. The gels were (**a**) stained with Coomassie blue or (**b**) transferred to nitrocellulose membranes and immunoblotted with anti-CEACAM1 antibody, and the relative amount of CEACAM1 was determined by densitometric quantification of the signals in the Western blots. *M* molecular weight markers. [Reproduced from Lee et al. (9) with permission from Elsevier].

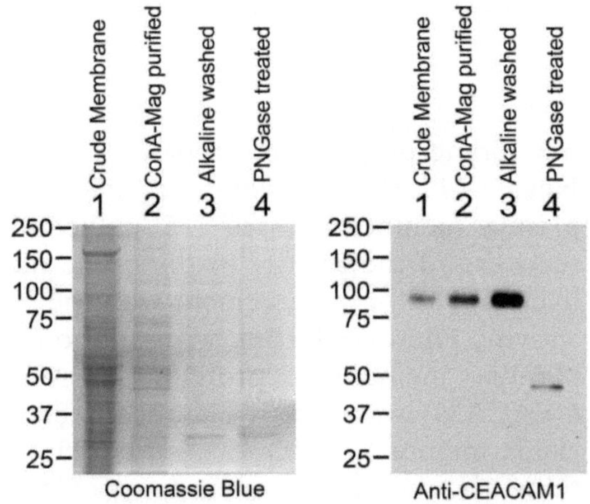

Fig. 3. Enrichment of transmembrane protein CEACAM1 by alkaline wash of ConA magnetic bead-purified membranes. Proteins from crude membranes, ConA magnetic beads, alkaline wash, and PNGaseF digestion were loaded onto SDS-PAGE gels. The gels were stained with Coomassie blue or transferred to nitrocellulose membranes and western blot with CEACAM1 antibody. Compared to the crude membrane, the transmembrane protein CEACAM1 was enriched about 6- and 26-fold after purified through ConA magnetic beads and alkaline wash, respectively. PNGase digestion reduced the apparent molecular weight of CEACAM1 to around 50 kDa.

Table 2
LC-ESI-MS/MS identification in crude liver plasma membranes

Membrane protein	Integral	Membr. assoc.	Identification
Cadherin	Yes	–	No
Dipeptidyl peptidase IV	Yes	–	Yes
CEACAM1	Yes	–	No
Cytochrome p450	Yes	–	Yes
Asialoglycoprotein receptor	Yes	–	No
LYRIC	Yes	–	No
Transferrin receptor	Yes	–	No
Integrin beta	Yes	–	No
CD 147	Yes	–	No
HSP 70	–	Yes	Yes
HSP 71	–	Yes	Yes
HSPA 78	–	Yes	Yes
HSP 90	–	Yes	Yes
Annexin A1	–	Yes	Yes
Annexin A3	–	Yes	Yes
Annexin A6	–	Yes	Yes
Annexin A7	–	Yes	Yes
Annexin A11	–	Yes	Yes

Deglycosylation of the total transmembrane proteins by PNGaseF yielded the expected size of deglycosylated CEACAM1 protein (~50 kDa). Deglycosylation of proteins should improve the protein identification by mass spectrometry. These results suggest that highly purified plasma membrane from tissue or cells can be obtained by using the ConA magnetic bead purification method and alkaline wash. As shown in Tables 2 and 3, such membrane preparation is suitable for the identification of integral and membrane-associated protein as potential biomarker candidates for diagnostic and prognostic purposes.

Table 3
LC-ESI-MS/MS identification of some liver plasma membrane proteins after isolation with magnetic beads with immobilized concanavalin A

Membrane protein	Integral	Membr. assoc.	Identification
Cadherin	Yes	–	Yes
Dipeptidyl peptidase IV	Yes	–	Yes
CEACAM1	Yes	–	Yes
Cytochrome p450	Yes	–	Yes
Asialoglycoprotein receptor	Yes	–	Yes
LYRIC	Yes	–	Yes
Transferrin receptor	Yes	–	Yes
Integrin beta	Yes	–	Yes
CD 147	Yes	–	Yes
HSP 71 kDa	–	Yes	Yes[a]
HSPA 78	–	Yes	Yes[a]
HSP 70 kDa	–	Yes	Yes
HSP 90 kDa	–	Yes	Yes
Annexin A1	–	Yes	No
Annexin A3	–	Yes	Yes[a]
Annexin A6	–	Yes	Yes[a]
Annexin A7	–	Yes	No
Annexin A11	–	Yes	Yes[a]

[a]Very low signal

References

1. Boone CW, Ford LE, Bond HE, Stuart DC, Lorenz D (1969) Isolation of plasma membrane fragments from HeLa cells. J Cell Biol 41:378–392
2. Lawson EL, Clifton JG, Huang F, Li X, Hixson DC, Josic D (2006) Use of magnetic beads with immobilized monoclonal antibodies for isolation of highly pure plasma membranes. Electrophoresis 27:2747–2758
3. Bhattacharyya L, Ceccarini C, Lorenzoni P, Brewer CF (1987) Concanavalin A interactions with asparagine-linked glycopeptides. Bivalency of high mannose and bisected hybrid type glycopeptides. J Biol Chem 262: 1288–1293
4. Neville DMJ (1968) Isolation of an organ specific protein antigen from cell surface membrane of rat liver. Biochim Biophys Acta 154:540–552
5. Cheung PH, Luo W, Qiu Y, Zhang X, Earley K, Millirons P et al (1993) Structure and function of C-CAM1. The first immunoglobulin domain is required for intercellular adhesion. J Biol Chem 268:24303–24310
6. Clifton J, Huang F, Rucevic M, Cao L, Hixson D, Josic Dj (2011) Protease inhibitors as possible pitfalls in proteomic analyses of complex biological samples. J Proteomics 74: 935–941
7. Eng JK, McCormack AL, Yates JR III (1994) An approach to correlate tandem mass spectral data

of peptides with amino acid sequence in a protein database. J Am Soc Mass Spectrom 5:976–989

8. Elias JE, Gygi SP (2007) Target-decoy search strategy for increased confidence in large-scale protein identification by mass spectrometry. Nat Methods 4:207–214

9. Lee YC, Block G, Chen H, Folch-Puy E, Foronjy R, Jalili R et al (2008) One-step isolation of plasma membrane proteins using magnetic beads with immobilized concanavalin A. Protein Expr Purif 62:223–229

Chapter 4

Analysis of the Cattle Liver Proteome by High-Sensitive Liquid Chromatography Coupled with Mass Spectrometry Method

Anna Maria Timperio, Gian Maria D'Amici, and Lello Zolla

Abstract

The present chapter describes methods for the separation and identification of proteins in liver metabolism through a comparison of the protein expression profiles of the two breeds taken into account as a model: Holstein Friesian and Chianina cattle. The liver has received special attention, containing as it does, enzymes involved in energy generation, carbohydrate, lipid, amino acid, and xenobiotic metabolism, as well as proteins involved in polypeptide synthesis, folding, and cell structure. The first step in the procedure is the preparation of purified protein fractions from liver tissues, followed by sample preparation for 2-DE analysis in order to identify proteins which could be differentially expressed in the livers of the two breeds and relate them to different liver functions. Data can be then statistically elaborated with cluster analysis, which stressed the up-/on-regulation trend of these proteins. Quantitative data can be used to perform a two-way hierarchical cluster analysis of the 39 differentially expressed protein spots, either up- or on-regulated in Chianina versus Holstein Friesian liver samples. Thus, spots from 2-DE maps can be carefully excised from the gel and subjected to in-gel trypsin digestion and analyzed by tandem mass spectrometry in their contents.

Key words: Liver, Two-dimensional gel electrophoresis, Mass spectrometry, Fused silica capillary

1. Introduction

The liver plays a major role in metabolism and has a number of functions in the body, including glycogen storage, decomposition of red blood cells, plasma protein synthesis, hormone production, and detoxification. The liver's highly specialized tissues regulate a wide variety of high-volume biochemical reactions, including the synthesis and breakdown of small and complex molecules, many of

Djuro Josic and Douglas C. Hixson (eds.), *Liver Proteomics: Methods and Protocols*, Methods in Molecular Biology, vol. 909, DOI 10.1007/978-1-61779-959-4_4, © Springer Science+Business Media New York 2012

which are necessary for normal vital functions (1). It plays critical roles in synthesizing molecules that are utilized elsewhere to support homeostasis, in converting molecules of one type to another, and in regulating energy balances. Thus, it is a general opinion that liver, being in contact with external food, plays a main role in addressing chemical substrate to other issue of body. To this regard, a recent study provides evidence on the global dynamic changes in the liver protein repertoire associated with distinct populations (2), and as some important metabolic differences have arisen upon modification of relatively few genes/proteins (or gene/protein networks), over thousand years of human breeding selection. In particular, changes in liver metabolism may be associated with different biological and productive aptitudes. Single enzymatic activities or genetic anomalies have been associated with specific bovine physiological and pathological conditions. Thus, there is interest in performing systematic studies to evaluate the entire protein repertoire in bovine tissues and fluids during certain specific phenomena and reveal unique profiles for different cattle breeds, despite the relatively short phylogenetic distance. In this way, it is possible to reveal important protein actors underlying those extreme traits that have been selected for in these breeds.

Hereby, it is of interest to present robust protocols about separation and identification of the differentially expressed proteins and gene products through an integrated proteomics and transcriptomics (microarray) approach (3).

Nowadays, significant advances in key technologies of proteome analysis are allowing a positive achievement for a series of investigations on the whole protein repertoire of different eukaryotes and prokaryotes (3–6). Indeed, proteomics offers unique possibilities for the investigation of changes in the levels and modifications of proteins (Fig. 1). However, a limited number of 2-DE studies for bovine tissues and biological fluids have been reported in the literature to date, mainly focused on cerebrospinal fluid, seminal plasma, mammary glands, pulmonary endothelial cells, chondrocytes, corneal lens, and milk (7–11). In parallel to genetic investigation, a series of structural studies have been performed on isolated bovine proteins, depending on their easy availability or their importance as main constituents of foods used in human diet (11).

1.1. Brief Introduction to MS-Based Proteomics

One critical goal of proteomics is qualitatively defining the proteome. There are several methods for submitting proteins to identification but the most powerful to date is mass spectrometry (MS). Proteins are sent into a pair of tandem MS devices (MS/MS). The proteins are sorted and groups of proteins of similar mass-to-charge ratio (m/z) are sent on to be ionized and characterized to determine the identity of each protein. This process is automated

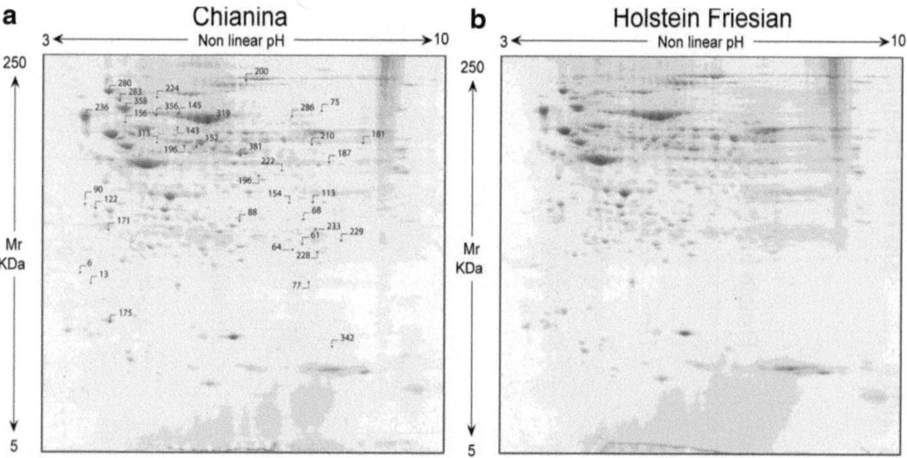

Fig. 1. 2-DE of liver extracts from Chianina and Holstein Friesian breeds. Each gel image has been elaborated with progenesis SamaSpots (Nonlinear Dynamics, NewCastle, UK) and represent an average of 18 gels (three technical replicates for six biological replicate samples), upon background subtraction. (Reproduced with permission from Elsevier Science).

so that thousands of proteins can be identified for each experiment. Mass spectrometry utilizing electrospray ionization (ESI) (12, 13) or matrix-assisted laser desorption/ionization (MALDI) (14, 15) as ionization techniques has been shown to be applicable to the mass spectrometric investigation of proteins and even protein complexes ranging in molecular mass from a few thousand to well over 200,000 Da. For peptides, tandem mass spectrometry (MS/MS) is currently the primary method. To extract meaningful sequence data by MS/MS, proteins are first digested into smaller peptides, usually by the enzyme trypsin. These peptides are then separated by reverse-phase HPLC online with the mass spectrometer and are transferred to the gas phase by ESI. Inside the mass spectrometer individual peptide ions are selected in a serial fashion for collision-induced dissociation (CID) so that a tandem mass spectrum contains fragments specific to a single parent peptide. Several methods have been employed to determine peptides corresponding to tandem mass spectra. For example, the SEQUEST (16), Mascot (17), and Sonar (18) applications compare a tandem mass spectrum against those expected for all possible peptides of identical mass within a given mass tolerance obtained from a sequence database. De novo sequencing methods derive a peptide sequence from tandem mass spectra without using a database, and are especially valuable for samples from organisms with polymorphic mutations or unsequenced genomes (19). Once peptides have been satisfactorily assigned to individual tandem mass spectra, they are used to infer the original protein contents of the sample (Table 1).

Table 1
Proteins differentially expressed from Chianina and Holstein Friesian breeds. (Reproduced with permission from Elsevier Science)

No.	SSP	Mw, kDa	p*I*	No. of peptides identified	Mascot score	NCBI accession number	Protein ID (*Bos taurus*)	Protein function	Gene
Chianina up									
1.	210	61,992	8.03	22	1,040	gi|74354891	GLUD1 protein	Dehydrogenase	GLUD1
2.	113	33,115	6.78	10	407	gi|230290	Chain rhodanese	Transferase	1RHD
3.	286	61,774	7.55	6	313	gi|89574195	Succinate dehydrogenase subunit A	Dehydrogenase	SDHA
4.	75	79,870	6.75	13	562	gi|2501351	Serotransferrin precursor (transferrin) (siderophylin) (beta-1-metal-binding globulin)	Metal-binding globulin	TF
5.	77	57,664	8.45	4	165	gi|229299	Catalase	Oxidoreductase	CAT
6.	313	53,959	5.50	3	94	XP_001787596.1	Similar to cytosolic beta-glucosidase, partial	Hydrolase	GBA
7.	233	34,410	7.71	10	429	gi|78369248	Hypothetical protein LOC509274	Hydrolase	HAGH
8.	222	43,302	6.44	22	1,152	gi|76639590	Similar to betaureidopropionase	Hydrolase	UEP1
9.	224	71,423	5.49	6	109	gi|123644	Heat shock cognate 71 kDa protein (heat shock 70 kDa protein 8)	Chaperonine	HSPA8
10.	228	40,045	8.20	3	98	gi|29135271	Ornithine carbamoyltransferase	Transferase	OTC

11.	779	44,809	8.82	3	63	gi|77735757	Acyl-coenzyme A dehydrogenase, C-2–C-3 short chain	De-hydrogenase	ACADS
12.	342	43,302	6.44	9	206	gi|76639590	Similar to beta-ureidopropionase	Hydrolase	BUP1
Chinnina on									
13.	61	26,901	6.45	12	677	gi|62888856	Triosephosphate isomerase	Isomerase	TP1
14.	64	25,810	6.45	12	574	gi|76613071	Similar to class mu glutathione S-transferase isoform 1	Transferase	GSTM1
15.	68	27,294	8.45	3	169	gi|27805907	Hydroxyacyl-coenzyme A dehydrogenase, type II hydroxyacyl-coenzyme A	Oxydoreductase	HADH
16.	187	46,673	7.14	13	629	gi|27806925	Argininosuccinate synthetase	Ligase	ASS
17.	181	46,943	8.31	6	258	gi|11549790	Acyl-coenzyme A dehydrogenase, C-4–C-l 2 straight chain	Dehydrogenase	ACADM
18.	196	46,491	6.49	3	122	gi|119913765	Similar to fumarylacetoacetate (FAA) hydrolase	Hydrolase	FAH
19.	154	55,426	6.24	7	280	gi|27806321	aldehyde dehydrogenase 1 family, member A1	Oxydoreductase	ALDH1
Holstein Friesian up									
1.	358	72,470	5.07	23	977	gi|115495027	Heat shock 70 kDa protein 5	Chaperonine	HSPA5
2.	143	33,287	5.39	8	391	gi|77736137	Chaperonin containing TCP1, subunit 5 (Epsilon)	Chaperonine	CCT5
3.	145	62,861	5.62	5	172	gi|119910189	Similar to esterase 31	Chaperonine	Es31

(continued)

Table 1 (continued)

No.	SSP	Mw, kDa	p*I*	No. of peptides identified	Mascot score	NCBI accession number	Protein ID (*Bos taurus*)	Protein function	Gene
4.	236	46,524	4.31	14	532	gi|631545	Calreticulin, brain isoform 1	Chaperonine	CALR
5.	280	92,654	4.76	38	1,770	gi|27807263	Tumor rejection antigen (gp96) 1	Chaperonine	TRA1
6.	175	11,044	5.15	5	209	gi|353817	Cytochrome b5	Hemoprotein	CYB5A
7.	200	165,833	6.28	33	1,668	gi|76659916	Similar to carbamoyl phosphate synthetase 1 isoform 1	Transferase	CAD
8.	156	42,084	5.22	9	370	gi|116242946	Actin, cytoplasmic 1 (beta-actin)	Structural protein	ACTB
9.	319	76,197	5.60	8	342	gi|2914360	Chain, bovine annexin Vi (calcium bound)	Lipid binding	1AVC
10.	122	24,902	4.54	2	75	gi|58760396	Eukaryotic translation elongation factor 1 beta 2-like	Transferase	EF1B
11.	356	71,424	5.37	26	1,051	gi|146231704	Heat shock 70 kDa protein 8	Chaperonine	HSPA8
12.	171	27,946	4.80	18	764	gi|2852383	14-3-3 protein beta	Oxydoreductase	YWHAB
13.	283	85,077	4.93	28	668	gi|60592792	Heat shock 90 kDa protein 1, alpha	Chaperonine	HSPCA
Holstein Friesian on									
14.	152	53,078	5.60	3	94	gi|78042564	Hypothetical protein LOC512626	Peptidase	CNDP2

<table>
<tr><td>15.</td><td>90</td><td>48,120</td><td>5.88</td><td>5</td><td>223</td><td>gi|77735583</td><td>S-adenosylhomocysteine hydrolase</td><td>Hydrolase</td><td>AHCY</td></tr>
<tr><td>16.</td><td>196</td><td>47,116</td><td>5.49</td><td>6</td><td>197</td><td>gi|6980816</td><td>Chain C, the crystal structure of modified bovine fibrinogen</td><td>Structural protein</td><td>FGA</td></tr>
<tr><td>17.</td><td>6</td><td>18,813</td><td>4.33</td><td>2</td><td>87</td><td>gi|51493666</td><td>Cytosolic prostaglandin E synthase</td><td>Chaperonine</td><td>cPGES</td></tr>
<tr><td>18.</td><td>381</td><td>48,120</td><td>5.88</td><td>17</td><td>843</td><td>gi|77735583</td><td>S-adenosyhomocysteine hydrolase</td><td>Hydrolase</td><td>AHCY</td></tr>
<tr><td>19.</td><td>88</td><td>35,062</td><td>6.00</td><td>13</td><td>583</td><td>gi|124249242</td><td>Hypothetical protein LOC783020</td><td>Transferase</td><td>SULTIC4</td></tr>
<tr><td>20.</td><td>13</td><td>23,356</td><td>4.51</td><td>4</td><td>257</td><td>gi|62460528</td><td>Hypothetical protein LOC513312</td><td>Structural protein</td><td>NACA</td></tr>
</table>

2. Materials (*See Notes 1–3*)

2.1. Sample Collection

Liver tissue 20 mg per sample.

2.2. Protein Precipitation

Tri-*n*-butyl phosphate/acetone/methanol (1:12:1) (Sigma Life Science, Italy).

2.3. Protein Concentration

Bradford assay (20) using BSA as a standard curve (Sigma Life Science Italy).

2.4. Protein Solubilization

7 M urea, 2 M thiourea, 4% (w/v) CHAPS, 0.8% (w/v) Carrier ampholyte pH 3–10, 40 mM Tris–HCl, 5 mM TBP, 0.1 mM EDTA pH 8.5, Protease inhibitor cocktail, 2 mM PMSF (all products are supplied by Pharmacia Biotech Sweden.

2.5. IPG Rehydratation

IEF strip gels pH 3–10 NL (Pharmacia Biotech Sweden), 8 M urea, 3. 2% CHAPS, 2% IPG buffer pH 3–10 NL, 0.3% DTT, and trace amounts of bromophenol blue.

2.6. Equilibration Stock Solution

10% Tris–HCl solution pH 6.8, 36% urea, 30% glycerol, 1% SDS, 25–100 mg DTT/10 ml equilibration, 0.25 g iodoacetamide plus a few grains of bromophenol blue/10 ml solution for the second step (all products are supplied by Sigma Life Science Italy.

2.7. Staining Procedure

Coomassie Brilliant Blue G-250 stain (Sigma Life Science Italy).

2.8. Image Analysis

LabScan software 3.01, Progenesis SameSpots software v.2.0.2733.19819 software package.

2.9. Cluster Analysis

PermutMatrix graphical interface, Pearson's distance, and Ward's algorithm.

2.10. In-Gel Digestion

Termomixer (24 spots), bidistiller water bidistilled, original Eppendorf, 100 mM buffer NH_4HCO_3, acetronitrile 100%, trypsin, iodoacetamide, ammonium bicarbonate, dithioerythritol, and formic acid (FA).

2.11. Setup for RP-HPLC and Nano-HPLC–ESI-MS

For RP-HPLC setup use trifluoroacetic acid (TFA) and formic acid, acetonitrile specifically for HPLC-MS, water bidistilled, analytical reverse column (microbore C4, 8, 18), capillary column home made. For mass spectrometry setup use tune-mix for tuning of the instrument and that is performed by direct infusion of a solution of 10 ng/μl lysozyme in water–acetonitrile (80:20 v/v) containing 0.05% trifluoroacetic trypsin digestion of bovine serum albumin (BSA), (all reagents for HPLC-MS were purchased from Sigma Aldrich).

2.12. Homemade Capillary Column

Fused silica capillary (FSC) column (ID 75 μm, OD 375 μm, and ID 50 μm) manufactured by VitroCom, stationary phase: ReproSil-Pur C18-AQ 3 μm (Dr. Maisch GmbH), methanol for HPLC-MS, isopropanol, water bidistiller, and preparation of kasil frits (kasil = potassium silicate low backpressure robust frits).

2.13. Packing Microcapillary Columns

Isopropanol-methanol (2:1) GmbH, stationary phase: ReproSil-Pur C18-AQ 3 μm (Dr. Maisch), pneumatic packing vessel (homemade), sonicator, stereomicroscope, stir plate, and vortexer.

2.14. Mass Spectrometry

Mass spectrometer with any type of analyzer (linear quadrupole, quadrupole or linear ion trap, linear or reflectron time-of-flight (TOF) mass analyzer, and ion-cyclotron resonance mass analyzer) suitable for interfacing via ESI, mass range at least m/z 500–1,500, preferably m/z 500–3,000.

2.15. Data Analysis Search

Binary data file generated from an LC-MS/MS experiment computer running Windows XP or Linux.

3. Methods

3.1. Sample Collection

1. All animals used for the sample collection are treated according to International Guiding Principles for Biomedical Research Involving Animals.
2. In a commercial dairy farm 12 animals (6 per breed), which are of the same breed, age, and body score condition, must be selected.
3. Cows are fed the same diet.

3.2. Protein Precipitation, Protein Solubilization, and Protein Concentration

1. Perform sample preparation and solubilize by the SWISS-2D PAGE sample preparation procedure.
2. Frozen samples of liver tissue from the six Chianina and six Holstein Friesian (approximately 20 mg per sample).
3. Crush proteins in a mortar containing liquid nitrogen, and remove lipids from a desired volume of each sample with a cold mix of tri-*n*-butyl phosphate/acetone/methanol (1:12:1).
4. Centrifuge at 2,800 × *g*, for 20 min at 4 °C after incubation pellet at 4 °C for 90 min.
5. Air-dry pellet and resuspend samples in the focusing solution.
6. Incubate sample in this solution for 3 h at room temperature, under strong agitation.
7. To prevent over-alkylation, acrylamide was destroyed by adding an equimolar amount of DTE.

8. The protein concentration of each group was determined according to Bradford (20) using BSA as a standard curve. A total of 250 μl of the resulting protein solution was then used to rehydrate 13 cm long IPG 3–10 NL (Amersham Biosciences) for 8 h. IEF (see Notes 4 and 5).

3.3. Isoelectrofocusing (First Dimension)

1. Prepare the strip holder(s). Select the strip holder(s) corresponding to the IPG strip length chosen for the experiment (7, 11, 13, 18, or 24 cm).
2. Wash each holder with strip holder cleaning solution supplied to remove residual protein. Rinse thoroughly with double distilled water. Use a cotton swab or a lint-free tissue to dry the holder or allow it to air-dry. The holder must be completely dry before use.
3. Prepare the rehydration solution. The Protocol Guide contains a commonly used recipe. Required volumes per strip are: 125 μL for 7 cm IPG strip; 200 μL for 11 cm IPG strip; 300 μL for 17 cm IPG strip, 315 μL for 18 cm IPG strip and 450 μL for 24 cm IPG strip.
4. Pipet the appropriate volume of rehydration solution into each holder. Deliver the solution slowly at a central point in the strip holder channel away from the sample application wells. Remove any larger bubbles. Important: to ensure complete sample uptake, do not apply excess rehydration solution.
5. Place the IPG strip. Remove the protective cover from the IPG strip. Position it with the gel side down and the pointed (anodic) end of the strip directed toward the pointed end of the strip holder. Position the strip onto the solution. To help coat the entire strip, gently lift and lower the strip and slide it back and forth along the surface of the solution, tilting the strip holder slightly as needed to assure complete and even wetting. Finally, lower the cathodic (square) end of the strip into the channel, making sure that the IPG gel contacts the strip holder electrodes at each end. (The gel can be visually identified once the rehydration solution begins to dye the gel.) Be careful not to trap bubbles under the strip. Important: handle the holders with care, as they are brittle and fragile.
6. Apply Immobiline DryStrip Cover Fluid to minimize evaporation and urea crystallization. Pipet the fluid drop wise into one end of the strip holder until one half of the strip is covered. Then pipet the fluid drop wise into the other end of the strip holder, adding fluid until the entire IPG strip is covered.
7. Place the cover on the holder. Pressure blocks on the underside of the cover assure that the strip maintains good contact with the electrodes as the gel rehydrates.
8. Rehydration can proceed on the bench top or on the IPGphor unit platform. Ensure that the holder is on a level surface.

A minimum of 10 h is required for rehydration; overnight is recommended. Alternatively, the rehydration period can be programmed as the first step of an IPGphor protocol. This is especially convenient if temperature control during rehydration is a concern, or if a low voltage is applied during rehydration.

3.4. SDS-PAGE (Second Dimension)

1. Equilibrate IPG strips for 30 min in a solution containing 6 M urea, 2% (w/v) SDS, 20% (v/v) glycerol, and 375 mM Tris–HCl (pH 8.8), with gentle agitation.
2. Lay the IPG strips on a 5–16% T gradient SDS-PAGE gel with 0.5% (w/v) agarose.
3. Put in the cathode buffer (192 mM glycine, 0.1% w/v SDS, and Tris–HCl to pH 8.3).
4. Put in the anode buffer 375 mM Tris–HCl, pH 8.8.

3.5. Coomassie Brilliant Blue Staining (for Mini-Gels)

1. Fix gel in 100 ml 46% v/v methanol and 7% v/v acetic acid for 1 h.
2. Stain gel in 100 ml 46% v/v methanol, 7% v/v acetic acid, and 0.1% v/v Coomassie Brilliant Blue R-250 (filter this before use) for 1 h.
3. Destain gel in 100 ml 5% v/v methanol and 7.5% v/v acetic acid for 24 h. Replace if needed.
4. Store the gel in 1–2% v/v acetic acid in clean sealed sample tubes at 4 °C.

3.6. In-Gel Digestion

3.6.1. Gel Washing Step

Excise carefully from the gel spots from 2-DE maps, cut the gel band or spot, slice it and collect pieces on a lean PCR or a 500 μl tube, spin down and remove excess of water using a gel loader tip, add *150 μl of 50% v/v ethanol* and 50% v/v *50 mM AB* (pH 8.4), make sure the gel is well soaked and incubate 20 min at room temperature, spin down, remove liquid and repeat previous step once, spin down, remove liquid, and dry the gel in *Speed Vac* (around 20 min).

3.6.2. Reduction and Alkylation (If Necessary)

Cover the gel piece (25–30 μl) with freshly prepared 10 mM DTE and 50 mM AB (pH 8.4), vortex briefly, spin down and incubate 1 h at 37°C protecting from light, remove DTE solution, and add 25–30 μl of freshly prepared 55 M IAA and 50 mM AB (pH 8.4). Vortex briefly, spin down, and incubate 45 min at 37°C protecting from light, remove IAA solution, and wash 2–3 times with 50 mM AB (pH 8.4). Dehydrate the gel by adding 100 μl of 50% ethanol and 50% 50 mM AB (pH 8.4), and incubate at room temperature for 20 min, spin down, remove the liquid, and Speed Vac to complete gel dryness (around 30 min). Store sample tubes on ice.

3.6.3. Tryptic Digestion

Cover the gel piece (25–30 μl) with a trypsin solution made of 12.5 ng/μl in 50 mM A (pH 8.4), 10 mM calcium chloride, vortex shortly, spin down and incubate 10 min on ice, check liquid volume and add enough 50 mM AB (pH 8.4) solution to cover the gel piece, vortex shortly, spin down and incubate again for 10 min on ice, check liquid volume again, add 50 mM AB (pH 8.4) if necessary, seal PCR tube and incubate at 37 °C over night, dry the supernatant by vacuum centrifugation which contains tryptic peptides, and redissolve the peptide mixtures in 10 μl of 5% v/v FA. Prior to mass spectrometric analysis (see Notes 6–10).

3.7. Preparation Capillary Frits

1. Mix 88 μl Kasil 1 and 16 μl formamide.
2. Vortex thoroughly and suck up sufficient mixture with a piece of fused silica.
3. Place in an oven 60–110 °C for an hour or two (or overnight on the table) and flush out with methanol or isopropanol.
4. Cut of the excess frit if it is too long; 2–3 mm of frit is enough.
5. Frits at both ends can be made by packing the column to the desired length and cut off the empty capillary column at the end or at the start.
6. Dip the open end of the packed column shortly in Kasil mix and leave for polymerization.

3.8. Packing Fused Silica Microcapillaries Columns

1. Place a microstir bar into a 2 ml glass vial.
2. Add methanol and isopropanol (2:1) and 5 mg of RP packing material to the glass vial (see Note 11).
3. Vortex to resuspend the packing material and sonicate for 5 min to avoid aggregation of the particles.
4. Transfer the slurry to a pneumatic loading vessel and place the loading vessel on a stir plate.
5. Turn on the stir plate to keep the packing material suspended. Secure the lid to the loading vessel by tightening the bolts that attach the lid to the base (see Note 12).
6. Insert FSC column into the loading bomb for packing.
7. Pull the column up so that the capillary rests just off the bottom of the glass vial and stir bar.
8. Apply pressure to the loading vessel by first setting the regulator on the high-pressure helium gas cylinder to 100 bar, and then open the three-way valve. If the column is long enough, place the fritted end of the column under a stereomicroscope to observe the packing.
9. When the column has been packed, slowly turn off the pressure to the vessel at the three-way valve. Slowly releasing the pressure prevents the packing material from unpacking.

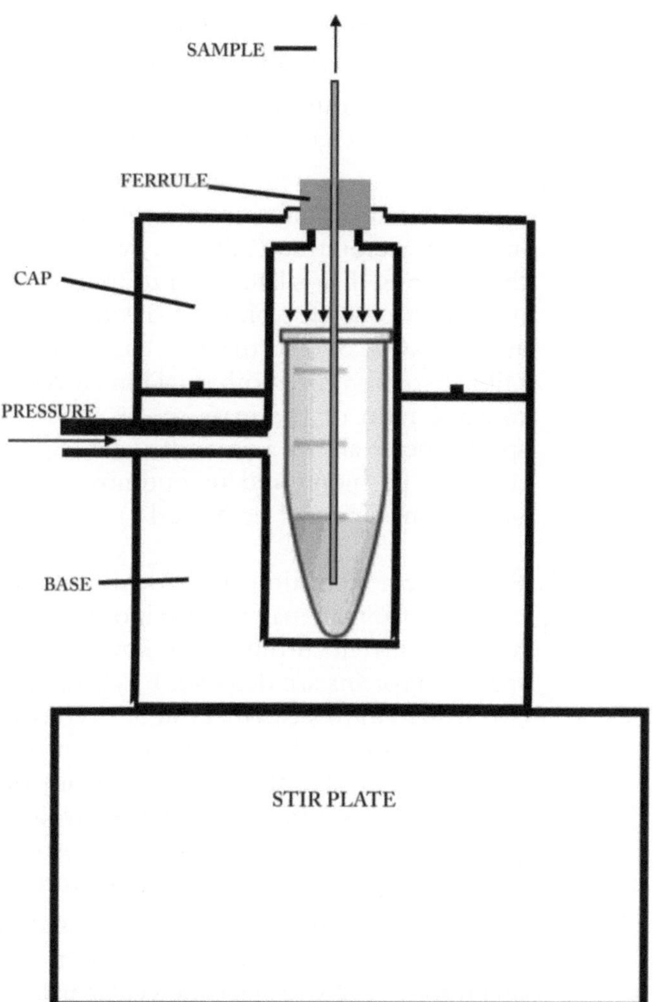

Scheme 1. Pneumatic loading device, "loading bomb," for packing microcapillary ESC HPLC columns.

10. Replace the vial containing the slurry with a 1.5-ml microcentrifuge tube filled with 5% acetonitrile, 0.1% v/v formic acid.
11. Wash the column for 10 min using the loading vessel.
12. Store the column in 5% v/v acetonitrile and 0.1% v/v formic acid until ready to use. Columns can be stored indefinitely at room temperature in this solution. Before using for any biological samples, run a blank HPLC gradient across the column to condition it. Conditioning the column is important to firmly pack the resin (see Scheme 1) (see Note 13).

3.9. Setup for RP-HPLC–ESI-MS Analysis

1. Run two blank gradients of 5–95% acetonitrile to remove strongly adsorbed compounds from the column.
2. Equilibrate the column with the starting eluents employed for gradient elution, typically water/0.05% TFA acetonitrile

0.05% TFA 90:10 for the butyl-silica column and water 0.1% v/v TFA for microcapillary column.

3. Calibrate and tune the mass spectrometer according to the manufacturer's recommendations (see Note 14).
4. In the case of proteins fine tuning of the instrument is performed by direct infusion of a solution of 10 ng/μl lysozyme in water–acetonitrile (80:20, v/v) containing 0.05% TFA at the flow rate that is later used to transfer the proteins into the mass spectrometer, typically 2–10 μl/min. The 7+ charge state at *m/z* 2,044.590 or the 10+ charge state at *m/z* 1,431.5 is utilized to maximize both total ion current stability and signal-to-noise ratio of the mass spectrometric signal. If the mass spectrometer allows a second signal, e.g., at *m/z* 1,590.461, this may be monitored to optimize ion transmission over a broader mass range (see Note 15).

3.10. Protein Detection and Identification

1. For protein detection and identification, mass spectra are extracted from all major and minor peaks in the reconstructed total ion current chromatogram by averaging about 5–10 mass spectra. Proteins are detected by characteristic series of multiply charged ion species that can be deconvoluted into the intact molecular mass of the uncharged protein by well-known deconvolution algorithms (12). In many cases, more than one protein mass is obtained after deconvolution, indicating that either different proteins or different isoforms of one protein are coeluting from the column (see Note 16).
2. Protein identification is performed based on comparison with elution patterns and intact molecular masses published in the literature (21–25). Intact molecular masses can be calculated by extracting the protein sequence from the publicly available protein databases and using the "peptide mass" tool from the ExPASY Proteomics Server of the Swiss Institute of Bioinformatics http://www.expasy.org. Measured masses may significantly deviate from the theoretical masses, either because of unknown exact cleavage position of the signal peptide, sequence variants (26), or unknown posttranslational modifications (27). Nevertheless, elution position and approximate mass range in many cases suffice to identify the protein correctly.
3. In the case of peptides a complex mixture of peptides is produced by Trypsin digestion of BSA that is reduced and alkylated with Iodoacetamide (CAM modified). This peptide mixture can be used to test a matrix-assisted laser desorption/ionization time-of-flight (MALDI-TOF) or ESI mass spectrometer (TOF, Q-TOF or Ion Trap). In the case of Nano LC-ESI MS/MS, 100 fmol of the peptide mixture is subjected to nano-reverse-phase liquid chromatography on a NanoLC

C18 Ion Trap and developed with water to acetonitrile gradient with both solvents containing 0.1% formic acid. The electrospray MS was operated in a data-dependent mode in which each full MS scan (m/z 300–1,800) is followed by three MS/MS scans, in which the three most abundant peptide molecular ions are dynamically selected for CID using a normalized collision energy of 35%. The temperature of the heated capillary and electrospray voltage is 160 °C and 2.0 kV, respectively.

4. In MS/MS data, 250 spectra are selected for analysis and a score of 1,200 or greater must contain. In the case of MALDI-TOF MS, 0.1–1 μl of each of the peptide mixture (0.5–5 pmol) is mixed with 1 μl of α-cyano-4-hydroxycinnamic acid matrix solution, air-dried, and subjected to MALDI-TOF MS analysis. The spectra obtained must contain more than 15 resolved peaks which match the theoretical peaks (see Note 17).

3.11. Data Analysis Format

1. Converting native mass spectrometry data files to an XLM.

In tandem mass spectrometric sequencing of a peptide, information about the peptide sequence is contained in the product ion or MS/MS spectrum. Low-energy fragmentation of ionized tryptic peptides occurs primarily at the amide bonds along the peptide backbone, generating a series of fragmentation or product ions. The sequence of the peptide is determined by comparing the differences among the masses of the product ions with the known masses of the individual amino acid residues (see Notes 18 and 19).

3.12. Results Evaluation

1. Select criteria for initially accepting database search results as correct.
2. Order the list of identified proteins by the number of peptides and tandem mass spectra identifying a protein in the sample. The higher the number of peptides significantly matching a protein, the more likely the protein is identified accurately.
3. Recognize proteins that are identified by unique peptides. In higher eukaryotic organisms, extensive gene duplication and alternative RNA splicing has resulted in a large number of proteins with similar sequences. Protein databases may contain multiple sequences for a protein that represent alternative forms that differ by only a single amino acid. Because peptides are frequently shared by multiple proteins in a database, the assembly of identified peptides into a list of proteins can drastically overstate the number of proteins in samples. It is important to recognize which proteins are identified by unique peptides and which proteins are indistinguishable because only shared peptides are identified. For proteins that share only peptide sequences, the identified proteins should be reported as a protein group.

4. Identify proteins or protein groups that are indistinguishable. Protein databases often contain large numbers of proteins with similar sequences. For example, human protein databases may contain many serum albumin and immunoglobin proteins that differ only by a few amino acid residues. These proteins will be indistinguishable because there are no unique peptides allowing the investigator to determine which protein produced the peptides. In these cases, reporting the smallest numbers of proteins necessary to explain the observed peptides is encouraged (28) (see Notes 20–22).
5. For proteins or modified peptides identified by a single spectrum, manual evaluation of the spectra and the matching sequence is required.

4. Notes

1. High-quality chemicals and all solutions should be prepared in bidistilled water having a resistivity of 18.2 MΩ cm and total organic content of less than 5 ppb.
2. Samples should be lysed before submission. Precipitated samples must be solubilized in lysis solution and clarified by centrifugation if necessary. The solution or supernatant will be analyzed. Contaminants (Table 2) should be removed from the sample prior to submission.
3. In case protein extraction contains components not compatible with 2D electrophoresis, remove these components prior to submission, e.g., by precipitating and redissolving in lysis solution.
4. Precautions before running the gel—optimum thickness for gels is 1 mm. It is desirable to use the narrowest gel lane width possible to achieve protein-to-gel volume ratio as high as possible. Commercial gels are available such as from Invitrogen and Bio-Rad. Homemade gels: polymerize acrylamide gels overnight to minimize cross-linking of unpolymerized acrylamide to proteins during electrophoresis.
5. Precautions after running the gel—use clean dishes and freshly made staining solutions for staining the gels to prevent contaminations from keratin, dust, saliva, and any other proteins you are using in your lab. Please be sure to follow the Coomassie Brilliant Blue Staining Protocol because many staining protocols, although perhaps slightly more sensitive or convenient than this one, may well render your proteins impervious to MS analysis. In general, chemical cross-linkers and strong oxidiz-

Table 2
List of contaminants

Contaminants		
Not acceptable		**Acceptable**
Reducing agents (e.g., DTT, TBP)	Buffers other than Tris–HCl	Urea
Salts (e.g., NaCl)	Ionic detergents (e.g., SDS)	Water
Nucleic acids	Polysaccharides	Tris base
Lipids	Phenolic compounds	Nonionic or zwitterionic
Insoluble material		Detergents (e.g., CHAPS)

ing or reducing agents should be avoided. After staining, gels may be stored in 1–2% acetic acid.

6. Any air bubbles that may be trapped beneath the strip should be removed.
7. When you acquire an image of your gel, please take special care not to allow your gel to contact any contaminated surfaces during the process.
8. Major precautions should always be taken during gel handling processes to avoid keratins contaminations which can severely interfere with the MS detection of your analyzed sample.
9. Wash all experiment devices (gel cutting surface, razor blades, solution bottles, instruments such as Speed Vac and centrifuge chambers, and internal and external sample tube surfaces) with an acidified mixture of aqueous and organic solvents.
10. Wear powder-free gloves and avoid touching organic solvent with them to minimize polymer contamination. Tie off your hair and work under laminar flow hood. Cut the gel under a pure deionized water film.
11. The thickness of the slurry will determine how the column will pack.
12. Always wear safety glasses when packing FSC column. The column is being packed under very high pressure. Improperly seated columns can be ejected from the loading vessel at high velocity.
13. Conditioning the column is important to firmly pack the resin. Multiple blank HPLC gradients may be required to obtain a reproducible baseline.
14. The tuning procedure has been shown to be essential to be able to detect the proteins/peptides with good signal-to-noise

ratios (26). Optimal tuning conditions generally differ significantly between instruments. Sometimes, tuning parameters need to be set in a significantly different manner from those suggested by automatic tuning routines in order to achieve high sensitivity and spectrum quality, especially for high-molecular mass proteins.

15. Monitoring the retention time (RT), signal intensity, and resolution of the albumin peptide (and background noise) is used to check the performance of the chromatography and mass spectrometry system prior to running unknown samples.
16. Deconvolution algorithms are usually implemented in the data analysis software components of the mass spectrometer.
17. The acquired data file can be searched against a protein database containing the albumin sequence to confirm the MS/MS. If this value or protein identification is not observed, a number of steps can be taken, including cleaning the ESI source checking the HPLC flow rates, and recalibrating and retuning the mass spectrometry.
18. Most mass spectrometer vendors encode the data files in a proprietary binary format. The *m/z* and intensity values from the precursor (MS) and product (MS/MS) spectra need to be extracted from these native files into a text format that can be read by the database search programs. XML is a simple, flexible text format derived from the standardized general mark up language (SGML), originally designed for electronic publishing. Bruker XMASS and flexAnalysis, for example, save peak lists in a simple XML format. For each peak, Mascot takes the *m/z* value from the ‹mass› element and the intensity from the ‹absi› element.
19. Most mass spectrometer vendors offer a commercially licensed database search program for processing the data with their mass spectrometers. Three of the most widely used search algorithms are Sequest, Mascot, and X!Tandem (16, 17, 28). All three programs can be used to analyze native data from any mass spectrometer once the data file has been converted to an XML or an ASCII format.
20. The database search is unbiased. All protein and peptide sequences in the database are typically compared with each MS/MS spectrum. The search engine did not make any assumptions about what proteins might be in the sample. This is probably the most important and powerful aspect of the approach. Unexpected proteins are identified that investigators would not have predicted to be in the sample.
21. Only peptides and proteins in the protein database will be listed in the output. If a protein or peptide sequence is not in the database, it will not be identified. The search algorithms do

not perform de novo sequencing of the MS/MS spectra and derive peptide sequences. The programs only match peptide sequences in a database to the MS/MS spectra using the parameters in the setup file.

22. The LC-MS/MS and MALDI-MS approaches are biased toward fragmenting peptides from abundant proteins. Since the selection process for fragmentation of precursor ions is typically based on ion intensity, the more-intense ions will be selected for fragmentation and the less-intense ions will be ignored and possibly escape fragmentation. While peptides ionize at different efficiencies, on average, the more abundant a peptide or protein is, the more likely it will generate peptides with strong precursor ion signals. Low-abundance peptides and proteins could escape detection.

References

1. Anthea M, Hopkins J, McLaughlin CW, Johnson S et al (1993) Human biology and health. Prentice Hall, Englewood Cliffs, NJ, USA
2. Sherwood RA, Parsons TS (1977) The vertebrate body. Holt-Saunders International, Philadelphia, PA
3. Pandey A, Mann M (2000) Proteomics to study genes and genomes. Nature 405:837–846
4. Li J, Sarah M, Assmann SM (2000) Mass spectrometry. An essential tool in proteome analysis. Plant Physiol 123:807–810
5. Wilm M, Shevchenko A, Houthaeve T, Breit S et al (1996) Femtomole sequencing of proteins from polyacrylamide gels by nano-electrospray mass spectrometry. Nature 379:466–469
6. Jobim MIM, Obersta ER, Salbegob CG, Souzab DO, Waldc VB et al (2004) Two-dimensional polyacrylamide gel electrophoresis of bovine seminal plasma proteins and their relation with semen freezability. Biol Sci 61:255–266
7. Vassallo JC, Kurpakus MA (1996) Nuclear matrix proteins of bovine corneal and conjunctival epithelium. Curr Eye Res 15:899–904
8. Havea P, Carr AJ, Creamer LK (2004) The roles of disulphide and non-covalent bonding in the functional properties of heat-induced whey protein gels. J Dairy Res 71:330–339
9. Wheeler TT, Broadhurst MK, Rajan GH, Wilkins RJ (1997) Differences in the abundance of nuclear proteins in the bovine mammary gland throughout the lactation and gestation cycles. J Dairy Sci 80:2011–2019
10. Timperio AM, D'Alessandro A, Pariset L, D'Amici GM, Valentini A, Zolla L (2009) Comparative proteomics and transcriptomics analyses of livers from two different Bos taurus breeds: "Chianina and Holstein Friesian". J Proteomics 73:309–322
11. Cole RB (1997) Electrospray mass spectrometry: fundamentals, instrumentation and applications. Wiley, New York
12. Gomez SM, Nishio JN, Faull KF, Whitelegge JP (2002) The chloroplast grana proteome defined by intact mass measurements from liquid chromatography mass spectrometry. Mol Cell Proteomics 1:46–59
13. Karas M, Hillenkamp F (1988) Laser desorption ionization of proteins with molecular masses exceeding 10,000 Daltons. Anal Chem 60:2299–2301
14. Szabò I, Seraglia R, Rigoni F, Traldi P, Giacometti GM (2001) Determination of photosystem II subunits by matrix-assisted laser desorption/ionization mass spectrometry. J Biol Chem 276:13784–13790
15. Eng JK, McCormack AL, Yates JR III (1994) An approach to correlate tandem mass spectral data of peptides with amino acid sequences in a protein database. J Am Soc Mass Spectrom 5:976–989
16. Perkins DN, Pappin DJ, Creasy DM, Cottrell JS (1999) Probability-based protein identification by searching sequence databases using mass spectrometry data. Electrophoresis 20:3551–3567
17. Field HI, Fenyo D, Beavis RC (2002) Radars, a bioinformatics solution that automates proteome mass spectral analysis, optimizes protein identification, and archives data in a relational database. Proteomics 2:36–47

18. Taylor JA, Johnson RS (1997) Sequence database searches via de novo peptide sequencing by tandem mass spectrometry. Rapid Commun Mass Spec 11:1067–1075
19. Goodlett DR, Keller A, Watts JD et al (2001) Differential stable isotope labeling of peptides for quantitation and de novo sequencing. Rapid Commun Mass Spec 15:1214–1221
20. Bradford MM (1976) A rapid and sensitive method for the quantitation of microgram quantities of protein utilizing the principle of protein–dye binding. Anal Biochem 72:248–254
21. Mann M, Meng CK, Fenn JB (1989) Interpreting mass spectra of multiply charged ions. Anal Chem 61:1702–1708
22. Timperio AM, Huber CG, Zolla L (2004) Separation and identification of the light harvesting proteins contained in grana and stroma thylakoid membrane fractions. J Chromatogr A 1040:73–81
23. Zolla L, Rinalducci S, Timperio AM, Huber CG, Righetti PG (2004) Intact mass measurements for unequivocal identification of hydrophobic photosynthetic photosystems I and II antenna proteins. Electrophoresis 25: 1353–1366
24. Huber CG, Walcher W, Timperio AM, Troiani S, Porceddu E, Zolla L (2004) Multidimensional proteomic analysis of photosynthetic membrane proteins by liquid extraction-ultracentrifugation-liquid chromatography-mass. Proteomics 4:909–3920
25. Huber CG, Timperio A-M, Zolla L (2001) Isoforms of photosystem II antenna proteins in different plant species revealed by liquid chromatography–electrospray ionization mass spectrometry. J Biol Chem 276:45755–45761
26. Rinalducci S, Larsen MR, Mohammed S, Zolla L (2006) Novel protein phosphorylation site identification in spinach stroma membranes by titanium dioxide microcolumns and tandem mass spectrometry. J Proteome Res 5: 973–982
27. Carr S, Aebersold R, Baldwin M, Burlingame A, Clauser K, Nesvizhskii A (2004) Working group on publication guidelines for peptide and protein identification data. The need for guidelines in publication of peptide and protein identification data. Mol Cell Proteomics 3:531–533
28. Craig R, Beavis RC (2004) TANDEM: matching proteins with tandem mass spectra. Bioinformatics 20:1466–1467

Chapter 5

Producing a One-Dimensional Proteomic Map for Human Liver Cytochromes P450

Natalia A. Petushkova and Andrey V. Lisitsa

Abstract

In this chapter we explore the inducible cytochrome P450 (CYP) forms as an example of membrane proteins analysis that relies on sodium dodecyl sulfate polyacrylamide gel electrophoresis (SDS-PAGE) fractionation with subsequent mass spectrometric (MS) identification. The approach involves cutting an SDS-PAGE gel lane into thin slices and identifying proteins in each slice by MS with the aim of obtaining detailed information on proteins of interest. A one-dimensional proteomic map showing the distribution of selected CYP isoforms across 40 slices was constructed using mass spectra obtained from each slice. Our protocol proved to be efficient enough to obtain a comprehensive profile of drug-metabolizing enzymes in the human liver. In addition to human tissues, the approach described should be applicable to the characterization of membrane proteins in other eukaryotic species.

Key words: Human liver microsomal ghosts, Cytochrome P450, MALDI-TOF mass spectrometry, 1DE-proteomic map, Protein abundance index

1. Introduction

Membrane proteins constitute one-third of all body proteins and play a crucial role in signaling, transport processes, and oxidative reactions. The superfamily of cytochromes P450 (CYPs) is of special importance (1, 2). The unique nature of CYPs, their involvement in the metabolism of a wide range of endogenous and exogenous substrates, and the existence of multiple CYP forms with overlapping substrate specificities make these enzymes an attractive subject for basic and applied studies (3, 4). In humans, about 60 different CYPs have so far been identified. Among these more than 20 are expressed by human liver with 70% of the CYP

Djuro Josic and Douglas C. Hixson (eds.), *Liver Proteomics: Methods and Protocols*, Methods in Molecular Biology, vol. 909, DOI 10.1007/978-1-61779-959-4_5, © Springer Science+Business Media New York 2012

content represented by 11 CYPs (1A2, 2A6, 2B6, 2C8/9/18/19, 2D6, 2E1, and 3A4/5) (5). There are two main MS-based approaches to CYPs' identification: peptide mass fingerprinting (PMF) and high-throughput liquid chromatography coupled with mass spectrometry (LC-MS/MS) (6). Using tandem one-dimensional electrophoresis (1DE) and PMF, Galeva and Alterman (7) identified 7 CYPs in rat liver microsomes, while Nisar et al. (8), using LC-MS/MS and working with the same object, reported identification of 24 CYPs. The same (i.e., LC-MS/MS) approach was employed by Lane et al. (9) for identification of 14 CYPs in a human liver microsomal fraction.

Proteomic protocols commonly incorporate two-dimensional electrophoresis (2DE) for protein separation; however, 2DE is not an effective technique for profiling membrane proteins because they are poorly solubilized by the sample buffer used for isoelectric focusing, and they have difficulty entering the gel due to their size and hydrophobicity (10). Membrane proteins are also extremely heterogeneous with respect to charge and are distributed over a wide pH range during isoelectric focusing. In addition, many proteins are present at low copy number in the cell, a characteristic that restricts their detection in a gel piece (11). An alternative approach to identify membrane protein relies on size separation using sodium dodecyl sulfate polyacrylamide gel electrophoresis (SDS-PAGE) and an MS strategy for the direct analysis of complex protein mixtures (11). SDS-PAGE provides the best coverage and is adopted as the reproducible method (12), but its resolution capacity is lower than that of 2DE because of complexity of a protein mixture used for analysis (13). To harvest the wealth of mass spectrometric information obtained during protein separation by SDS-PAGE, the following approach has been proposed: a gel lane is serially sectioned into the sequence of slices, and protein profiles refer to tissue types, slice numbers, or slice positioning the gel lane (13). A variety of approaches are now available for cutting an SDS-PAGE gel into slices with a width of 1.3–3 mm (14–17). In most cases each slice corresponds to a protein band with a known apparent molecular weight. Recently, we have reported that it is technically possible to cut the gel into much thinner slices (about 0.2 mm thick) (18, 19).

Application of MS for proteomics does not rely on a single technique for all purposes but rather on a collection of methodologies, a choice being determined by which is better suited for a given sample (20). Among the available mass spectrometric methods for identification of proteins separated by SDS-PAGE or 2DE, MALDI-TOF MS and electrospray tandem mass spectrometry (ESI MS/MS) are commonly utilized. These two approaches for protein identification have been found to be complementary in the analysis of a single band from SDS-PAGE with MALDI-TOF/PMF producing unique peptide molecular weight matches that were not identified by μLC-MS/MS (21). Reliable protein identification also depends on several parameters including the search engine, the accuracy of fragment mass determination, the number of masses submitted for query, the mass distribution of query masses, the number of masses matching between sample and database protein, the size of the sequence database, and the kind and number of modifications considered. One of the most important factors influencing successful protein identification by PMF is a choice of the corresponding number of matching *m/z* values (22). Berndt et al. (22) have shown that with a mass accuracy of 0.02%, 9–10 matching masses belonging to the 23 required masses are needed for 99% correct identification of proteins up to 500 kDa using a database comparable in size to UniProtKB/Swiss-Prot. The authors calculated these 23 fragments for false-positive protein identification as a function of database size and mass accuracy. They selected this number of fragments representing a hypothetical random protein from different proteins in the database and queried these artificially generated mass fingerprints against the databases. Over many such searches, the likelihood of false-positive identification is calculated as the fraction of searches leading to the identification of a real protein.

In this chapter we describe how to plan, design, and conduct experiments for identifying and distinguishing between highly homologous CYP forms in a complex protein mixture. In the first stage, a four-step ultracentrifugation protocol is employed to yield high-purified human liver microsomal fraction (ghosts, HLMGs) which are treated with 0.15% ionic detergent sodium cholate to remove intravesicular-localized soluble proteins. After incubation with a solution composed of equal volumes of 100 mM sodium citrate and 100 mM sodium pyrophosphate to remove ribosomes and proteins adsorbed onto the surface of microsomal vesicles, the microsomal suspension is centrifuged through 0.4 M sucrose. Obtained HLMGs (microsomal vesicles devoid of their intravesicular soluble proteins, ribosomes, and adsorbed surface proteins) are separated by SDS-PAGE; then the region that corresponded to the molecular masses of the CYPs is cut into 40 slices; each slice after in-gel trypsinolysis is analyzed by MALDI-TOF mass spectrometry. Data on the matched *m/z* values and the corresponding (identified) proteins are presented as a one-dimensional proteomic map (1D-PM). The mass spectrometric data obtained from several adjacent gel slices are pooled to increase the sequence coverage of CYP 3A4, CYP 2A13, and CYP 2A6 by 10–15%.

2. Materials

2.1. Tissue Homogenization and Human Liver Microsomal Ghosts Purification

1. 150 mM Potassium chloride (KCl) in water, pH 7.4 (see Note 1). Keep at 4°C.
2. Tissue homogenization buffer I: 1 mM ethylenediaminetetraacetic acid tetrasodium salt (EDTA), 1 mM dithiothreitol (DTT), 0.1 mM phenylmethyl-sulfonyl fluoride (PMSF), 150 mM KCl. Keep at 4°C.
3. Homogenization buffer II: 100 mM potassium phosphate buffer, pH 7.4, containing 1 mM EDTA, 1 mM DTT. Keep at 4°C.
4. 3% Sodium cholate in water. Keep at 4°C.
5. Resuspension buffer III: Homogenization buffer 2, containing 1 mM EDTA and ribonuclease (2 mg/mL). Keep at 4°C.
6. Buffer solution IV: 100 mM sodium pyrophosphate and 100 mM sodium citrate in water, pH 8.2 (see Note 1). Keep at 4°C.
7. Sucrose solution: 0.4 M sucrose in homogenization buffer II. Keep at 4°C.
8. Storage buffer V: 100 mM potassium phosphate buffer, pH 7.4, containing 1 mM EDTA, 1 mM DTT, 20% glycerol for −80°C storage of purified human liver microsomal membranes (ghosts, HLMGs). Keep at 4°C.
9. The following Beckman centrifuges, rotors, and centrifuge tubes: J2-21 centrifuge, JA-25.50 rotor, 36 mL 25.4 × 89 mm tubes, and Optima L-90K ultracentrifuge, Ti90 rotor, *Thinwall* open-top 5/8 × 3 *in* tubes.
10. Protein standard: Bovine serum albumin (BSA) 0.5 mg/mL in storage buffer.
11. Bradford solution: 25 mg Coomassie, 12.5 mL ethanol, and 25 mL phosphoric acid in 212.5 mL water.

2.2. SDS-Polyacrylamide Gel Electrophoresis

The SDS-PAGE method is that originally described by Laemmli (23). All reagents are from Bio-Rad Laboratories (Russian Federation).

1. SDS-PAGE equipment: The mini format electrophoresis apparatus Mini-Protean 3 Cell (Bio-Rad Laboratories) for 6–10 cm length gels.
2. Hand cast gels of variable acrylamide percentage.
3. Standard Laemmli loading buffer and running buffer.
4. Molecular weight markers: Precision Plus Protein Dual Color Standards.
5. Standard Solution for Coomassie Blue R-250 staining.

2.3. Gel Slicing of the Region of Interest (37–65 kDa) and In-gel Digestion of Each Slice

1. Thermoelectric Peltier Cooler; glass plate (10×7×0.15 cm); power unit; a manual microtome R100, profile C; magnifier (2×, 3×, d. 90–100 mm), and lamp for illumination of gel (see Note 2).
2. Destained solution: Acetonitrile and 100 mM ammonium bicarbonate in water 1:1 (v/v).
3. Modified trypsin (Promega) (see Note 3).
4. Stop solution: 0.7% trifluoroacetic acid (TFA).

2.4. MALDI-TOF Peptide Mass Fingerprinting

1. MALDI Target: MTP 600/384 Anchor Chip™ from Bruker Daltonics.
2. MALDI-Matrix: 2.5-dihydroxybenzoic acid in acetonitrile and 0.7% TFA 1:1 (v/v). Matrix solution should be prepared fresh each day.
3. MALDI-TOF mass spectrometer: Ultraflex II (Bruker Daltonics), accelerating voltage of 25 kV and 135 ns.
4. Mass calibration: Peptide calibration standard II (cat. 222570) (Bruker Daltonics):
   Angiotensin II, Angiotensin I, Substance P, Bombesin, ACTH clip 1–17, ACTH clip 18–39, Somatostatin 28, Bradykinin Fragment 1–7, Renin Substrate Tetradecapeptide porcine; covered mass range: ~700–3,200 Da. Calibration standards were prepared by adding 125 μL of 0.1% TFA to vials.

2.5. Data Interpretation

Data interpretation requires local installation of Mascot search engine (Matrix Sciences, UK) and an access to the proprietary package ZOOMer to produce the 1D-PMs (see Note 4). ZOOMer is a Web-based application free for academic use. Access is provided at http://projects.ibmh.msk.su/zoomer/site. For obtaining the access credentials contact authors by e-mail.

3. Methods

3.1. Human Liver Microsomal Ghosts Purification

A four-step protocol for obtaining purified preparations of HLMGs was first described by Karuzina and Archakov (24). All preparation procedures are performed at 0–4°C (see Note 5).

1. Take 10–20 g from each sample of human liver (see Note 6). Wash these samples twice in ice-cold 1.15% KCl solution, pH 7.4, containing 0.1 mM PMSF (1 g liver/4–5 mL of KCl); absorb residual liquid with filter paper before weighing the liver specimen (see Note 7).
2. Finely chop the liver samples with clippers and homogenize using a glass Potter–Elvehjem homogenizer with a Teflon

pestle (Sartorius) on ice in tissue homogenization buffer I at a w/v of 1:2.

3. Filter the crude homogenate through three layers of gauze to remove pieces of adipose, fatty, and connective tissues. Transfer the filtrate to 36-mL 25.4 × 89 mm centrifuge tubes and centrifuge in J2-21 centrifuge using a JA-25.50 rotor at 11,000 rpm at 4°C for 20 min.
4. Filter the supernatant through two layers of gauze, transfer the filtrate to *Thinwall* open-top 5/8 × 3 *in* centrifuge tubes, and centrifuge in Optima L-90K using a Ti90 rotor at 40,000 rpm (105,000 × *g*) at 4°C for 70 min.
5. Suspend the crude microsomal pellet using a glass Potter–Elvehjem homogenizer (10 mL volume) with a Teflon pestle on ice in homogenization buffer II at a w/v of 1:1 and incubate with 0.15% Na-cholate on ice for 30 min, with subsequent ultracentrifugation at 40,000 rpm (105,000 × *g*) at 4°C for 70 min in four *Thinwall* open-top 5/8 × 3 *in* tubes.
6. Overlay the pellets in the tubes with two volumes of homogenization buffer II each and store at 4°C overnight.
7. Remove the buffer overlay, resuspend the microsomal fraction in each tube in two volumes of ice-cold (4°C) resuspension buffer III, pool the partially purified membrane fraction from each tube in a 50 mL glass tube, and incubate on a rotary platform shaker for 30 min at 4°C. Add a volume of ice-cold buffer solution IV equivalent to 1/3 the total volume of the pooled microsomal pellet and incubate on a rotary platform shaker for 20 min at 4°C.
8. Divide the final suspension into four equal aliquots and carefully overlay each onto 0.4 M of ice-cold sucrose solution (ratio 1.2:1) in four *Thinwall* open-top 5/8 × 3 *in* tubes. Centrifuge at 40,000 rpm (105,000 × *g*) at 4°C for 70 min.
9. Remove the upper layer carefully using a pipette, resuspend the pellet in ice-cold resuspension buffer III at a w/v of 1:1, pool the suspension from each tube into one 50 mL glass tube, and incubate on a rotary platform shaker for 30 min at 4°C. Repeat step 8.
10. Using a syringe, carefully remove the upper layer from each tube. Resuspend the high-purified HLMGs in each tube in ice-cold storage buffer and pool HLMGs suspensions together.
11. Determine the protein concentration using the Bradford assay (25) (see Note 8).
12. Aliquot HLMGs suspension into tubes is suitable for storage, snap freeze in liquid nitrogen and store at −80°C (see Note 9).

3.2. SDS-Polyacrylamide Gel Electrophoresis of HLMGs

1. For SDS-PAGE separation of HLMGs' proteins Mini Protean III (Bio-Rad Laboratories, Russian Federation) was used. SDS-PAGE separation should be performed according to the manufacturer's instructions. Refer to the user's manual for general details.
2. Deposit 10–15 μL HLMGs suspension on each gel lane and 7 μL on the last lane of molecular weight markers: Precision Plus Protein Dual Color Standards (see Note 10).
3. Apply power to the Mini-Protean 3 cell and begin electrophoresis at 150 V. Run time is approximately 60 min; upon the exit of bromophenol blue into the electrode buffer the electrophoretic separation is considered as complete.
4. Gel on one of the glass plates is placed into a plastic bath (volume, 1,000 mL) with 250 mL of the dye – 0.1% Coomassie brilliant-blue G-250 for 1 h (or for the night) at room temperature. Remove excess dye by use of 250 mL of a solution, prepared from 400 mL ethanol, 100 mL acetic acid, and 500 mL water. The washing of the gel from Coomassie is carried out on a rotary platform shaker for 3 h with a fivefold substitution of the washing mixture.
5. The obtained gel (size, 70 × 100 × 1 mm) is placed for 15 min into distilled water, and then it is taken by hands in latex gloves without powdering and placed on the glass plate for scanning analysis (Fig. 1).

3.3. Gel Slicing and In-Gel Digestion

1. Put on the cap, mask, and unpowdered latex gloves. Wipe gloves, Eppendorf adjustable-volume pipette, and working surfaces of the table with 70% ethanol (to avoid contamination of MS samples with human keratin which can be a major problem during proteomic analysis).
2. Preparation of gel slices:
   (a) Assemble the device for cutting SDS-PAGE gel lane (see Note 2).
   (b) Place a gel lane containing a 110 μg sample of HLMGs resolved by SDS-PAGE onto a water-soaked glass plate cooled by a Peltier element.
   (c) Find out on SDS-PAGE gel lane the upper limit of the region as determined from protein standards (Precision Plus Protein Standards, Bio-Rad Laboratories) corresponding to the MW of CYPs (65 kDa). Discard a gel piece positioned above this area. Cut the 1-cm length on the remainder gel lane into forty 0.25-mm slices using a manual microtome R100, profile C. Divide the slices sequentially with the microtome being positioned perpendicularly to the gel lane. Control accuracy of the slicing using a magnifier and graph paper (with tenth divisions)

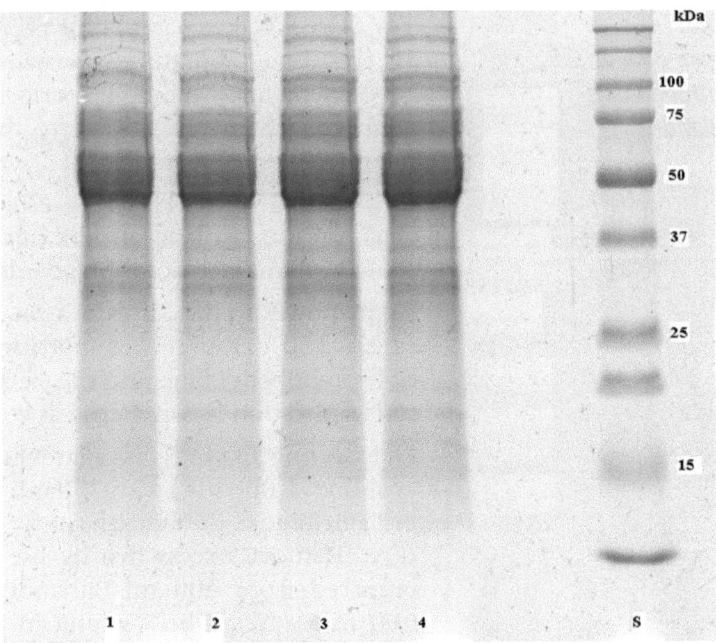

Fig. 1. SDS-PAGE separated proteins of HLMGs sample: *lanes 1 and 2*—80 μg of protein; *lanes 3 and 4*—100 μg of protein. *Lane S*—molecular weight markers. The protein profiles are highly reproducible between lanes with equal amount of loaded protein in the wells. Clear differences are seen between different amounts of loading protein: intensity of staining of protein bands with greater amount of protein (*lanes 3 and 4*) is higher than that of bands with lesser amount of protein (*lanes 1 and 2*) in this gel. Depending on electrophoretic conditions, 17–21 protein bands were observed after Coomassie blue staining of SDS-PAGE gels. Most intensively colored was the 45–65 kDa mass range corresponding to weights of cytochrome P450, where approximately 2–3 major (wide) bands along with several minor bands were revealed.

placed under glass plate. The lower limit of cutting will correspond to the lowest MW of CYPs ~37 kDa (by from protein standards). Cut each slice into three equal fragments (parts). Using the plastic forceps transfer middle piece into a 200 μL Eppendorf tube (see Note 11).

3. Destain each gel piece by incubation at 50°C for 45–60 min in 100 μL of Coomassie brilliant-blue G-250 destaining solution. Remove the destaining solution, add 100 μL of fresh destain, and incubate as before at 50°C. Remove the destaining solution.
4. Dehydrate gel pieces by adding 150 μL of acetonitrile and incubating for 15 min at room temperature. Remove the acetonitrile and carefully dry the gel pieces at 37°C in uncapped Eppendorf tubes.
5. Add to the gel pieces the calculated amount of trypsin solution (see Note 12). Tap the tubes and incubate at 0°C for 45–60 min

to allow the trypsin to penetrate into the gel. Digest the gel slices on a shaking incubator for 18 h at 37°C. Terminate the reaction by addition of 7.8–8.8 μL of stopping solution (item 2, Subheading 2.3) and place the Eppendorfs in the shaking incubator for a 2-h storage at 25°C.

The extracted tryptic peptides solutions (40 samples) are now ready for spotting onto MALDI target and measurement.

3.4. MALDI-TOF Peptide Mass Fingerprinting

1. Clean the MALDI target (item 1, Subheading 2.4) by sonication for 7 min in 250 mL of 50% ethanol. Remove the ethanol and repeat the cleaning process two more times. Dry the target at room temperature.
2. Spot the proteolytic peptides' mixture by adding 1 μL of tryptic peptides extracted from each slice to 1 μL of MALDI-Matrix solution (item 2, Subheading 2.4) on a MALDI target and dry on air. The extracted tryptic peptides obtained from the slices' fragments should be placed on the MALDI target in three replicates. The correspondence between slice number and target position should be thoroughly recorded.
3. Calibrate the instrument (item 3, Subheading 2.4) in reflectron mode using molecular weights standards (item 4, Subheading 2.4).
4. Perform measurements in the reflection/delayed extraction mode at an accelerating voltage of 25 kV and a 135-ns delay when using the Ultraflex II (Bruker Daltonics). Typically, each mass spectrum is the sum of 100 laser shots. Peak detection is performed using the SNAP algorithm (FlexAnalysis 2.0, Bruker Daltonics) and the resulting mass spectra are internally calibrated by trypsin autolysis peaks (m/z 842.5094 Da and 2,211.1046 Da).

3.5. Data Interpretation

1. Perform PMF identification of proteins for each mass spectrum which has been obtained across all the samples, slices, and replicates. Use Mascot search with the following parameters: (1) taxonomy: *Homo sapiens* (human); (2) trypsin cleavage allowing up to 1 missed cleavage; (3) variable modification including oxidation (M), propionamide (C), (4) peptide tolerance 90 ppm; and (5) SwissProt as a protein sequence database. The searches can be automated by use of Mascot batch mode or by exploiting special functions of LIMS, such as ProteinScape (Bruker Daltonics, Germany). Extract and merge resulting protein identification into the single list, and remove duplicate identifications. Prepare appropriate amino acid sequences for the listed proteins in the FASTA-format file (say, a < *source.fasta* > file), using, for example, "Retrieve" section of UniProtKB (www.uniprot.org). For convenience you can

assign human-readable names to the proteins (instead of accession numbers).

2. Define the protein which you would like to investigate for the presence of isoforms or single amino acid polymorphisms (SAPs). Further we shall describe the workflow for the latter case; however, for the case of deciphering isoforms the similar steps should be followed. Prepare a separate FASTA-file (further referred to as a <*polymorphism.fasta*> file), where amino acid sequences should be modified to carry at least one SAP per tryptic peptide for the recruited protein. Information about putative SAPs can be taken from "Natural variation" section of UniProt record.

3. Generate, with a tolerance of around 180 ppm, a consensus peak list by selecting from replicated peaks from mass spectra of the same slice. For example, if each tryptic mixture from one slice was spotted onto the MALDI target in five replicates, select those m/z values, which occur at least in three of these replicates. Average m/z values across replicates to produce consensus peak list. Note that for consensus peak list creation, as well as for the following steps, it is necessary to maintain the mapping of the MALDI target, which attributes each spot on the target to the proper slice.

4. Produce for each sample a table, whose rows are protein names from the <*source.fasta*> file and whose columns indicate slice numbers. When proteins in the table are sorted by increasing molecular weights, such a table is called a 1D-PM as the maximum values in the rows correspond to the location of each protein along the SDS-PAGE gel lane (Fig. 2a). Input into each cell of the table the protein abundance index (PAI), i.e., the number of consensus peak list values which match (with a tolerance of 90 ppm) to the peptides for the protein in a row. Calculate average PAI values for each row and SDs. By the visual inspection of the 1D-PM verify that the highest (in the row) PAI values occur approximately along the diagonal of the table. Such distribution of PAI values indicates that all previous steps are correct.

5. Find on the 1D-PM the row for the protein that was earlier defined in step 2 of Subheading 3.5 and determine those slices, where PAI exceeds mean value by 1 SD. If several of such slices are adjacent to each other notify them as a "protein abundance bandwidth" (BW). Collecting m/z values from the BW enables to enrich sequence coverage by 5–15% (see Note 13).

6. Record m/z values from consensus spectra, corresponding to the bandwidth slices, and store them to a third <*sampleN.bandwidth.mz*> file.

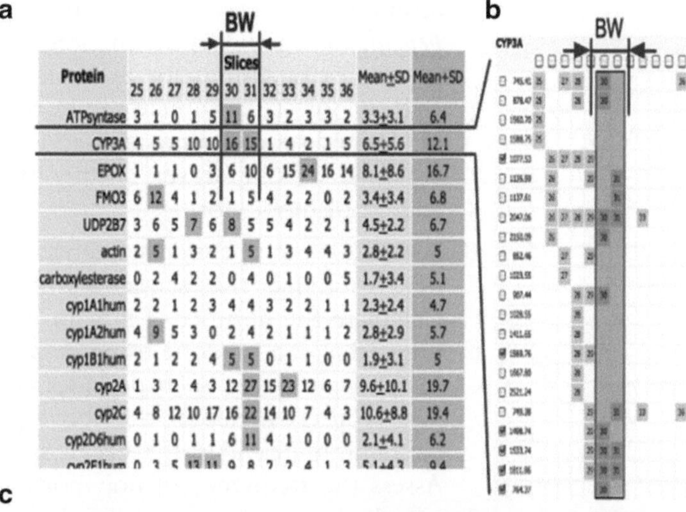

Protein	25	26	27	28	29	30	31	32	33	34	35	36	Mean±SD	Mean+SD
ATPsyntase	3	1	0	1	5	11	6	3	2	3	3	2	3.3±3.1	6.4
CYP3A	4	5	5	10	10	16	15	1	4	2	1	5	6.5±5.6	12.1
EPOX	1	1	1	0	3	6	10	6	15	24	16	14	8.1±8.6	16.7
FMO3	6	12	4	1	2	1	5	4	2	2	0	2	3.4±3.4	6.8
UDP2B7	3	6	5	7	6	8	5	5	4	2	2	1	4.5±2.2	6.7
actin	2	5	1	3	2	1	5	1	3	4	4	3	2.8±2.2	5
carboxylesterase	0	2	4	2	2	0	4	0	1	0	0	5	1.7±3.4	5.1
cyp1A1hum	2	2	1	2	3	4	4	3	2	2	1	1	2.3±2.4	4.7
cyp1A2hum	4	9	5	3	0	2	4	2	1	1	1	2	2.8±2.9	5.7
cyp1B1hum	2	1	2	2	4	5	5	0	1	1	0	0	1.9±3.1	5
cyp2A	1	3	2	4	3	12	27	15	23	12	6	7	9.6±10.1	19.7
cyp2C	4	8	12	10	17	16	22	14	10	7	4	3	10.6±8.8	19.4
cyp2D6hum	0	1	0	1	1	6	11	4	1	0	0	0	2.1±4.1	6.2
cyp2E1hum	0	3	5	13	11	9	8	7	7	4	1	3	5.1±4.3	9.4

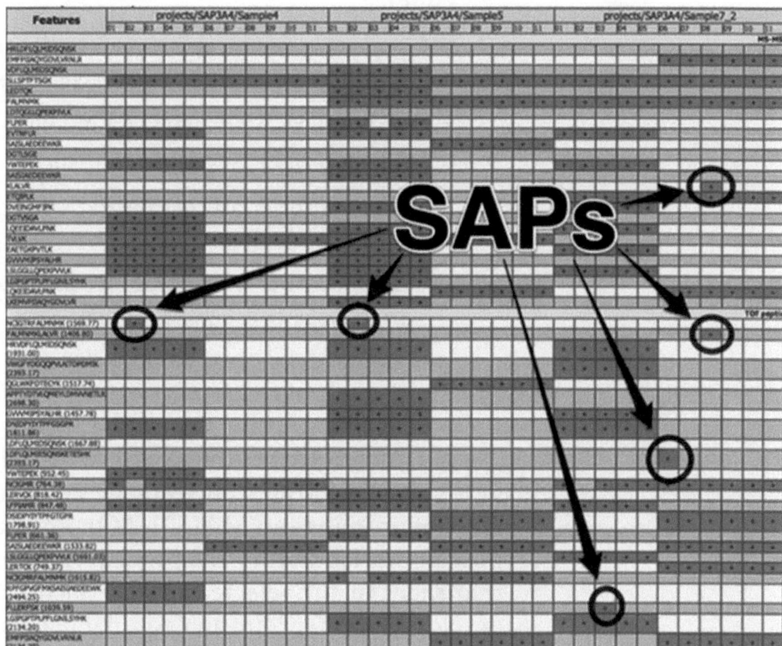

Fig. 2. The report generated by Web-package ZOOMer based on processing of MS of SDS-PAGE slices from several HLMGs samples: (**a**) view of the 1D-PM with a bandwidth (BW) shown for CYP3A subfamily of proteins; (**b**) view of the *m/z* values across the selected bandwidth; (**c**) cross-sample view: 18 polymorphic variants (columns headed as 01–18) are screened across for three HLMGs samples (Sample 4, Sample 5, and Sample 7_2). (*) sign indicates the presence of the peptide (row name) in the given polymorphic sequence (column number). SAPs are observed as situations, when peptide matches single polymorphic sequence.

7. Match each m/z value from the <*sampleN.bandwidth.mz*> to the m/z values, obtained by the theoretical tryptic digestion of the predefined set of sequences stored in the <*polymorphism.fasta*>. Select the matching peptides within the 90 ppm tolerance

interval and store their sequences to a fourth <*sampleN.peptides.mz*> file. Get the theoretical peak lists for polymorphic sequences by using, for example, MS-Digest program from ExPaSy.org. To automate the matching of the peak list to the predefined set of amino acid sequences, use VDigest program, available at our server (http://projects.ibmh.msk.su/msa/vdigest.pl?mode=form).

8. Search for each peptide sequence from the <*sampleN.peptides.mz*> matching the sequences in the <*polymorphism.fasta*>. Calculate the number of hits for each peptide and select peptides (that can be done using *grep* utility, for example), which match only once. Such peptides contain the site of the SAP (see Note 14). Repeat the above procedures to collect SAP-containing peptides for every sample you have processed. Assess the frequency of polymorphism occurrence across the samples.
9. Prepare the final report for each sample by including the following information (Fig. 2, see Note 15):
   - Tabular view of the 1D-PM, with shaded bandwidth cells
   - The number of bandwidth slices
   - The number of *m/z* values in the consensus spectra of bandwidth slices
   - The number of peptides from the <*polymorphism.fasta*> file matching these *m/z* values
   - List of SAP-specific peptides and frequency of occurrence of each of them

3.6. Conclusion

This chapter describes the workflow aimed at identification of CYPs in a complex protein mixture by combining SDS-PAGE separations of proteins and MALDI-TOF PMF approaches. For this purpose in our laboratory the procedure for obtaining and processing thin slices in the selected region on the SDS-PAGE gel lane was developed. For spectra analysis, database search, and data interpretation we used the proprietary Web-based information management system ZOOMer, which enables to consolidate and visualize a large massif of MS data (above 400 mass spectra per one experiment), collected from the series of SDS-PAGE slices. We have proposed a way of visually (spatial) displaying the results of protein identification in the form of the so-called 1D-PM, which allows you to anchor each protein with its position on SDS-PAGE gel lane. By analyzing the 1D-PMs experimental information on MS-based protein identification may be processed to encounter specificity of peptide masses across proteins and their isoforms; isoform-specific masses and corresponding peptide sequences; frequency of occurrence of specific peptides across slices; and frequency of occurrence of specific peptides across samples. The protocol described in this chapter may well be applied for screening of HLMGs to enrich coverage of the membrane proteins.

4. Notes

1. The pH of the solutions is adjusted by using 0.15 M hydrochloric acid (HCl).
2. Install an assembly for cutting the 1-cm gel length into slices as shown in Fig. 3. To avoid ripping of the gel, it should be slightly frozen. For this purpose we used Peltier cooler to maintain a temperature of −9°C. The gel lane was placed onto a glass plate (10 × 7 × 0.15 cm) and for convenience; the gel was additionally illuminated during slicing. For a more precise cutting, a magnifier was helpful.
3. Prepare a stock solution of modified trypsin (on ice). For 20 μg add 100 μL of enzyme resuspension buffer (provided by Promega). Stock solution is divided into 5-μL aliquots for storage at −20°C. Prior to running the reaction of tryptic hydrolysis, the stock solution (5 μL) is diluted eightfold with

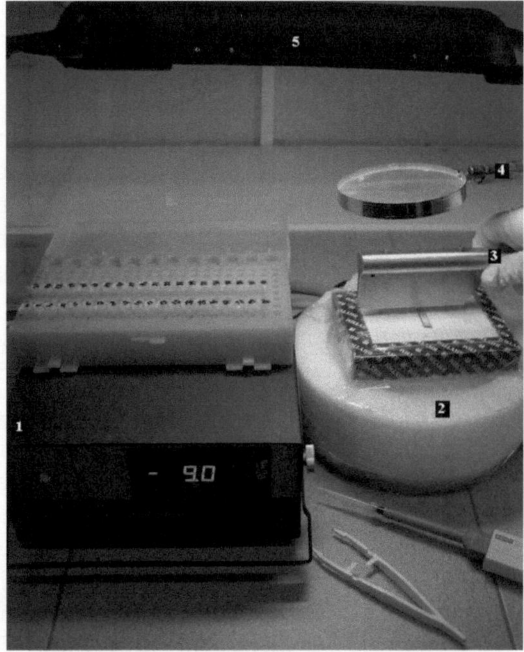

Fig. 3. An assembly for cutting SDS-PAGE gel slices; 1—power unit; 2—thermoelectric (a Peltier) cooler with glass plate (10 × 7 × 0.15 cm) on it; 3—a manual microtome R100 profile C; 4—magnifier (2×, d. 90–100 mm); 5—lamp for illumination of gel.

10 μL of 100 mM ammonium hydrocarbonate diluted with 25 μL of H_2O (dd, Merck).

4. We have combined and evaluated at least 6–12 MALDI-TOF mass spectra per one slice. "Peaklist.xml" files, containing information about *m/z* value and intensity of mass-spectrometric peaks, are loaded into the proprietary ZOOMer software suite (Subheading 2.5) which contains several modules and tools for mass spectrum interpretation and protein sequence analysis. The program which is written in Active State Perl scripting language optimized for Apache 2.0 can be easily modified or extended by modules of specific need. ZOOMer was used (1) to compute consensus peak lists for MALDI-TOF data and (2) to construct 1D-PM by performing theoretical digest of a selected protein sequence with subsequent comparison of resulting tryptic peptides with observed peak lists. Essentially, each of the identified proteins is anchored to a particular position in the gel. By anchoring the protein to a certain bandwidth it is possible to enrich the collection of identified sequence fragments.

5. It is well known that protein identification via MS depends on sensitivity for ion detection. Ion formation is largely dependent on sample preparation and its presentation to a mass spectrometer. Besides, the preparation of MS sample for proteomic researches is an important step as there are many factors that may impair protein separation and, hence, impede the interpretation of the data obtained (26). The protocol presented in this chapter has been employed in our lab and works well with MS-based analysis. It provides, in sufficient amounts, "enriched" HLMGs although some cross-contamination between microsomal membranes and the liver cells' endomembrane material was observed. In the past decade, several authors have reported the presence of microsomal CYPs in cellular organelles other than the endoplasmic reticulum, in particular, in plasma membranes and mitochondria. However, levels detected in these compartments were significantly lower than those observed in the endoplasmic reticulum (27).

6. Tissues for the present study were obtained from the resected and discarded liver tissue masses taken from patients undergoing hepatic surgery. Samples of morphological normal liver were taken from regions distant from the metastatic foci of colon cancer. Samples were immediately packed in ice and transported, within 30 min after partial hepatectomy, to the laboratory where they underwent for further processing. Usually, the procedure for microsomal fraction purification requires 12–13 h, so during microsome preparation the protease inhibitor PMSF was added to avoid proteolytic degradation.

7. The protocol presented in this chapter has been originally developed to obtain a pure microsomal fraction from rat or rabbit liver. Removal of blood from liver tissue was achieved by inserting a plastic cannula into an incision made in the portal vein and perfusing the liver with cold 1.15% KCl, pH 7.4. After perfusion, the liver was removed and the procedure of isolating liver microsomes and their ghosts was initiated. Isolation of HLMGs from human liver samples was hampered by the inability to remove blood by liver perfusion. For a human liver great care should be exercised to wash the specimen from blood on the liver surface.
8. Determine the protein concentration using the Bradford assay with a calibration curve generated with 0.5 mg/mL BSA stock solution (item 1, Subheading 2.2). For this purpose, mix BSA aliquots (in duplicate) containing 0–60 μg of protein (0–60 μL of BSA stock solution) with 100 to 40 μL of storage buffer and 5 mL of Bradford reagent. To determine the protein concentration of purified HLMGs (in duplicate) mix 1–2 μL of sample with 99 to 98 μL of storage buffer, add 5 mL of Bradford reagent, vortex, and incubate at room temperature for 5 min. Using a spectrophotometer, measure the absorbance of the standards and samples at 595 nm. Estimate the protein concentration of the HLMGs samples from the calibration curve.
9. The typical yield of purified HLMGs is 4.2 ± 0.6 mg from 1 g of human liver tissue. For storage at −80°C, resuspend the final pellet of purified HLMGs in 2 mL of storage buffer and divide resultant suspension into 100-μL aliquots.
10. HLMGs suspension is mixed with sample buffer containing deionized water, 0.5 M Tris-HCl buffer, pH 6.8, glycerol, 10% SDS solution, β-mercaptoethanol and bromophenol blue by the following proportion: 110 μg of HLMGs' protein in 10 μL of resultant homogenate protein. Samples are vortexed at room temperature and heated at 96°C for 4 min.
11. As a result of slicing, 40 samples for subsequent trypsinolysis are obtained. The number on a tube designed for trypsinolysis corresponds to the slice number. As a rule, gel slicing proceeds from greater to lesser MW values and, accordingly, slice numbering goes in descending direction: sample #1 contains a piece of slice from the 65 kDa region, while in sample #40 one will find a piece of slice from the 37 kDa region.
12. Calculation of the amount of trypsin for a piece of the gel slice. The calculation involves two parameters: the amount of loaded protein and the volume of the gel piece taken for trypsinolysis, according to the following equations:
    (a) The volume of the gel piece taken for trypsinolysis was calculated as follows:

Table 1
Distribution of *m/z* values of CYP2E1 peptides across gel slices

No	MW Da	Slice number	Accompanying proteins[a]
1	911.46	11 12 13 14 15 16 17 19 20 39 40	CES
2	966.56	11 14	
3	1,120.56	14 15 16 17 18 19 20 28	
4	1,198.64	14 15 16 17 18	
5	1,542.74	14 16 18	CYP4A11/22
6	1,694.91	15	
7	1,845.84	15 16 17	
8	837.47	16 17 18 19 20 23 25	CYP2C8
9	1459.7	16	CYP2C181
10	1,087.58	17 18 19 20 25	CYP2B6,CYP4A11/22
11	1,294.65	17 18 23 25	
12	1,483.69	17 18 19 20 21 23 26	
13	1,246.64	18	
14	1,517.77	19 31	
15	2,562.27	19	
16	991.6	29	EPOX
17	999.63	30	
18	1,009.53	31	
19	1,031.54	31 34 35	
20	1,665.81	31 34 35	CYP2F1
21	980.49	36 38 40	CYP4A11/22
22	1,730.85	39	
23	858.49	40	
24	907.42	40	

Marked in black are the slices denoting putative localization of CYP2E1 (the slices numbered from 15 to 19), Marked in grey are the slices not belonging to the CYP2E1 bandwidth
CYP cytochrome P450, *CES* liver carboxylesterase 1, *EPOX* epoxide hydrolase
[a]The proteins, which do not belong to the CYP2E1 but which contain peptides with the indicated mass

$V = a \times b \times c$, where V is the gel piece volume, mm^3 (a is the length of the gel piece; b is its thickness, and c is its height; all measured in mm).

(b) The required amount of trypsin was determined by the following proportion: 10.5 μL* of trypsin solution on 1.0-mm^3 the gel piece.

*The calculation assumes a uniform distribution of protein along the CYPs region on the gel lane. In cases of more or less intensive staining of protein bands, the amount of trypsin varies in the range ± 0.5 μL. In most cases 5.8–6.8 μL of trypsin solution were added to a piece of the gel slice.

13. Table 1 shows CYP2E1-specific m/z values and their location along numbered slices from no. 15 to no. 19. A total of 24 m/z values whose MH^+ (monoisotopic atom mass) was equal to or higher than 1 kDa were detected. From the table it can be seen that the most abundant slice (MA-slice, i.e., the slice in which the maximum PAI value is observed) for CYP2E1 is slice no. 17, for which 8 m/z values were observed. In the neighboring slices (nos. 15, 16, 18, and 19) there are several m/z values (1,542.74, 1,694.91, 1,459.70, 1,246.64, 1,517.77, and 2,562.27) that were attributed to CYP2E1 but not observed

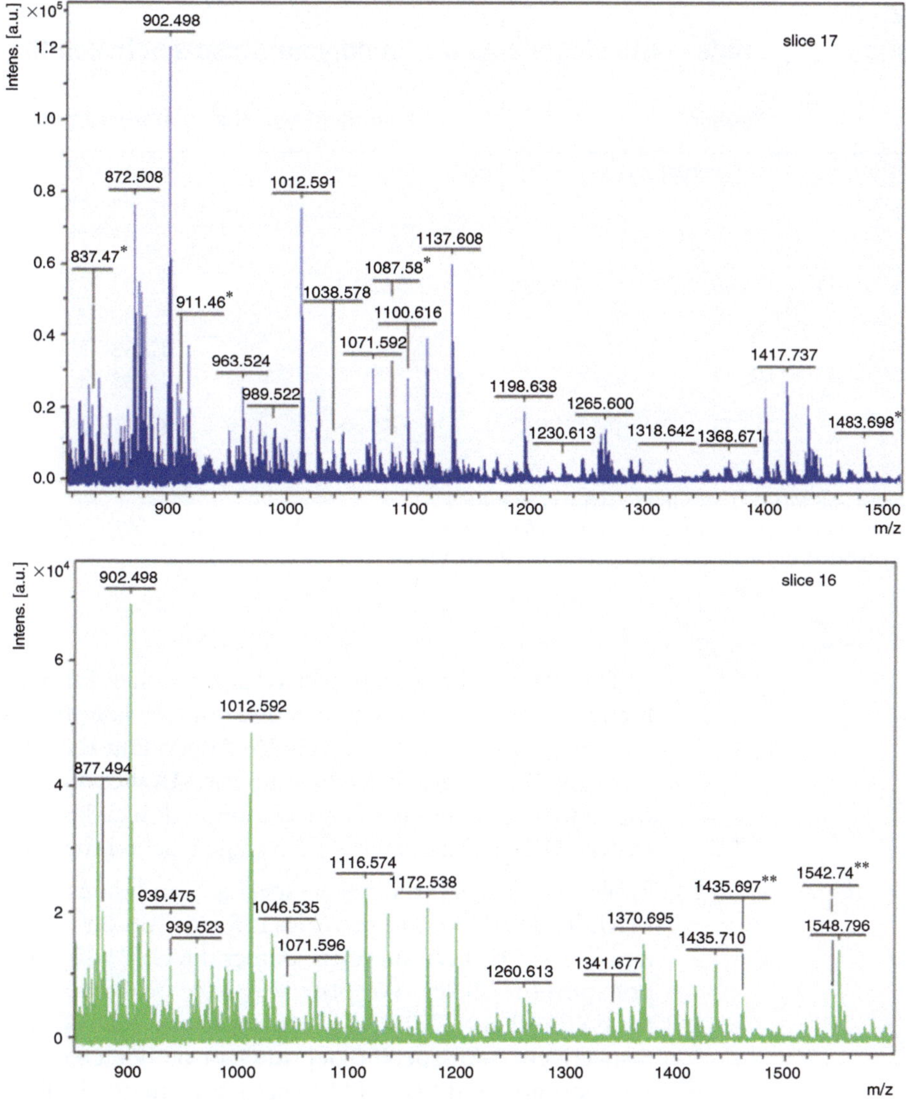

Fig. 4. MALDI-TOF spectra for slice nos. 16 and 17. Peaks, corresponding to peptides of CYP2E1, are indicated by asterisk (*).**Peaks matching two peptides of CYP2E1, which are present in slice no. 16, but absent in slice no. 17.

in the MA slice (Fig. 4). Appending the data for slice nos. 15, 16, 18, and 19 to the data for the MA slice no. 17 increased the sequence coverage of CYP2E1 by approximately 14%, which totally reached 41%. In a similar way compiling MS data from neighboring slice nos. 14–19 enhanced the sequence coverage of CYP1A2 by approximately 7%. Earlier we observed the same trend with respect to CYP 2A6 family proteins for which the enhancement of the sequence coverage constituted, on average, 12 ± 5% (18). No enrichment was observed for those proteins that resided in a single slice or nonsequential slices, namely, CYP2D6, CYP4F2, and CYP2F1.

Table 2
Distribution of peptides with single amino acid polymorphisms between the slices

Allele	Peptide	Theoretical MH+ (Da)	Observed *m/z*	Slices
CYP1A2 R281W	[278]NSV**W**DITGALFKHSK[292]	1,702.93	1,702.86[a]	17
			1,702.87[b]	18
			1,702.87[b]	19
CYP3A4 G56D	[55]**D**FCMFDMECHKK[68]	1,533.81	1,533.72[b]	17
			1,533.73[b]	18
			1,533.72[b]	19
			1,533.74[a]	20
EST1 S75N	[64]FTPPQPAEPW**N**FVK[79]	1,657.88	1,657.82[a]	11
			1,657.82[a]	14

CYP cytochrome P450, *EST1* liver carboxylesterase 1
[a]*m/z* was revealed in the most abundant (MA) slice within protein bandwidth
[b]*m/z* was revealed in the slice(s) neighboring the MA slice

1D-PM can be also produced for another HLMGs' membrane protein, e.g. for arylacetamide deacetylase; this enzyme is identified in four slice nos. 33–36. Appending the MS data for slice nos. 33, 35, and 36 to those for the MA slice no. 34 increased the sequence coverage of arylacetamide deacetylase by approximately 21%; and its sequence coverage reached the 56%.

14. Table 2 presents the allele-specific candidates revealed by MALDI-TOF MS and provides information on the ability of a rapid MALDI-TOF survey to detect the CYP1A2 and CYP3A4 polymorphic alleles. We observed *m/z* peaks that are specific for CYP1A2 and CYP3A4 polymorphisms, R281W and G56D, respectively. SAP-specific peptide can be detected not only in the most abundant slice (MA slice), e.g., in no. 17 for CYP1A2 (R281W), but also in the neighboring slices nos. 18 and 19. SAP-specific peptide G56D for CYP3A4 is observed in the MA slice no. 20 and in the neighboring slice nos. 17–19.

Table 2 also presents a SAP-specific peptide for liver carboxylesterase 1 (EST1). Two landmark proteins are used to determine the range of CYPs' molecular masses. The upper boundary was determined by EST1, which is known to be involved in the detoxification of xenobiotics and the activation of ester and amide prodrugs. This enzyme was usually detected in several slices, forming a wide bandwidth from slice no. 7 to slice no. 14. A total of six SAPs can be identified for this protein. Table 2 presents an *m/z* peak (1,657.82 Da) that is specific for EST1 polymorphism S75N, and can be detected in slices nos. 11 and 14.

15. The steps of data interpretation described in this chapter should be automated by a skilled bioinformatician as a series of scripts, specific to the MS data formats used in a particular laboratory. For those working with the FlexAnalysis/DataAnalysis software suit from Bruker Daltonics we provide such scripts in the ZOOMer package, enabling to process the MS data and to visualize this data as the 1D-PM. ZOOMer is also helpful in selecting bandwidth m/z values and for observing the occurrence of SAPs (or protein isoforms) across several samples (see Fig. 2c).

ZOOMer software also can be applied to utilize the data obtained by high-throughput LC-MS/MS. In this case it is necessary to process the html-format file where Mascot reports are stored for each slice. In our experiments proteins from the slice are identified by database MS/MS search using Mascot under the following parameters: accuracy of determination of mass peptide ions ± 1.5 Da, the parent ion mass tolerance 200 ppm, the fragment mass tolerance 0.8 Da, and the possibility of methionine oxidation and modification of cysteine residues by acrylamide. The 1D-PM should indicate the number of MS/MS-identified peptides that match a given protein in the averaged peak list of a slice.

On average, the peak list for PMF searching included 96 ± 23 m/z peaks (for positive identification of CYP forms we used 16 ± 4 m/z peaks), whereas the MS/MS search report contained from 4 to 17 peptide sequences. 1D-PMs obtained using PMF and MS/MS data were basically similar (18).

Acknowledgments

The authors would like to thank Dr. Alexey Chernobrovkin for technical assistance and Cryptome Research Ltd. (http://www.cryptome.ru/) for supporting and developing ZOOMer software. This work was supported by a contract of Ministry of Education and Science of the Russian Federation (No. 2010–01.648.12.3004) as a part of the "Development of methodology for estimation of a safety of nanomaterials" Project.

References

1. Archakov AI, Bachmanova GI (1990) Cytochrome P450 and active oxygen. Taylor and Francis, London
2. Lewis DFV (2001) Guide to cytochromes P450, Structure and Function. Taylor and Francis, London and New York
3. McFadyen MC, Melvin WT, Murray GJ (2004) Cytochrome P450 enzymes: novel options for cancer therapeutics Molecular. Cancer Therapy 3:363–371
4. Zhang H, Davis CD, Sinz MW, Rodrigues AD (2007) Cytochrome P450 reaction-phenotyping:

an industrial perspective. Expert Opin Drug Metab Toxicol 3(5):667–687

5. Anzenbacher P, Anzenbacherova E (2001) Cytochromes P450 and metabolism of xenobiotics. Cell Mol Life Sci 58:737–747
6. Wang Y, Al-Gazzar A, Seibert C, Sharif A (2006) Proteomic analysis of cytochromes P450: a mass spectrometry approach. Biochem Soc Trans 34:1246–1251
7. Galeva N, Alterman M (2002) Comparison of one-dimensional and two dimensional gel electrophoresis as a separation tool for proteomic analysis of rat liver microsomes. Proteomics 2:713–722
8. Nisar S, Lane CS, Wilderspin AF et al (2004) A proteomic approach to the identification of cytochrome P450 isoforms in male and female rat liver by nanoscale liquid chromatography-electrospray ionization-tandem mass spectrometry. Drug Metab Dispos 32:382–386
9. Lane CS, Nisar S, Griffiths WJ et al (2004) Identification of cytochrome P450 enzymes in human colorectal metastases and the surrounding liver: a proteomic approach. Eur J Cancer 40:2127–2134
10. Molloy MP, Herbert BR, Slade MB et al (2000) Proteomic analysis of the Escherichia coli outer membrane. Eur J Biochem 267:2871–2881
11. Simpson RJ, Connolly LM, Eddes JS et al (2000) Proteomic analysis of the human colon carcinoma cell line (LIM 1215): development of a membrane protein database. Electrophoresis 21(9):1707–1732
12. Cottingham K (2010) 1DE proves its worth … again. J Proteome Res 9(4):1636
13. Supek F, Peharec P, Krsnik-Rasol M, Smuc T (2008) Enhanced analytical power of SDS-PAGE using machine learning algorithms. Proteomics 8:28–31
14. Thiede B, Treumann A, Kretschmer A et al (2005) Shotgun proteome analysis of protein cleavage in apoptotic cells. Proteomics 5:2123–2130
15. He N, Botelho JM, McNall RJ et al (2007) Proteomic analysis of cast cuticles from Anopheles gambiae by tandem mass spectrometry. Insect Biochem Mol Biol 37:135–146
16. Poetsch A, Schlüsener D, Florizone Ch et al (2008) Improved identification of membrane proteins by MALDI-TOF MS/MS using vacuum sublimated matrix spots on an ultraphobic chip surface. J Biomol Tech 19:129–138
17. Rezaul K, Wu L, Mayya V et al (2005) A systematic characterization of mitochondrial proteome from human T leukemia cells. Mol Cell Proteomics 4(2):169–181
18. Lisitsa AV, Petushkova NA, Thiele H et al (2010) Application of slicing of one-dimensional gels with subsequent slice-by-slice mass spectrometry for the proteomic profiling of human liver cytochromes P450. J Proteome Res 9(1):95–103
19. Petushkova NA, Larina OV, Chernobrovkin AL et al (2012) Optimization of the SDS-PAGE gel slicing approach for identification of human liver microsomal proteins via MALDI-TOF mass spectrometry. J Proteomics Bioinform 5:40–49 (http://www.omicsonline.org/ArchiveJPB/February2012.php)
20. Han X, Aslanian A, Yates JR III (2008) Mass spectrometry for proteomics. Curr Opin Chem Biol 12(5):483–490
21. Lim H, Eng J, Yates JR 3rd et al (2003) Identification of 2D-gel proteins: a comparison of MALDI/TOF peptide mass mapping to μLC-ESI tandem mass spectrometry. J Am Soc Mass Spectrom 14:957–970
22. Berndt P, Hobohm U, Langen H (1999) Reliable automatic protein identification from matrix-assisted laser desorption/ionization mass spectrometric peptide fingerprints. Electrophoresis 20:3521–3526
23. Laemmli UK (1970) Cleavage of structural proteins during the assembly of the head of bacteriophage T4. Nature 227(5259):680–685
24. Karuzina II, Archakov AI (1994) Hydrogen peroxide-mediated inactivation of microsomal cytochrome P450 during monooxygenase reactions. Free Radic Biol Med 17(6):557–567
25. Bradford MM (1976) A rapid and sensitive method for the quantitation of microgram quantities of protein utilizing the principle of protein-dye binding. Anal Biochem 72:248–254
26. Govorun VM, Archkov AI (2002) Proteomic technologies in modern biomedical science. Biochemistry (Mosc) 67:1109–1123
27. Neve EP, Ingelman-Sundberg M (2010) Cytochrome P450 proteins: retention and distribution from the endoplasmic reticulum. Curr Opin Drug Discov Devel 13(1):78–85

Chapter 6

Assessing Heterogeneity of Peroxisomes: Isolation of Two Subpopulations from Rat Liver

Markus Islinger, Afsaneh Abdolzade-Bavil, Sven Liebler, Gerhardt Weber, and Alfred Völkl

Abstract

Peroxisomes exhibit a heterogeneous morphological appearance in rat liver tissue. In this respect, the isolation and subsequent biochemical characterization of peroxisome species from different subcellular prefractions should help to solve the question of whether peroxisomes indeed diverge into functionally specialized subgroups in one tissue. As a means to address this question, we provide a detailed separation protocol for the isolation of peroxisomes from both the light (LM-Po) and the heavy (HM-Po) mitochondrial prefraction for their subsequent comparative analysis. Both isolation strategies rely on centrifugation in individually adapted Optiprep gradients. In case of the heavy mitochondrial fraction, free flow electrophoresis is appended as an additional separation step to yield peroxisomes of sufficient purity. In view of their morphology, peroxisomes isolated from both fractions are surrounded by a continuous single membrane and contain a gray-opaque inner matrix. However, beyond this overall similar appearance, HM-Po exhibit a smaller average diameter, float at lower density, and show a more negative average membrane charge when compared to LM-Po.

Key words: Peroxisomes, Liver, Enzymes, Mitochondria, Fractionation, Free flow electrophoresis

1. Introduction

Peroxisomes can be found in nearly all eukaryotic cells and tissues of multicellular organisms. Therefore they had to adapt to a plethora of cellular physiologies and metabolisms, resulting in a considerably heterogeneous cellular compartment (1). However, peroxisomes do not only vary from species to species or tissues, respectively, but also exhibit differential appearance in a single organ. Rat liver peroxisomes

Djuro Josic and Douglas C. Hixson (eds.), *Liver Proteomics: Methods and Protocols*, Methods in Molecular Biology, vol. 909, DOI 10.1007/978-1-61779-959-4_6, © Springer Science+Business Media New York 2012

represent morphologically heterogeneous organelles (2) implicating the existence of biogenetically or metabolically different subpopulations. For their biochemical characterization, peroxisomes are traditionally purified from the so-called light mitochondrial fraction, which sediments at a centrifugal force of 37,000 × g_{max}. However, peroxisomal enzyme activities can be found in considerable amounts in all fractions of the classical scheme for subcellular differential centrifugation (3, 4). The isolation of largely pure peroxisomes from these nonclassical sources is therefore a prerequisite for a detailed comparative analysis, which could clarify whether peroxisomes show significant functional variability in a single tissue.

Here we present isolation protocols for peroxisomes found in the heavy mitochondrial (HM-Po) and light mitochondrial fraction (LM-Po) of rat liver, which yield peroxisomes of sufficient purity for subsequent biochemical investigations. Whereas highly pure peroxisomes from the light mitochondrial fraction can be isolated by traditional gradient centrifugation, such an approach is not sufficient for the purification of HM-Po. Therefore we introduced free flow electrophoresis (FFE) as an additional purification step, thereby adding another physical parameter to the purification procedure. Unlike centrifugation techniques, separation by FFE is largely independent from organellar size because it mainly relies on different net charges on the organelles' surface. In FFE an electric field is applied perpendicular to a laminar buffer flow in a rectangular separation chamber. Thus, different analytes applied at the lower end of the separation chamber migrate in different velocities towards the anodic or cathodic end of the separation chamber, while they are drawn towards the instrument's outlet, where they are collected in 96 fractions. In principle, FFE can be run in three different modes: zonal-FFE, isoelectric focusing, and isotachophoresis, each with individual pros and cons for different biological compounds (5). Zonal-FFE is the most classic separation mode, only requiring a single separation buffer in the FFE-chamber and allowing a continuous separation at comparatively high flow rates. Therefore, organelles can be purified under gentle conditions but nevertheless high sample throughput. Due to these advantages, zonal-FFE has been used for the isolation of many different organelles, e.g., mitochondria, lysosomes, endosomes, Golgi vesicles, and peroxisomes (6–13). Even so, FFE still remains a technique largely overlooked in the field. Here we present a new low-ionic-strength separation system which by use of two buffer components allows a rapid shift to a stable, optimal pH and a separation at high voltage. Combining this separation technique with an Optiprep step gradient allows the purification of HM-Po suitable for a subsequent intensive biochemical characterization. Unlike the LM-Po, HM-Po are gained from the heavy mitochondrial fraction pelleted at a centrifugal force of only 2,700 × g_{max} and thus may comprise peroxisomes with altered physical but also metabolic functions.

2. Materials

2.1. Generation of Heavy and Light Mitochondrial Fractions

1. Motor-driven Potter–Elvehjem tissue grinder with loose-fitting pestle (clearance 0.1–0.15 mm).
2. Homogenization buffer (HB): 250 mM sucrose, 5 mM MOPS, 1 mM ethylenediaminotetraacetic acid (EDTA), 2 mM phenylmethylsulfonylfluoride (PMSF), 1 mM dithiothreitol (DTT), 1 mM ε-aminocaproic acid. Adjust pH to 7.4 with KOH.
3. Gradient Buffer (GB): 5 mM MOPS, 1 mM EDTA, 2 mM PMSF, 1 mM DTT, 1 mM ε-aminocaproic acid. Adjust pH to 7.4 with KOH.
4. Refrigerated centrifuge with a fixed angle rotor, e.g., Beckman Avanti with a J-25 or a JA-20 rotor.
5. Animals: Rats and mice approximately 8 weeks of age (220–225 g and 15–25 g body weight, respectively) (see Note 1).

2.2. Purification of LM-Po from the Light Mitochondrial Fraction

1. Refractometer for the preparation of the density gradients.
2. Optiprep: 60% (w/v) iodixanol in water (Axis Shield, Rodeløkka, Sweden).
3. Ultracentrifuge with a fixed-angle rotor and appropriate centrifugation tubes: Beckman Optima LE-80K with a Beckman VTi 50 vertical-type rotor and Quick-seal polyallomer tubes 25 × 89 mm.

2.3. Purification of HM-Po from the Heavy Mitochondrial Fraction

1. Prepare an Optiprep-step gradient as described in Subheading 3.2, steps 1 and 2.
2. Ultracentrifuge with a swing-out rotor and appropriate centrifugation tubes: Beckman Optima LE-80K with a Beckman SW32 Ti rotor and Quick-seal polyallomer tubes 25 × 89 mm.
3. Reagents for the Optiprep step gradient: Dilution Buffer (DB): 8% sucrose, 1 mM EGTA, 20 mM tricine, pH 7.8; 50% Optiprep solution in 1.3% sucrose, 1 mM EGTA, 20 mM tricine, pH 7.8.
4. FFE system, e.g., Octopus (Dr. Weber GmbH), Pro Team FFE (Tecan), BD FFE System (BD Diagnostics).
5. Anodic electrolyte stabilizer: 250 mM sucrose, 50 mM MES, 100 mM H_2SO_4, 250 mM morpholinoethanol, 0.1% HPMC.

6. Cathodic electrolyte stabilizer: 250 mM sucrose, 150 mM NaOH, 30 mM Tris, 200 mM TAPS, 0.1% HPMC.
7. Separation buffer: 250 mM sucrose, 7 mM 3-[[2-hydroxy-1,1-bis(hydroxymethyl)ethyl]amino]-1-propanesulfonic acid (TAPS), 14 mM morpholinoethanol, 0.1% HPMC, pH 7.8.
8. Counterflow: 250 mM sucrose, 0.1% HPMC.
9. Electrode anode circuit: 100 mM H_2SO_4, 250 mM morpholinoethanol.
10. Electrode cathode circuit: 150 mM NaOH, 200 mM TAPS.
11. Pump tubing (Tygon standard; Ismatec): Media, seven tubes of 0.64 mm ID (E1–E7); counterflow, one tube of 0.86 ID, one tube of 1.42 ID, one tube of 0.86 ID; sample, one tube of 0.64 mm ID.
12. 96-well plates and 96-deep well plates (capacity 4 mL/well).

2.4. Enzyme Assays for Purity Validation

1. General equipment:

   TVBE-Buffer: 1 mM $NaHCO_3$, 1 mM EDTA, 0.1% (w/v) ethanol, 0.01% (w/v) Triton X-100, pH adjusted to 7.6 with HCl.

2. Catalase assay

   Catalase substrate: Dilute 10 mL imidazole solution pH 7.0 in 100 mL of 1 mg/mL bovine serum albumin solution. Add 35 μL 30% H_2O_2.

   TiOSO4-solution: Boil up 6.8 g titanoxide sulfate-1-hydrate in 1 L 1 M H_2SO_4 (see Note 2). Thereafter filter through two layers of folded paper filters. Bring to 1.5 L with 1 M H_2SO_4.

3. Succinate dehydrogenase assay

   Reaction buffer: Prepare fresh by mixing the following stock solutions (volumes are given for one reaction): 100 μL 50 mM potassium phosphate buffer, pH 7.4, 50 μL 1% Triton WR-1339, 50 μL 2.5 mg/mL Nitrotetrazolium Blue chloride (NBT, prepare from a 50 mg/mL ethanol stock solution), 50 μL 100 mM sodium succinate.

   Stop-solution: 2% SDS in H_2O.

4. Acid phosphatase assay

   Acetate buffer: 180 mM sodium acetate, bring to pH 5.0 with acetic acid.

   Substrate solution: 16 mM *p*-nitrophenyl-phosphate.

5. Esterase assay

   Reaction buffer: 20 mM potassium phosphate, 0.1% Triton X-100, 10 mM EDTA, pH 7.4.

   Substrate solution: Dissolve 32.6 mg *o*-nitrophenyl-acetate in 1 mL methanol.

3. Methods

After differential centrifugation of liver tissue, activities of peroxisomal marker enzymes can be found in all particulate subfractions in varying amounts (4). Since peroxisomes show also a heterogeneous ultrastructural morphology in the same tissue or even in single cells (14–17), it can be assumed that their distribution over a broad density range is not just caused by insufficient separation but may result from true physically and potentially functionally distinct subfractions. However, existing protocols for peroxisome isolation use the light mitochondrial fraction as starting material, which means that peroxisomes sedimenting at higher or lower densities have not been characterized in detail.

To overcome this limitation, we present a fractionation scheme which enables the isolation of peroxisomes from both the light and the heavy mitochondrial fraction and from the same starting material, thus enabling further comparative analyses.

Starting with a liver homogenate, heavy and light mitochondrial fractions are prepared by differential centrifugation. Both fractions are further purified by individual Optiprep gradients (Fig. 2b) where peroxisomes allocate near the bottom of the gradient. As peroxisomes are only a minor constituent of the heavy mitochondrial fraction, obtaining organellar fractions of high purity was not feasible with only a single gradient centrifugation step. Therefore, we added free flow zone electrophoresis to increase the purity of the peroxisome-enriched fraction of the Optiprep gradient. In FFE analytes are separated according to their electrophoretic mobility. For organelles this is mainly determined by the overall surface charge of their surface membrane since opposing frictional forces are comparably low in aqueous separation media (for a detailed description of the principles of FFE see 18). Presumably due to the negative charges of the phosphate-head groups of the membrane lipids, organelles generally migrate to the anodic side of the instrument. Thus, when the sample is applied next to the cathode of the instrument, the distance in the separation chamber accessible for separation is maximized. Nevertheless, peroxisomes do not separate well from other organelles in the mode of zone electrophoresis that use common FFE-buffer systems (19), making it necessary to introduce a new separation buffer to accomplish this task. Separation capacity in FFE as in other electrophoretic systems is mainly hindered by the ionic strength of the separation medium leading to higher currents with rising ion concentrations and thus restricting the maximum charge applicable for separation. To minimize the ion concentrations we therefore devised a separation medium consisting of two buffer components of opposing pI (TAPS, morpholinoethanol), thus stabilizing the solution at a physiologic pH even when applied in the low molarity range.

In their morphological appearance peroxisomes from both LM-Po and HM-Po fractions do not significantly differ from those found in the starting material and show an electron dense, opaque inner matrix surrounded by a well-preserved single membrane (Fig. 1a). With respect to purity of the fractions, the mitochondria, lysosomes, and microsomes are efficiently removed in stepwise fractions, as shown by both microscopical observation and the enrichment or decrease of characteristic marker enzymes, respectively (Fig. 1a, c). Although the grade of contamination found in HM-Po is still higher than in LM-Po, this fraction proved to be adequate for a detailed biochemical and mass spectrometrical comparison to LM-Po (5).

Indeed comparison of both fractions revealed significant differences not only in density, but also in size, net surface charge (Table 1), and quantitative distribution of individual proteins. These initial findings may provide the basis for further studies directed towards unraveling the functional significance of the heterogeneity of peroxisomes in liver.

3.1. Preparation of Heavy and Light Mitochondrial Fractions

1. All steps should be performed at a temperature of 4°C.
2. Anesthetize rats (see Note 3), open the body cavity, and rapidly excise the liver (see Note 4). After determining its weight, cut the liver into pieces of approximately 2.5 mm^3 and suspend in ice-cold Homogenization Buffer at a ratio of 3 mL/g tissue.
3. Homogenize the liver pieces using a motor-driven Potter–Elvehjem tissue grinder and a loose-fitting pestle with exactly one stroke of 1,000 rpm in 2 min.
4. Centrifuge the homogenate at $600 \times g_{max}$ for 10 min, 4°C, to sediment nuclei and cellular debris.
5. Store the supernatant on ice and resuspend the pellet in at least 10 mL of HB per liver. Homogenize the pellet for a second time using a stroke of 1,000 rpm in 1 min. Centrifuge again at $600 \times g_{max}$.
6. Combine both supernatants and centrifuge at $2{,}700 \times g_{max}$ for 10 min to pellet the heavy mitochondrial fraction.
7. Decant the supernatant and save it on ice. Manually resuspend the heavy mitochondrial pellet in HB manually with a glass rod. Avoid disturbing the blood at the bottom of the pellet. Centrifuge at $2{,}700 \times g_{max}$ for a second time and resuspend the pellet as described in Subheading 3.3.2.
8. Combine the supernatants and centrifuge at $37{,}000 \times g_{max}$ for 20 min. This light mitochondrial fraction contains a mixture of mainly mitochondria, microsomes, and peroxisomes.

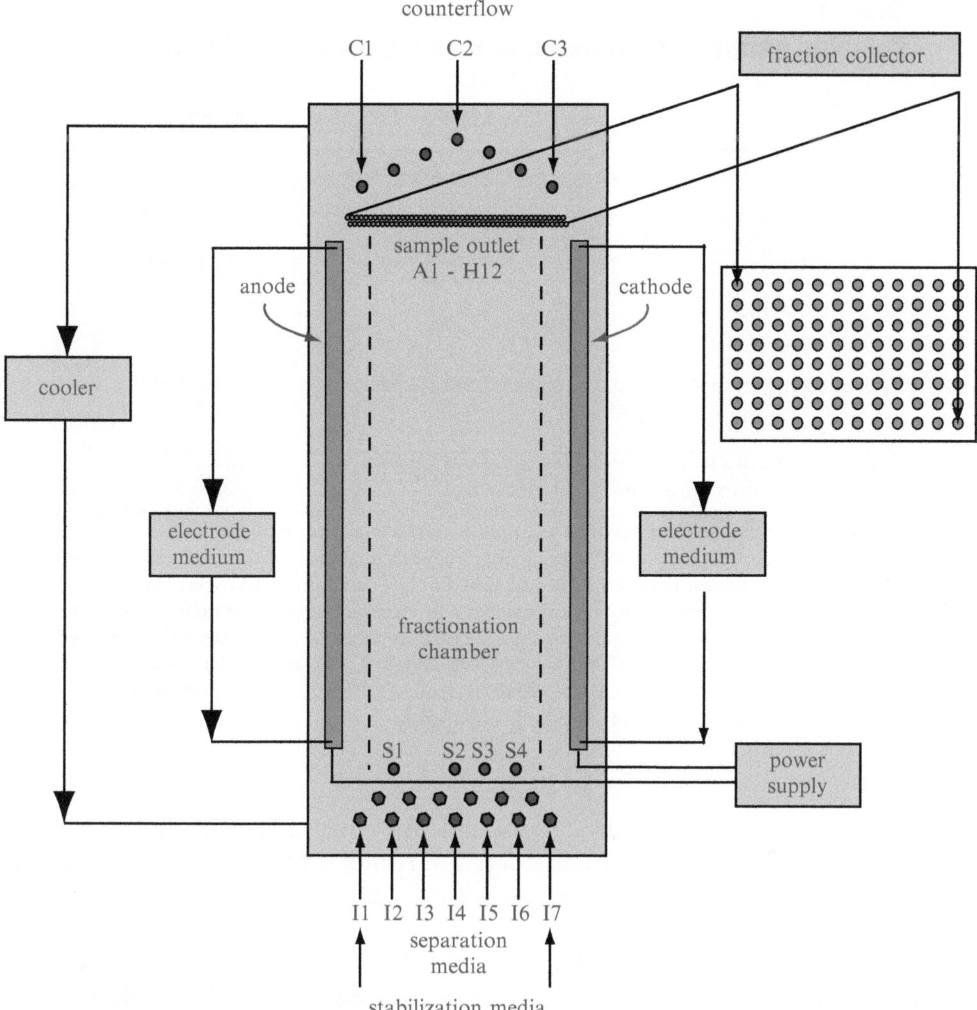

Fig. 1. Schematic overview of the Free Flow Electrophoresis system based on the BD instrument. The abbreviations used for the various buffer inlets correspond to the terms used in the text section. In brief: counterflow inlets C1–3 are operated with counterflow medium, media inlets I1 and I7 with anodic and cathodic stabilization buffer, respectively, and inlet I3 I5 with separation medium all delivered by the same peristaltic pump. Samples instead are applied by an independently working pump via the anodic sample inlet (S4) (with kind permission from Springer Science + Business Media: Methods in Molecular Biology vol. 432: Organelle Proteomics (2008). Islinger M., Weber G., Free Flow Isoelectric Focusing—A method for the separation of both hydrophilic and hydrophobic proteins of rat liver peroxisomes. Humana Press Inc., Totowa, NJ, pp. 199–215, Fig. 3).

9. Aspirate the supernatant including the reddish fluffy layer on top of the pellet mainly comprising microsomes.
10. Resuspend the light mitochondrial pellet with a glass rod, add HB drop-wise until a homogenous suspension is gained, and then adjust with HB to a volume of approximately 2 mL/g. Centrifuge at 37,000 × g_{max} for 15 min.

Table 1
Selected parameters differing between LM-Po and HM-Po

	LM-Po	HM-Po
Size	470.3 ± 104.6 nm	369.5 ± 103.9 nm
Mobility in FFE	0.47 mm/V h	0.38 mm/V h
Buoyant density	1.20 g/mL	1.18 g/mL
Ratio of enzyme activities of ACOX1/OXDA	3.9	2.2
Ratio staining intensity of ACOX1/OXDA on immunoblots	1.06 ± 0.53	0.79 ± 0.39
Ratio staining intensity of ACBD3/PXMP2 on immunoblots	1.79 ± 0.60	0.26 ± 0.17

Size is given as the mean diameter of peroxisome cross sections determined by electron microscopy using Metamorph analysis software. Selected peroxisomal proteins, which were found to differ in their abundance between LM- and HM-Po by quantitative mass spectrometry (5), were validated by enzyme activity measurements and immunoblotting. Western Blots (three independent experiments) were quantified using Quantity One (Bio-Rad). Diverging ratios of peroxisomal matrix enzymes (ACOX1—Acyl-CoA oxidase 1, OXDA—D-aminoacid oxidase) as well as membrane proteins (ACBD3—PMP70, Pxmp2—PMP22) proof a different enrichment of individual proteins in both organelle populations

11. Carefully, resuspend the final pellet comprising the light mitochondrial fraction (crude LM-Po fraction) as described in step 9 in HB at 1–2 mL/g.

3.2. Isolation of Peroxisomes from the Light Mitochondrial Fraction

1. Prepare dilutions of 1.26, 1.22, 1.19, 1.15, and 1.12 g/cm^3 from the 60% Optiprep (1.32 g/cm^3) stock solution with GB using a refractometer and the formula: $\rho = 3.350 \times$ refractive index – 3.462.
2. Layer sequentially 4, 3, 6, 7, and 10 mL of the Optiprep dilutions in a decreasing order of concentration (1.26–1.12 g/cm^3), in a 40-mL centrifugation tube.
3. Freeze the discontinuous gradient in liquid nitrogen and store at –80°C (see Note 5).
4. Thaw the gradient at room temperature immediately before use in a metallic stand.
5. Layer 5 mL of the light mitochondrial fraction on top of the Optiprep gradient. Overlay with GB and seal the tubes (see Note 6).
6. Centrifuge in a vertical type rotor (e.g., VTi 50) at an integrated force of 1,256 × 10^6 g min (g_{max} = 33,000) with slow acceleration/deceleration, 4 °C.

7. Puncture the tubes with a syringe and remove the peroxisomes banding at 1.20 g/cm^3 (Fig. 1b).
8. Remove the Optiprep by diluting the fraction 1:4 with HB, pellet the organelles at 37,000 × g_{max}, and dilute the pellet in HB to a final protein concentration of 1–2 mg/mL (usually ~ 0.25–0.5 mL per rat).

3.3. Isolation of Peroxisomes from the Heavy Mitochondrial Fraction

1. Prepare gradient media of 32, 28, 25, and 20% Optiprep by diluting the 50% Optiprep stock solution with DB.
2. Resuspend the heavy mitochondrial pellet produced in step 7 of Subheading 3.1 in the 25% Optiprep solution. Adjust the organelle suspension to an exact concentration of 25% Optiprep using a refractometer.
3. Layer 3 mL of 32% and 5 mL of 28% Optiprep into a 25 × 89 mm polyallomer centrifugation tube. Overlay with 20 mL of the heavy mitochondrial fraction suspended in 25% Optiprep (see Note 7). Top-load a 4 mL layer of 20% Optiprep buffer. Balance the weight of opposing tubes with HB.
4. Centrifuge for 2 h at 50,000 × g_{av}, 4°C, in an SW32 Ti rotor. The fraction enriched in peroxisomes accumulates at the interface of the 32 and 28% Optiprep steps (Fig. 1b). Collect fractions by aspiration with a pipette beginning at the top of the gradient.
5. Assemble the chamber as described in the operating manual of your FFE system. Use the 0.5 mm spacer designed for zonal FFE of organelles. Start cooling and wait until the chamber has reached 10°C.
6. Fill the separation chamber at the inclined position with water, remove any trapped air bubbles, and check for leaks. Bring the chamber to the horizontal position and perform a stripe test to insure laminar flow conditions. To do this, feed inlets I2, I4, and I6 (Fig. 2) with a 1:100 dilution of Trisodium (4,5-dihydroxy-3-(*p*-sulfophenyl)-2,7-)naphthalenedisulfonic acid (SPADNS) in water and run media pump at 40 rpm. The SPADNS solution should flow in parallel, even-colored stripes through the separation chamber. If this is not the case, clean and reassemble the system.
7. Change to the separation media according to the following instructions (Fig. 2): hang inlets I2–I6 into separation medium, I1 into anodic stabilization medium, and I7 into cathodic stabilization medium. Place counterflow tubings C1–3 into counterflow medium. Fill the electrolyte anode and cathode circuit with liquid circuit (+ve) and liquid circuit (–ve), respectively. Equilibrate the chamber for at least two transition times (~20 min).

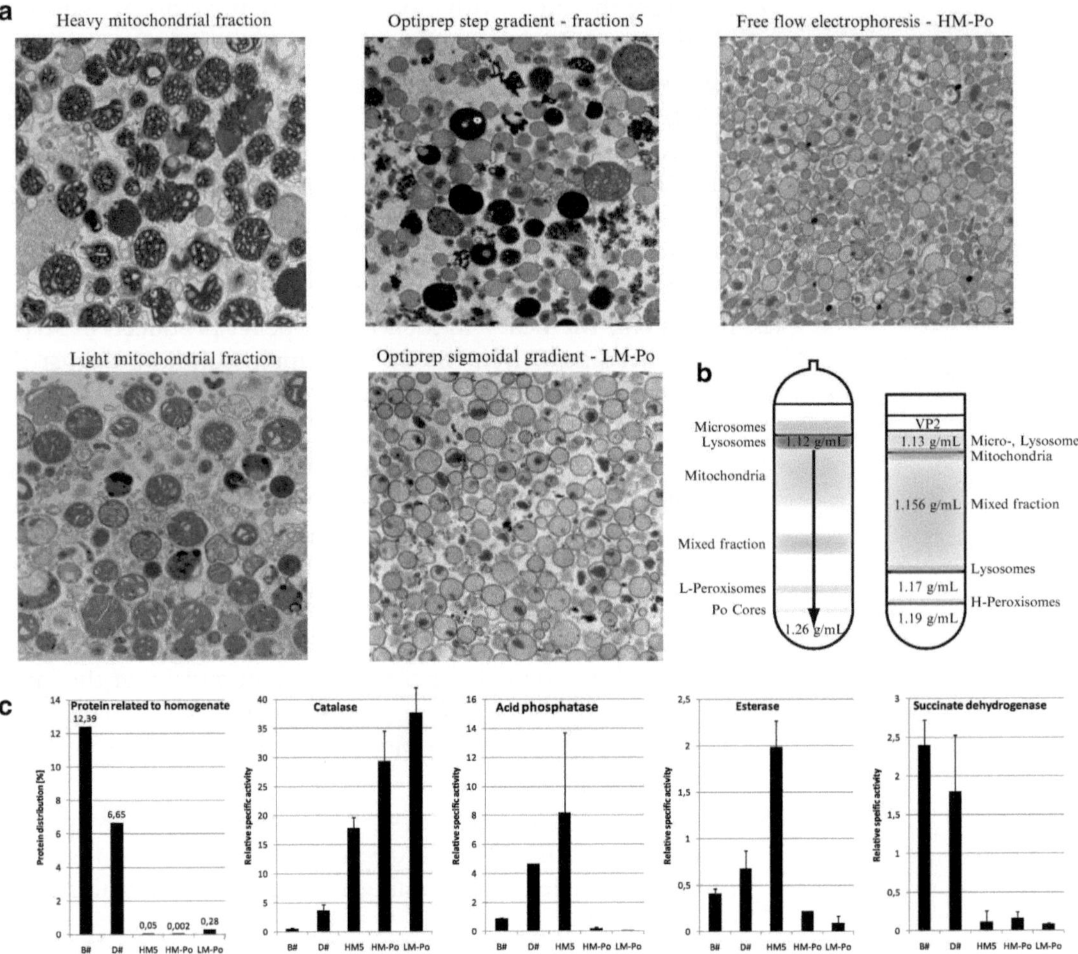

Fig. 2. Overview of the fractions obtained during the isolation procedure: (**a**) Ultrastructural appearance of the main organellar fractions analyzed. For the electron microscopic examination the organelle fractions were fixed with 2% glutaraldehyde, pelleted, and mounted on 1% agar. Then the pellets were cut into pieces of approx. 1 mm^2 dehydrated according to standard protocols and embedded in Epon. Note that both the heavy and the light mitochondrial fractions are dominated by a high proportion of the eponymous organelle. After gradient centrifugation, the LM-Po fraction generated from the light mitochondrial fraction shows a homogenous, even distribution of peroxisomes with only minimal contamination. In comparison, a single centrifugation was not sufficient in the case of the heavy mitochondrial fraction, where mitochondria were also drastically reduced but lysosomes and microsomes that copurify in part with peroxisomes are still present. After FFE, however, the fraction peaking in catalase activity shows also highly enriched, well-preserved peroxisomes. (**b**) Scheme of the Optiprep gradients used for peroxisome isolation. Because of the leakiness of peroxisomes for small molecules, gradient media are taken up by the organelles leading to a significant increase in their buoyant density. Therefore peroxisomes can be found as the organelle with the highest density located near the bottom of the gradients. Note, however, that peroxisomes of the heavy mitochondrial fraction exert a lower density than those of the light mitochondrial fraction. (**c**) Distribution of organellar marker enzyme activities in selected fractions mostly significant for peroxisome isolation. Whereas catalase as a marker for peroxisomes is enriched around 30–40 times in the peroxisome fractions obtained from light and heavy mitochondrial fractions, the enzyme markers for mitochondria (succinate dehydrogenase), lysosomes (acid phosphatase), and microsomes (esterase) are drastically decreased.

8. After changing the medium, carry out a performance test. For this purpose, switch on high voltage (850 V, I_{max}: 45 mA, P_{max}: 60 W) and set the media pump at 300 mL/h. Dilute the BD FFE system p*I*-Marker 1:2 in separation medium and equilibrate with 0.8% HPMC stock solution and solid sucrose to reach a final concentration of 0.1% and 250 mM, respectively. Apply the mixture via the right sample inlet (S4) at a flow rate of 800 μL/h. Begin to collect samples when colored drops are visible at the collecting tube outlets (5 min) as described under Subheading 3.2, step 5. The p*I* mix should distribute into five distinguishable peaks.
9. Spin down the peroxisome fraction HM5 from the Optiprep gradient at a velocity of 37,000 × g_{max}, 4°C for 15 min. Suspend the pellet with a glass rod by drop-wise addition of separation buffer until a final protein concentration of 1–2 mg/mL is reached.
10. Switch on high voltage and media flow as described for the performance test. Apply the organelle suspension via inlet S4 at a flow rate of 800–1,500 μL/h (see Note 8). Begin to collect the sample approximately 5 min after the organelle suspension reaches the separation chamber. First use a standard 96-well plate to monitor the separation profile at 420 nm, and then use a 4 mL 96-well plate for further collection, since the probe is significantly diluted with separation buffer during the run. Monitor the enrichment of peroxisomes in individual fractions using the microplate catalase assay described under Subheading 3.4.

3.4. Measurement of Catalase Activity (in Microtiter Plates) (20)

1. Withdraw 5 μL of sample directly from the microtiter plate collected from the FFE separation. If needed, dilute the samples in TVBE buffer to a concentration, which avoids a complete discoloring of the reaction solution.
2. Mix the 5 μL sample with an equal volume of 2% Triton X-100 solution and incubate for 5 min on ice.
3. Add 100 μL of catalase substrate and incubate for another 15 min on ice.
4. Add 100 μL $TiOSO_4$ solution and incubate for 10 min at room temperature.
5. Use the following formula to calculate the catalase activity:

$$\text{Catalase activity (BU/mL)} = \frac{\textit{Incubation volume}}{\textit{Sample volume} \times \textit{Incubation time} \times 50} \times \textit{Sample dilution} \times \log \frac{(\textit{OD blank})}{\textit{OD sample}}.$$

3.5. Measurement of Succinate Dehydrogenase Activity (21)

1. Pool and pellet samples as described above and suspend in 50–100 μL HB.
2. Dilute the samples in TVBE buffer to a concentration that gives a color change in the linear range (begin with 1:100).
3. For a 1.5 mL reaction volume, mix 250 μL reaction buffer and 250 μL sample solution and incubate for 30 min at 37°C. For the blank replace sample by TVBE buffer.
4. Terminate the reaction by adding 1 mL SDS stop solution.
5. Measure the extinction at a wavelength of 630 nm and calculate the enzyme activity according to the following formula:

$$\text{Enzyme activity (U/mL)} = (\text{OD}_{\text{sample}} - \text{OD}_{\text{blank}}) \times 111.1 \times \left(\frac{\text{Reaction volume}}{\text{Sample volume}} \times \text{Reaction volume} \right) \times \text{Dilution.}$$

3.6. Measurement of Acid Phosphatase Activity (22)

1. Mix 0.4 mL acetate buffer with 0.4 mL substrate solution.
2. Add 50 μL of sample or TVBE buffer as blank. Incubate for 20 min at 37°C.
3. Stop the reaction by adding 1.2 mL 0.25 M NaOH.
4. Measure the extinction at a wavelength of 630 nm and calculate the enzyme activity according to the following formula:

$$p\text{-nitrophenol turnover (mg/mL)} = (\text{OD}_{\text{sample}} - \text{OD}_{\text{blank}}) \times 0.0116.$$

5. The reaction volume can be downscaled to microtiter plate format by using one-tenth of each solution. The factor to calculate the nitrophenol concentration is then 0.0074.

3.7. Measurement of Unspecific Esterase Activity (23)

1. Mix 930 μL reaction buffer with 50 μL sample. Use TVBE buffer for the blank. Start the reaction by adding 20 μL substrate solution.
2. Measure the extinction at 420 nm directly after starting the reaction. Repeat the measurement every 30 s for a total period of 5 min.
3. Calculate the activity according to the following formula:

$$\text{Enzyme activity (U/mL)} = \frac{\Delta A/\text{min (Sample} - \text{blank)}}{3.06} \times \text{Dilution factor} \times \frac{\text{Reaction volume (mL)}}{\text{Sample volume (mL)}}.$$

4. Notes

1. Before sacrifice, the animals have to be fasted overnight; otherwise the excess amounts of glycogen in the liver tissue disturb the separation in the density gradients. One rat liver yields around 30 μg of purified HM-Po. To equal the amount of starting material approximately six mice livers have to be used per single rat liver. When animals of older age are used, for hitherto unknown reasons, the separation in the FFE gets less efficient.
2. Use an oversized beaker and submerge a glass rod to avoid superheating and concomitant boil over of the highly acidic solution. After boiling, the titan salt will not completely dissolve but the liquid will remain a slightly white-opaque color.
3. Anesthetize animals by having a properly trained operator with the appropriate license; perform an i.p. injection of Nembutal or chloralhydrate.
4. Perfusion of the liver via the portal vein prior to liver excision will efficiently remove blood from the tissue. Alternatively, remaining agglutinated blood, not separated from the sample in the 600 × *g* run, can be removed from the heavy mitochondrial pellet by cautiously aspirating the organellar layer situated on top of the blood pellet.
5. Even when Optiprep gradients are used right after their preparation, the freezing–thawing process is a prerequisite for a functional density gradient. During the thawing process the still frozen, solid core floats upwards and consequently transforms the step gradient into a gradient of slightly sigmoidal shape.
6. To ensure a proper separation of a crude peroxisomal fraction, do not apply more than 10 g liver tissue on top of the density gradient (best use one gradient per rat liver).
7. For optimal results and to minimize contaminations in the peroxisome-enriched HM5-fraction of the gradient centrifugation do not apply more than the equivalent of one rat liver (approx. 6 g tissue) on a single-step gradient.
8. If you sense sample precipitation during the electrophoresis, reduce the sample amount injected into the separation chamber. Alternatively, you can supplement the separation buffer with 5 mM NaCl to increase its ionic strength. As this will result in higher currents, the separation capacity will be slightly reduced.

References

1. Islinger M, Cardoso MJ, Schrader M (2010) Be different – the diversity of peroxisomes in the animal kingdom. Biochim Biophys Acta 1803:881–897
2. Baumgart E (1997) Application of in situ hybridization, cytochemical and immunocytochemical techniques for the investigation of peroxisomes. A review including novel data. Robert Feulgen Prize Lecture 1997. Histochem Cell Biol 108:185–210
3. Leighton F, Poole B, Beaufay H, Baudhuin P, Coffey JW, Fowler S, De Duve C (1968) The large-scale separation of peroxisomes, mitochondria, and lysosomes from the livers of rats injected with triton WR-1339Improved isolation procedures, automated analysis, biochemical and morphological properties of fractions. J Cell Biol 37:482–513
4. Volkl A, Fahimi HD (1985) Isolation and characterization of peroxisomes from the liver of normal untreated rats. Eur J Biochem 149: 257–265
5. Islinger M, Li KW, Loos M, Liebler S, Angermuller S, Abdolzade A, Weber G, Eckerskorn C, Voelkl A (2010) Peroxisomes from the heavy mitochondrial fraction: isolation by zonal free flow electrophoresis and quantitative mass spectrometrical characterization. J Proteome Res 9:113–124
6. Marsh M, Kern H, Harms E, Schmid S, Mellman I, Helenius A (1988) Co-fractionation of BHK-21 cell endosomes and lysosomes by free-flow electrophoresis. Prog Clin Biol Res 270:21–33
7. Marsh M, Schmid S, Kern H, Harms E, Male P, Mellman I, Helenius A (1987) Rapid analytical and preparative isolation of functional endosomes by free flow electrophoresis. J Cell Biol 104:875–886
8. Morre DJ (1988) Free-flow electrophoresis: preparative applications to cell-free analyses of exocytotic membrane traffic. Prog Clin Biol Res 270:7–19
9. Paulik M, Nowack DD, Morre DJ (1988) Isolation of a vesicular intermediate in the cell-free transfer of membrane from transitional elements of the endoplasmic reticulum to Golgi apparatus cisternae of rat liver. J Biol Chem 263:17738–17748
10. Zischka H, Braun RJ, Marantidis EP, Buringer D, Bornhovd C, Hauck SM, Demmer O, Gloeckner CJ, Reichert AS, Madeo F, Ueffing M (2006) Differential analysis of Saccharomyces cerevisiae mitochondria by free flow electrophoresis. Mol Cell Proteomics 5:2185–2200
11. Zischka H, Weber G, Weber PJ, Posch A, Braun RJ, Buhringer D, Schneider U, Nissum M, Meitinger T, Ueffing M, Eckerskorn C (2003) Improved proteome analysis of Saccharomyces cerevisiae mitochondria by free-flow electrophoresis. Proteomics 3:906–916
12. Eubel H, Lee CP, Kuo J, Meyer EH, Taylor NL, Millar AH (2007) Free-flow electrophoresis for purification of plant mitochondria by surface charge. Plant J 52:583–594
13. Eubel H, Meyer EH, Taylor NL, Bussell JD, O'Toole N, Heazlewood JL, Castleden I, Small ID, Smith SM, Millar AH (2008) Novel proteins, putative membrane transporters, and an integrated metabolic network are revealed by quantitative proteomic analysis of Arabidopsis cell culture peroxisomes. Plant Physiol 148: 1809–1829
14. Cimini AM, Singh I, Farioli-Vecchioli S, Cristiano L, Ceru MP (1998) Presence of heterogeneous peroxisomal populations in the rat nervous tissue. Biochim Biophys Acta 1425:13–26
15. Schrader M, Baumgart E, Volkl A, Fahimi HD (1994) Heterogeneity of peroxisomes in human hepatoblastoma cell line HepG2, Evidence of distinct subpopulations. Eur J Cell Biol 64:281–294
16. Roels F, Cornelis A (1989) Heterogeneity of catalase staining in human hepatocellular peroxisomes. J Histochem Cytochem 37:331–337
17. Gorgas K, Krisans SK (1989) Zonal heterogeneity of peroxisome proliferation and morphology in rat liver after gemfibrozil treatment. J Lipid Res 30:1859–1875
18. Krivankova L, Bocek P (1998) Continuous free-flow electrophoresis. Electrophoresis 19:1064–1074
19. Volkl A, Mohr H, Weber G, Fahimi HD (1998) Isolation of peroxisome subpopulations from rat liver by means of immune free-flow electrophoresis. Electrophoresis 19:1140–1144
20. Baudhuin P (1974) Isolation of rat liver peroxisomes. Methods Enzymol 31:356–368
21. Nachlas MM, Margulies SI, Seligman AM (1960) A colorimetric method for the estimation of succinic dehydrogenase activity. J Biol Chem 235:499–503
22. Bergmeyer HU (1972) Standardization of enzyme assays. Clin Chem 18:1305–1311
23. Beaufay H, Amar-Costesec A, Feytmans E, Thines-Sempoux D, Wibo M, Robbi M, Berthet J (1974) Analytical study of microsomes and isolated subcellular membranes from rat liver. I. Biochemical methods. J Cell Biol 61:188–200

Chapter 7

Purification and Proteomic Analysis of Liver Membrane Skeletons

Jintang He, Yashu Liu, Qingsong Wang, and Jianguo Ji

Abstract

The detergent-resistant membrane skeletons play a critical role in cell shaping and signaling. The focus of the methods described in this chapter is first on the preparation of membrane skeletons from liver by multistep sucrose density gradient centrifugation, and then on the analysis of the protein components of membrane skeletons using proteomics techniques. Two proteomic analysis strategies are described. In the first strategy, membrane skeleton proteins are separated by two-dimensional gel electrophoresis and identified by matrix-assisted laser desorption/ionization tandem time-of-flight mass spectrometry. In the other strategy, proteins are separated by SDS-PAGE and identified by liquid chromatography-electrospray ionization Fourier transform ion cyclotron resonance mass spectrometry. The methods facilitate the understanding of the structure of membrane skeletons.

Key words: Liver, Plasma membrane, Membrane skeleton, Mass spectrometry, Proteomics

1. Introduction

Membrane skeletons refer to the detergent-resistant cytoskeletal networks located on the inner surface of plasma membranes. Membrane skeletons are found in various cells such as human red blood cells (1), neutrophils (2), and liver cells (3). They play a critical role not only in shaping the cell and determining tissue integrity, but also in signal transduction pathways (4, 5). The cytoskeletal proteins within these membrane skeleton structures connect tightly with each other, and the interactions between them provide resistance against disruption by the nonionic detergent Triton X-100. High-purity membrane skeletons can be prepared based on their detergent-resistant property.

Djuro Josic and Douglas C. Hixson (eds.), *Liver Proteomics: Methods and Protocols*, Methods in Molecular Biology, vol. 909, DOI 10.1007/978-1-61779-959-4_7, © Springer Science+Business Media New York 2012

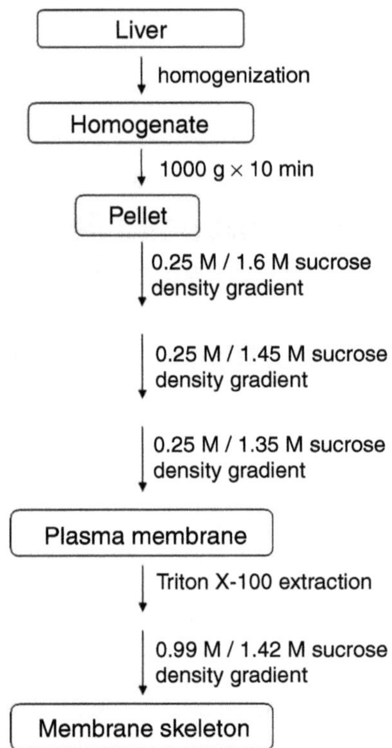

Fig. 1. Workflow of the method to prepare membrane skeletons from liver.

The parenchyma of the liver, the largest epithelial organ in the body, is rich in membrane skeleton-associated domains, especially various cell junctions (6–8). Spectrin-based cytoskeletal meshworks have also been found in liver (3, 9, 10). These specialized domains may greatly influence the growth and development of hepatocytes, the chief functional epithelial cells that contribute 80% of the liver mass. Mass spectrometry-based proteomics approaches have been employed to analyze membrane skeletons, and a lot of cytoskeletal proteins within these structures have been identified (2, 3, 11, 12). These studies will improve the understanding of the organization and function of membrane skeletons.

In this chapter, we describe an approach to prepare membrane skeletons from human or rat liver. We prepare plasma membranes and further purify membrane skeletons by Triton X-100 extraction. A diagram of the workflow of this method is shown in Fig. 1. We also describe proteomic analysis of membrane skeletons using the following two strategies: protein separation by two-dimensional gel electrophoresis and identification by matrix-assisted laser desorption/ionization tandem time-of-flight mass spectrometry (MALDI-TOF–TOF MS) and protein separation by SDS-PAGE and identification by liquid chromatography-electrospray ionization Fourier transform ion cyclotron resonance mass spectrometry (LC-ESI-FTICR MS).

2. Materials

2.1. Preparation of Liver Plasma Membranes

1. Homogenization buffer (HB): 0.25 M sucrose, 10 mM HEPES, 1 mM $MgCl_2$, 1 mM PMSF, pH 7.5.
2. High-density sucrose: 2.4 M sucrose, 10 mM HEPES, 1 mM $MgCl_2$, 1 mM PMSF, pH 7.5.
3. Scissors and tweezers (Thermo Fisher Scientific).
4. 40-ml Dounce homogenizer with a "loose" pestle and "tight" pestle (Wheaton, Millville, NY).
5. Himac CR21 centrifuge (Hitachi, Toyota, Japan).
6. Optima™ L-80 XP ultracentrifuge (Beckman Coulter, Fullerton, CA).
7. Himac CP 100MX ultracentrifuge (Hitachi, Toyota, Japan).
8. Abbe refractometer (Shanghai Optical Instruments Factory, Shanghai, China).
9. Microcentrifuge tubes (Eppendorf).

2.2. Preparation of Detergent-Resistant Membrane Skeletons

1. Triton X-100 extraction buffer: 1% Triton X-100, 25 mM Tris, 250 mM NaCl, 1 mM EGTA,1 mM ATP, 1 mM PMSF.
2. 7-ml Dounce homogenizer with a "loose" pestle and "tight" pestle (Wheaton, Millville, NY).
3. Himac CP 100MX ultracentrifuge (Hitachi, Toyota, Japan).

2.3. Evaluation of the Morphology and Purity of Plasma Membranes and Membrane Skeletons

2.3.1. Electron Microscopy

1. Glutaraldehyde, ethanol, and acetone (Sino-American Biotechnology, Beijing, China).
2. PBS, osmium tetroxide, uranyl acetate, and lead citrate (Sigma-Aldrich).
3. Epon 812 (Merck).
4. JEM-1010 transmission electron microscope (JEOL, Tokyo, Japan).

2.3.2. Western Blotting

1. Lysis buffer: 7 M urea, 2 M thiourea, 2% amidosulfobetaine-14 (ASB-14, a zwitterionic detergent), and 65 mM dithiothreitol (DTT).

2. PBST: 0.05% Tween-20 in PBS.
3. Blocking buffer: 1% BSA in PBST.
4. 2-D Quant Kit (GE Healthcare, Uppsala, Sweden).
5. Mini-PROTEAN 3 system (Bio-Rad, Hercules, CA).
6. Polyvinylidene difluoride (PVDF) membrane (Bio-Rad, Hercules, CA).
7. Trans-Blot SD Semi-Dry System (Bio-Rad, Hercules, CA).
8. Anti-flotillin-1 (Chemicon, Temecula, CA, USA). Flotillin-1 is a lipid raft-associated plasma membrane protein involved in signal transduction.
9. Anti-Oxphos complex I 39 kDa subunit (Molecular Probes, Eugene, OR, USA). Oxphos complex I 39 kDa subunit is a subunit of the NADH–ubiquinol oxidoreductase (oxidative phosphorylation complex I).
10. Anti-KDEL (StressGen, Victoria, Canada). The tetrapeptide KDEL is located at the carboxy-terminal sequences of luminal proteins that perform essential functions in the endoplasmic reticulum (ER).
11. Anti-Golgi 97 (Molecular Probes, Eugene, OR, USA). Golgi 97 is a 97 kDa protein located on the cytoplasmic face of the Golgi apparatus.
12. Anti-Na^+/K^+ ATPase $\alpha 1$ subunit (Abcam, Cambridge, MA, USA). Na^+/K^+ ATPase is an integral membrane protein responsible for maintaining the electrochemical gradients of Na^+ and K^+ across the plasma membrane. The $\alpha 1$ subunit of this enzyme is the catalytic subunit.
13. Peroxidase-conjugated goat anti-rabbit and anti-mouse IgG (H+L) (Millipore, Billerica, USA).
14. Immobilon Western Chemiluminescent HRP Substrate Kit (Millipore, Billerica, USA).

2.4. Protein Separation by Two-Dimensional Gel Electrophoresis and In-Gel Protein Digestion

1. Lysis buffer: 7 M urea, 2 M thiourea, 2% ASB-14, and 65 mM DTT.
2. Ampholytes, pH 3–10 (GE Healthcare).
3. IPG DryStrips, 13 cm, pH 3–10 (GE Healthcare).
4. The first-dimensional IPGphor™ II and the second-dimensional SDS-PAGE system (GE Healthcare).
5. Equilibration buffer 1: 50 mM Tris–HCl, 6 M urea, 2% SDS, 30% glycerol, 1% DTT.
6. Equilibration buffer 2: 50 mM Tris–HCl, 6 M urea, 2% SDS, 30% glycerol, 1% iodoacetamide.
7. Hand cast 10% SDS polyacrylamide gels: 0.0375 M Tris–HCl, pH 8.8, 10% acrylamide, 0.1% SDS, 0.03% TEMED, 0.05% ammonium persulfate.

8. ImageMaster Platinum™ software (GE Healthcare).
9. Acetonitrile, ammonium bicarbonate, TFA, and *n*-octyl-β-D-glucopyranoside (Sigma-Aldrich).
10. Sequencing grade modified trypsin (Promega, Madison, WI).

2.5. Protein Identification by MALDI-TOF–TOF MS

1. α-Cyano-4-hydroxycinnamic acid (CHCA), AnchorChip™ 600/384, standard peptide mixtures, Ultraflex TOF–TOF mass spectrometer, FlexControl™ 2.2 software, FlexAnalysis™ 2.2 software, and Biotool™ 2.2 software (Bruker Daltonics, Bremen, Germany).
2. Ammonium citrate, TFA (Sigma-Aldrich).

2.6. Protein Separation by SDS-PAGE and In-Gel Protein Digestion

1. SDS running buffer: 25 mM Tris–HCl, pH 8.3, 192 mM glycine, 0.1% SDS.
2. Brilliant Blue R staining solution (Sigma-Aldrich).
3. Ethanol, acetic acid (Sino-American Biotechnology, Beijing, China).
4. Ammonium bicarbonate, TFA (Sigma-Aldrich).
5. DTT, iodoacetamide (Merck).
6. SpeedVac Concentrator (Thermo Fisher Scientific).

2.7. Protein Identification by LC-ESI-FTICR MS

1. Nano-LC system, C18 reverse phase column (Micro-Tech Scientific, Vista, CA, USA).
2. 7-T Fourier transform ion-cyclotron resonance (FTICR) mass spectrometer Apex-Qe, ApexControl 1.0 software, DataAnalysis 3.4 software (Bruker Daltonics, Bremen, Germany).
3. Mascot 2.1.0 software (Matrix Science, London, UK).
4. Formic acid, acetonitrile (Sigma-Aldrich).

3. Methods

Membrane skeletons were prepared by differential centrifugation and sucrose gradient centrifugation. The workflow of this strategy is shown in Fig. 1.

3.1. Preparation of Liver Plasma Membranes

Plasma membranes were prepared by differential centrifugation and sucrose density gradient centrifugation as previously described (3, 13). Carry out all procedures at 4°C unless otherwise specified.

1. Place ~20 g of human or rat liver in ice-cold homogenization buffer. Remove connective tissues with scissors and tweezers.

2. Mince livers into small pieces with scissors and wash twice with HB to remove blood (see Note 1).
3. Resuspend the liver pieces with 5 volumes of ice-cold HB, transfer them to a 40-ml Dounce homogenizer, and homogenize with ten up-and-down strokes using the loose-fitting pestle (see Note 2).
4. Filter the homogenate through four layers of nylon gauze. Wash the residues with 1 volume of ice-cold HB and pool the filtrate.
5. Centrifuge the filtrate at 1,000 × g for 10 min in a Himac CR21 centrifuge. Resuspend the pellets in HB and pool them.
6. Add high-density sucrose (2.4 M sucrose, 10 mM HEPES 1 mM $MgCl_2$, 1 mM PMSF, pH 7.5) to obtain a concentration of 1.6 M (45.7% w/w sucrose) (see Note 3). Transfer 25 ml of the suspension into each SW32 rotor tube and overlay it with 13 ml of HB to fill the tube. Centrifuge at 70,900 × g (R_{av}) for 70 min in a Beckman Optima™ L-80 XP ultracentrifuge (see Note 4).
7. Collect the interface layer (mainly contains plasma membranes, mitochondria, and endoplasmic reticulums) in all tubes, resuspend it gently in 2 volumes of cold water and enough HB, and centrifuge at 1,200 × g for 10 min (see Note 5).
8. Resuspend the pellets in HB and adjust the concentration to 1.45 M (42% w/w sucrose) with high-density sucrose. Transfer the suspension into SW32 rotor tubes, overlay with HB to fill the tubes, and centrifuge at 68,400 × g (at R_{av}) for 60 min.
9. Collect the interface layer from 0.25 M/1.45 M sucrose, resuspend it in HB, and centrifuge at 17,600 × g for 10 min. Resuspend the pellets in HB and add high-density sucrose to bring the final concentration to 1.35 M (39.4% w/w sucrose). Transfer the suspension into P40ST rotor tubes, overlay with HB to fill the tubes, and centrifuge at 231,000 × g for 60 min in a Himac CP 100MX ultracentrifuge.
10. Collect the purified plasma membranes from the interface, resuspend in HB, and centrifuge at 14,500 × g for 15 min.
11. Aliquot the plasma membranes to 20 microcentrifuge tubes, wherein an aliquot is for electron microscopy analysis, an aliquot is used to determine the protein concentration and Western blotting analysis, and the other aliquots are used for preparation of membrane skeletons.

3.2. Preparation of Detergent-Resistant Membrane Skeletons

Membrane skeletons are prepared as previously described (2, 11). Carry out all procedures at 4°C unless otherwise specified.

1. Resuspend the plasma membranes to 1–2 mg of membrane proteins per ml of ice-cold Triton X-100 extraction buffer containing 1 mM PMSF (see Note 6).

2. Incubate the suspension on ice for 15 min.
3. Homogenize the plasma membranes with 15 strokes in a 7-ml Dounce glass homogenizer with a tight-fitting pestle. After a short pause, perform an additional 15 strokes.
4. Add 2.4 M sucrose in 10 mM HEPES (pH 7.5) to the homogenate to obtain a concentration of 1.42 M (41% w/w sucrose).
5. Transfer 5 ml of the suspension to each of the two P40ST rotor tubes, and gently overlay with 4 ml of 0.99 M (30% w/w) sucrose followed by 3 ml of 0.12 M (4% w/w) sucrose. Centrifuge at 240,000 × *g* for 16 h in a Himac CP 100MX ultracentrifuge.
6. Collect the bands around the 30%/41% sucrose interfaces with a blunt-ended plastic transfer pipet, and resuspend them in TEB. Centrifuge at 4°C for 1 h at 100,000 × *g*.
7. The pellets represent the membrane skeleton fraction. Store them in aliquots at –80°C.

3.3. Evaluation of the Morphology and Purity of Plasma Membranes and Membrane Skeletons

3.3.1. Electron Microscopy

1. Add an equal volume of 5% glutaraldehyde in PBS to a sample (plasma membranes or membrane skeletons). Mix gently.
2. Incubate the mixture overnight at 4°C, and centrifuge at 10,000 × *g* for 5 min.
3. Wash the pellet with PBS three times.
4. Postfix the pellet in 1% osmium tetroxide for 1 h, and wash the pellet with PBS three times.
5. Dehydrate through a graded series of ethanol (15 min each in 70, 80, 90, 95, and 100%) with a final 15-min dehydration in 100% acetone.
6. Embed the pellet in Epon 812 and examine ultrathin sections stained with uranyl acetate and lead citrate using a JEM-1010 transmission electron microscope.

3.3.2. Western Blotting

1. Lyse the plasma membranes or membrane skeletons with a lysis buffer containing 7 M urea, 2 M thiourea, 2% ASB-14, and 65 mM DTT (see Note 7).
2. Determine protein concentration using 2-D Quant Kit.
3. Separate 25 μg of proteins from different fractions by SDS-PAGE in a Mini-PROTEAN 3 system (see Fig. 2a).
4. Transfer proteins to a PVDF membrane using a Trans-Blot SD Semi-Dry System according to the manufacturer's instruction manual (see Note 8).
5. Wash the PVDF membrane once with PBST, and block the membrane with blocking buffer for 1–2 h at RT.

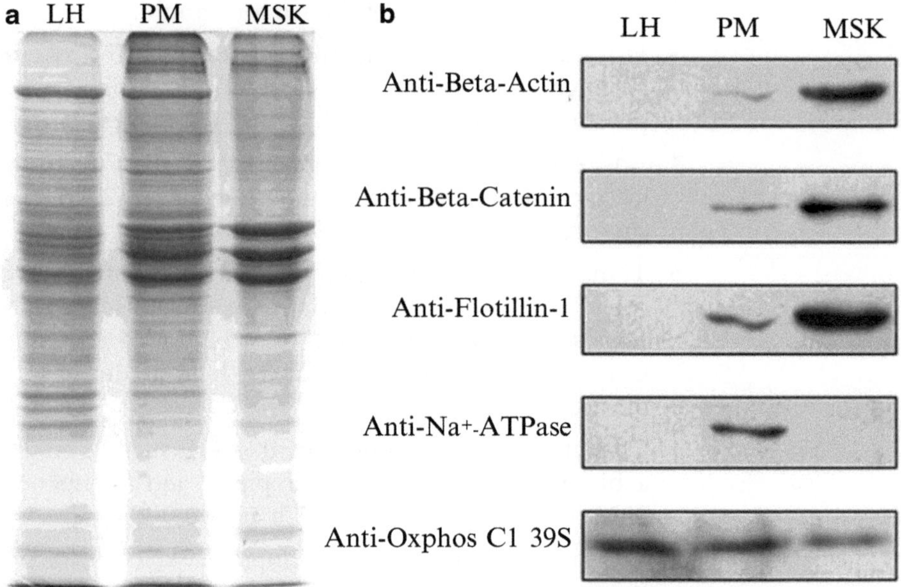

Fig. 2. Western blotting analysis of plasma membranes and membrane skeletons from rat livers. Twenty-five micrograms of proteins from liver homogenate (LH), plasma membrane (PM), and membrane skeleton (MSK) fractions were separated by 10% SDS-PAGE and stained by Coomassie Brilliant Blue R (**a**). Another replicate of samples was transferred to a PVDF membrane from the gel. The blots were probed with the anti-Beta-Actin, anti-Beta-Catenin, anti-Flotillin-1, anti-Na^+–K^+–ATPase, and anti-Oxphos C1 39S (**b**). Figure 2b is reproduced with permission from ref. (11).

6. Incubate the membrane with various primary antibodies overnight at 4°C or 4 h at RT. To evaluate the purity of plasma membranes or membrane skeletons, the following antibodies recognizing organelle markers may be used: anti-flotillin-1, anti-Oxphos complex I 39 kDa subunit, anti-KDEL, anti-Golgi 97, anti-Na^+/K^+ ATPase α1 subunit.
7. Wash the PVDF membrane 3 × 10 min with PBST.
8. Incubate the membrane with peroxidase-conjugated goat anti-rabbit or anti-mouse IgG (H + L) for 1 h at RT.
9. Wash the PVDF membrane 3 × 10 min with PBST.
10. Visualize the bound antibodies with Immobilon Western Chemiluminescent HRP Substrate Kit (see Fig. 2b).

3.4. Protein Separation by Two-Dimensional Gel Electrophoresis and In-Gel Protein Digestion

1. Resuspend the membrane skeleton fraction in 250 μl of lysis buffer containing 7 M urea, 2 M thiourea, 65 mM DTT, and 2% ASB-14. Solubilize proteins by vigorous vortexing and ultrasonication, and incubate at RT for 5 h.
2. Replenish the sample solution with 0.5% ampholytes (pH 3–10) before isoelectric focusing (IEF).
3. Load the sample on a pH 3–10 IPG DryStrip.

4. Conduct IEF using the IPGphor™ II (GE Healthcare, Sweden) according to the following procedure: 30 V×8 h; 50 V×4 h; 300 V×1 h; 1,000 V×1 h; 3,000 V×1 h; 5,000 V×1 h; and 8,000 V until reaching a total of 65,000 Vh.
5. Incubate the IPG strip in equilibration buffer 1 for 15 min at RT.
6. Discard equilibration buffer 1, add equilibration buffer 2, and incubate for 15 min at RT in the dark (see Note 9).
7. Perform the second-dimension separation using a 10% SDS polyacrylamide gel. Stain the gel by a procedure called "blue silver" (14).
8. Scan the gel at 300 dpi resolution, and analyze protein spots with the ImageMaster Platinum™ software.
9. Excise protein spots with scissor-sheared tips. Transfer the spots to V-bottom 96-well microplates (see Note 10).
10. Destain the gel spots with 50% acetonitrile/25 mM ammonium bicarbonate solution for 1 h.
11. Dehydrate gel pieces with 100% acetonitrile for 20 min.
12. Remove acetonitrile and air-dry gel pieces (see Note 11).
13. Add 2 μL of 5 μg/ml trypsin in 25 mM ammonium bicarbonate to each well, seal the microplate with parafilm, and re-swell gel spots at 4°C for 45 min.
14. Incubate at 37°C for 10 h.
15. After tryptic digestion, add 8 μL of the extraction solution consisting of 0.25% TFA and 5 mM *n*-octyl-β-D-glucopyranoside (see Note 12) to each well and shake at 37°C for 1 h to extract the peptides.

3.5. Protein Identification by MALDI-TOF–TOF MS

MALDI samples were prepared according to a thin layer method as described previously (15, 16).

1. Drag saturated CHCA in a solution containing 97% acetone and 0.003% TFA with an Eppendorf pipette tip (see Note 13) across the surface of AnchorChip™ 600/384 to form a thin layer of crystalline CHCA.
2. Deposit 4 μl of each tryptic digest onto a spot. Incubate for 3 min at RT.
3. Remove the remaining liquid with Eppendorf tips, and wash each spot with 6 μl of a solution containing 10 mM ammonium citrate and 0.3% TFA.
4. Deposit 1 μl of 0.1 pmol/μl standard peptide mixtures and wash them using the same method.
5. Obtain mass spectra on an Ultraflex MALDI-TOF–TOF mass spectrometer under the control of FlexControl™ 2.2 software.

6. Record MS spectra in the positive ion reflector mode in a mass range from 800 to 4,000 Da and the ion acceleration voltage was 25 kV.
7. Analyze MS spectra with FlexAnalysis™ 2.2 and Biotool™ 2.2 software, and automatically search the MS spectra against Swiss-Prot or IPI database (see Note 14) using the MASCOT 2.1.0 software. Set the main parameters as follows: S/N ≥ 3.0; Fixed modification: Carbamidomethyl (C); Variable modification: Oxidation (M); Maximum number of missing cleavages 1; MS tolerance ±100 ppm.
8. Perform TOF–TOF analysis in "LIFT" mode as follows: select three strongest peaks of each MS spectra as precursor ions which were accelerated in TOF1 at a voltage of 8 kV and fragmented by lifting the voltage to 19 kV. Analyze the MS/MS spectra by FlexAnalysis™ 2.2, Biotool™ 2.2, and MASCOT. Set the parameters as follows: S/N ≥ 3.0; Fixed modification: Carbamidomethyl (C); Variable modification: Oxidation (M); Maximum number of missing cleavages 1; MS tolerance ± 0.7 Da.
9. Table 1 shows the membrane skeleton proteins identified from human liver using the 2-D gel and MALDI-TOF–TOF MS strategy.

3.6. Protein Separation by SDS-PAGE and In-Gel Protein Digestion

1. Prepare 10% polyacrylamide gels (see Note 15) according to standard protocol.
2. Load 50 μg of membrane skeleton proteins and run gels in SDS running buffer.
3. Stain gels with Coomassie Brilliant Blue R staining solution for 2 h at RT.
4. Destain gels with a buffer containing 10% ethanol and 10% acetic acid.
5. Cut the entire gel lane into 12 pieces of equal size.
6. Destain the gel pieces with 50% acetonitrile/25 mM ammonium bicarbonate solution.
7. Reduce proteins by 20 mM DTT in 25 mM ammonium bicarbonate, and alkylate proteins by 50 mM iodoacetamide in 25 mM ammonium bicarbonate.
8. Digest the proteins with trypsin overnight at 37°C.
9. Extract tryptic peptides from the gel pieces with 50% acetonitrile and 5% formic acid, and then dry them in a SpeedVac Concentrator.

3.7. Protein Identification by LC-ESI-FTICR MS

The LC-ESI-FTICR analysis was performed essentially as previously described (17).

1. Dissolve peptides extracted from each gel band in 0.1% formic acid, and then separate them by a Nano-LC system equipped

Table 1
Membrane skeleton proteins identified from human liver by MALDI-TOF–TOF MS

No.	Accession[a]	Protein name	Score[b]	Pep[c]	MW	pI
1	PLEC1_HUMAN	Plectin 1	378	3	531,733	5.73
2	DESP_HUMAN	Desmoplakin	438	2	331,774	6.44
3	SPTA2_HUMAN	Alpha-fodrin	535	4	274,539	5.22
4	FLNB_HUMAN	Beta-filamin	74	0	278,195	5.49
5	SPTB2_HUMAN	Beta-fodrin	357	3	274,609	5.41
6	MYH14_HUMAN	Nonmuscle myosin heavy chain IIc	112	1	228,002	5.76
7	MYH9_HUMAN	Nonmuscle myosin heavy chain IIa	178	2	226,532	5.5
8	IPI00307829	Cingulin-like 1	207	0	149,079	5.51
9	CING_HUMAN	Cingulin	204	0	136,386	5.46
10	VINC_HUMAN	Vinculin	144	1	123,799	5.51
11	DSG2_HUMAN	Desmoglein-2 precursor	73	1	122,385	5.15
12	CO6A1_HUMAN	Collagen alpha 1(VI) chain	186	3	108,529	5.29
13	CTND1_HUMAN	Catenin delta-1 (p120 catenin)	221	3	108,170	5.86
14	ACTN4_HUMAN	Nonmuscle alpha-actinin 4	206	1	104,854	5.27
15	ACTN1_HUMAN	Nonmuscle alpha-actinin-1	389	4	103,058	5.25
16	DSC2_HUMAN	Desmocollin-2 precursor	360	4	99,962	5.91
17	CTN1_HUMAN	Alpha-1 catenin	143	1	100,071	5.95
18	CTNB1_HUMAN	Beta-catenin	336	4	85,497	5.53
19	GELS_HUMAN	Gelsolin precursor	84	0	85,698	5.9
20	PLAK_HUMAN	Junction plakoglobin	596	5	81,630	5.95
21	LAMA_HUMAN	Lamin A/C	162	1	74,139	6.57
22	LAM1_HUMAN	Lamin B1	218	1	66,408	5.11
23	COR1C_HUMAN	Coronin-1 C	132	1	53,249	6.65
24	K2C8_HUMAN	Type II cytoskeletal 8	389	3	53,704	5.52
25	TBA1_HUMAN	Tubulin alpha-1 chain	511	5	49,924	4.95
26	TBB2_HUMAN	Tubulin beta-2 chain	537	5	49,671	4.78
27	K1C18_HUMAN	Type I cytoskeletal 18	411	4	48,058	5.34
28	ARP3_HUMAN	Actin-related protein 3	182	1	47,371	5.61
29	ARP2_HUMAN	Actin-related protein 2	346	3	44,761	6.3
30	ACTG_HUMAN	Gamma-actin	318	3	41,793	5.31

(continued)

Table 1 (continued)

No.	Accession[a]	Protein name	Score[b]	Pep[c]	MW	pI
31	ACTB_HUMAN	Beta-actin	293	2	41,737	5.29
32	ARPC2_HUMAN	ARP2/3 complex 34 kDa	194	1	34,333	6.84
33	CAZA1_HUMAN	F-actin capping protein alpha-1	381	3	32,923	5.45
34	CAZA2_HUMAN	F-actin capping protein alpha-2	360	3	32,949	5.58
35	CAPZB_HUMAN	F-actin capping protein beta	217	2	31,350	5.36
36	MLRM_HUMAN	Nonsarcomeric myosin RLC	130	1	19,794	4.67
37	HSPB1_HUMAN	HSP 27	189	0	22,783	5.98
38	SRBS1_HUMAN	CAP/ponsin; SH3P12	220	2	142,455	6.4

Reproduced with permission from ref. (3)
[a]Swiss-Prot or IPI accession numbers
[b]MOWSE scores obtained by the combined search (PMF and LIFT data) using Mascot engine. For those proteins without LIFT data, the scores represent MOWSE scores by PMF search
[c]Numbers of peptides identified by MALDI-TOF–TOF MS/MS

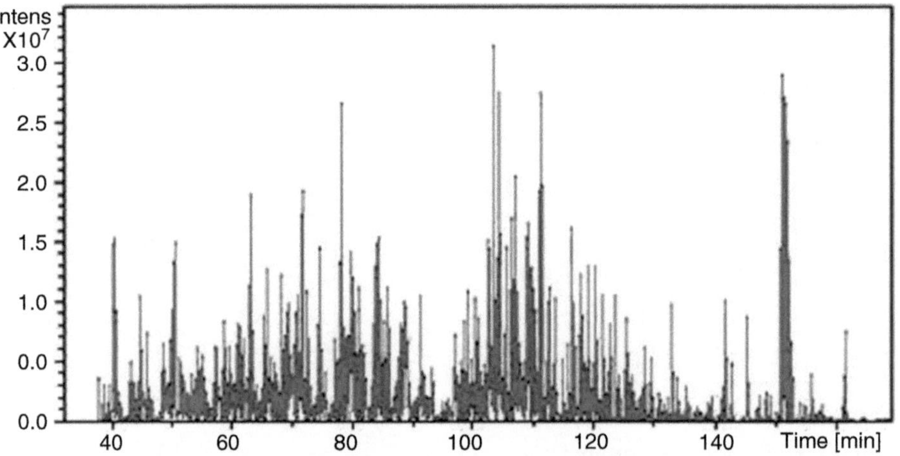

Fig. 3. A representative total ion chromatogram of the peptides from a gel slice analyzed by LC-ESI-FTICR MS.

with a C18 reverse-phase column. Elute the peptides using a 120-min reverse-phase gradients from 0 to 50% acetonitrile in 0.1% formic acid at a constant flow rate of 400 nl/min.

2. Record mass spectra on a 7-T FTICR mass spectrometer, Apex-Qe.

3. Acquire data in data-dependent mode using ApexControl 1.0 software. Select three strongest peaks of each MS acquisition for the following MS/MS analysis. A representative total ion chromatogram of the peptides from a gel slice is shown in Fig. 3.

4. Process the MS/MS spectra by DataAnalysis 3.4, and set S/N to ≥4.0.
5. Search the data against IPI database using the Mascot 2.1.0. Set the mass tolerances for MS and MS/MS to 5 ppm and 0.02 Da, respectively. Select the instrument setting for the Mascot search as "ESI-FTICR."
6. A total of 87 membrane skeleton proteins were identified from rat liver using the 1-D gel and LC-ESI-FTICR MS strategy in our previous work (11) (see Note 16).

4. Notes

1. For preparation of rat liver plasma membranes, use a perfusion method to remove blood from rats under anesthesia.
2. Homogenize liver gently to avoid formation of vesicular plasma membranes, which are indistinguishable from other intracellular vesicles.
3. Measure the density of the mixture using an Abbe refractometer.
4. Balance the tubes very carefully before ultracentrifugation.
5. A blunt-ended plastic transfer pipet can be used for manual collection of pellicles from the top. Alternatively, an automated gradient-collection device may also be used.
6. PMSF is not stable in aqueous solutions. Therefore, add PMSF to the extraction buffer immediately before use.
7. The solubility of membrane skeleton proteins is relatively low. Solubilize these proteins by repeat vortex and ultrasonication.
8. PVDF membrane is hydrophobic and should be prewet in methanol before use.
9. Equilibration buffer 1 is used to reduce the disulfide in proteins, while equilibration buffer 2 is used for alkylation. Iodoacetamide is light sensitive. Therefore, make sure to prepare equilibration buffer 2 immediately before use and perform alkylation in the dark.
10. Use V-bottom instead of U-bottom 96-well microplates to finish in-gel digestion with minimum volumes of various solutions.
11. Alternatively, use a SpeedVac to dry gel pieces quickly.
12. *n*-Octyl-β-D-glucopyranoside will facilitate the extraction of hydrophobic peptides from gels.
13. Make sure that the pipette tips will not introduce any polymer contamination. Eppendorf LoRetention pipette tips are a good choice.

14. Search against the protein sequence database corresponding to the appropriate species.
15. Most of membrane skeleton proteins have a high molecular weight. Therefore, the 10% polyacrylamide gel is a good choice to separate these proteins. Alternatively, the 4–15% gradient gel can also be used for the separation.
16. The 2-D gel-coupled MALDI-TOF–TOF MS strategy is complementary with the SDS-PAGE-coupled LC-ESI-FTICR MS strategy. As most of very-hydrophobic proteins can be separated by SDS-PAGE rather than 2-D gel electrophoresis, the latter strategy is able to identify more membrane skeleton proteins. Therefore, we prefer SDS-PAGE-coupled LC-ESI-FTICR MS if only one strategy can be used.

Acknowledgements

This work was supported by grants from National Key Basic Research Program of China (No. 2010CB912203 and 2010CB915504) and National Natural Science Foundation of China (No. 90919023 and 30970652).

References

1. Yu J, Fischman DA, Steck TL (1973) Selective solubilization of proteins and phospholipids from red blood cell membranes by nonionic detergents. J Supramol Struct 1:233–248
2. Nebl T, Pestonjamasp KN, Leszyk JD, Crowley JL, Oh SW, Luna EJ (2002) Proteomic analysis of a detergent-resistant membrane skeleton from neutrophil plasma membranes. J Biol Chem 277:43399–43409
3. He J, Liu Y, He S, Wang Q, Pu H, Ji J (2007) Proteomic analysis of a membrane skeleton fraction from human liver. J Proteome Res 6:3509–3518
4. Cowin P, Burke B (1996) Cytoskeleton-membrane interactions. Curr Opin Cell Biol 8:56–65
5. Luna EJ, Hitt AL (1992) Cytoskeleton–plasma membrane interactions. Science 258:955–964
6. Tsukita S (1989) Isolation of cell-to-cell adherens junctions from rat liver. J Cell Biol 108: 31–41
7. Farquhar MG, Palade GE (1963) Junctional complexes in various epithelia. J Cell Biol 17:375–412
8. Montesano R, Gabbiani G, Perrelet A, Orci L (1976) In vivo induction of tight junction proliferation in rat liver. J Cell Biol 68:793–798
9. Bennett V, Davis J, Fowler WE (1982) Brain spectrin, a membrane-associated protein related in structure and function to erythrocyte spectrin. Nature 299:126–131
10. Glenney JR Jr, Glenney P (1983) Fodrin is the general spectrin-like protein found in most cells whereas spectrin and the TW protein have a restricted distribution. Cell 34:503–512
11. Wang Q, He J, Meng L, Liu Y, Pu H, Ji J (2010) A proteomics analysis of rat liver membrane skeletons: the investigation of actin- and cytokeratin-based protein components. J Proteome Res 9:22–29
12. Blonder J, Terunuma A, Conrads TP, Chan KC, Yee C, Lucas DA et al (2004) A proteomic characterization of the plasma membrane of human epidermis by high-throughput mass spectrometry. J Invest Dermatol 123:691–699
13. He JT, Liu YS, He SZ, Wang QS, Pu H, Tong YP et al (2007) Comparison of two-dimensional gel electrophoresis based and shotgun strategies

in the study of plasma membrane proteome. Proteomics Clin Appl 1:231–241

14. Candiano G, Bruschi M, Musante L, Santucci L, Ghiggeri GM, Carnemolla B et al (2004) Blue silver: a very sensitive colloidal Coomassie G-250 staining for proteome analysis. Electrophoresis 25:1327–1333
15. Yang W, Liu P, Liu Y, Wang Q, Tong Y, Ji J (2006) Proteomic analysis of rat pheochromocytoma PC12 cells. Proteomics 6:2982–2990
16. Gobom J, Schuerenberg M, Mueller M, Theiss D, Lehrach H, Nordhoff E (2001) Alpha-cyano-4-hydroxycinnamic acid affinity sample preparation, a protocol for MALDI-MS peptide analysis in proteomics. Anal Chem 73:434–438
17. Liu Y, He J, Ji S, Wang Q, Pu H, Jiang T et al (2008) Comparative studies of early liver dysfunction in senescence-accelerated mouse using mitochondrial proteomics approaches. Mol Cell Proteomics 7:1737–1747

Chapter 8

Liver Plasma Membranes: An Effective Method to Analyze Membrane Proteome

Rui Cao and Songping Liang

Abstract

Plasma membrane proteins are critical for the maintenance of biological systems and represent important targets for the treatment of disease. The hydrophobicity and low abundance of plasma membrane proteins make them difficult to analyze. The protocols given here are the efficient isolation/digestion procedures for liver plasma membrane proteomic analysis. Both protocol for the isolation of plasma membranes and protocol for the in-gel digestion of gel-embedded plasma membrane proteins are presented. The later method allows the use of a high detergent concentration to achieve efficient solubilization of hydrophobic plasma membrane proteins while avoiding interference with the subsequent LC-MS/MS analysis.

Key words: Liver, Plasma membranes, In-gel digestion, Proteomics, Tandem mass spectrometry

1. Introduction

The liver is the largest organ in the body, and has main function in digestion and metabolism. It also has a myriad of additional functions, including the production of red blood cells during embryonic development, the production of various plasma proteins, and the detoxification of xenobiotics (1). Membranes are critical components of cellular structure. It has been reported that membrane proteins constitute roughly 25–30% of cellular proteins (2). Plasma membrane proteins play a crucial role in the fundamental biological process of the cell, including the exchange of the material and energy between the cell and its environment, cell–cell interactions, and signal transport. Furthermore, it has been reported that more than half of all membrane proteins are predicted to be pharmacological targets (3). Despite the importance of the plasma membrane

Djuro Josic and Douglas C. Hixson (eds.), *Liver Proteomics: Methods and Protocols*, Methods in Molecular Biology, vol. 909, DOI 10.1007/978-1-61779-959-4_8, © Springer Science+Business Media New York 2012

proteome, there is less understanding of this protein class than of any other in the cell because plasma membrane proteins are present in lower abundance. Plasma membrane preparations are also easily contaminated by other components such as mitochondrial and endoplasmic reticulum membranes (4).

Another challenge in studying membrane proteins is their hydrophobicity and poor water solubility. As a result, the identification of PM proteins are usually in low sequence coverage because of inefficient digestion. Thus, the high-throughput analysis of the plasma membrane proteins is more difficult than the high-throughput analysis of soluble proteins due to the requirement for efficient protein solubilization and the incompatibility of MS with ionic and nonionic detergents (5).

In the protocol presented in this chapter we try to address these issues. First, we describe a method for isolation of plasma membranes consisting of a traditional sucrose density centrifugation followed by a two-phase aqueous polymer partition (6). Second, the plasma membrane proteins that are dissolved in solution containing strong surfactants and detergents are directly incorporated into a polyacrylamide gel matrix. After formation of the gel, proteins in the gel section are digested with trypsin, and the resulting peptides are subjected to reversed-phase, high-performance liquid chromatography followed by electrospray ion-trap tandem mass analysis. The in-gel digestion-based shotgun method combines the advantages of in-gel and in-solution digestion for mass spectrometry-based proteomics, which allowed the use of a high detergent concentration to achieve efficient solubilization of very-hydrophobic membrane proteins while avoiding interference with the subsequent LC-MS/MS analysis (7, 8).

The analysis was demonstrated to be comprehensive with respect to the number of identified membrane proteins and the range of membrane protein p*I* and GRAVY values and the detection of hydrophilic as well as highly hydrophobic proteins.

The in-gel shotgun method is high throughput and comprehensive enough to analyze membrane samples.

2. Materials

2.1. Density Gradient Centrifugation

Due to the strong influence of ions on membrane partitioning in the two-phase systems, Mili-Q water (18.2 MΩ cm, TOC < 15 ppm) should be used throughout the experiments.

1. Sprague–Dawley rats with body weight between 250 and 300 g.

2. Homogenization buffer: 50 mM HEPES (pH 7.4), containing 1 mM $CaCl_2$ and 0.1 mM phenylmethylsulfonyl fluoride (see Note 1).
3. Homogenizer: Ultra Turrax T8 (IKA Labortechnik, Stanfen, Germany).
4. Beckman ultracentrifuge with an SW 28, or SW41, and Ti 70 rotor (Beckman Coulter, Brea, CA, USA) or similar.
5. 60, 50, 44, 42.3% (w/v) sucrose stock solutions, prepared with100 mM NaCl, 10 mM HEPES-NaOH, pH 7.4, containing 1 mM EDTA.

2.2. Two-Phase Partitioning

1. Dextran stock solution: Dextran T-500 (20% w/w) (see Note 2).
2. PEG stock solution: PEG 3350 (40% w/w).
3. 0.2 M potassium phosphate buffer, pH 7.2.
4. 1 M Sucrose stock solutions.
5. Resuspended buffer: 1 mM $NaHCO_3$.
6. Wash buffer 1: 0.1 M sodium carbonate (Na_2CO_3): dissolve 106 mg Na_2CO_3 in 10 mL ddH_2O. The pH of the solution should be approximately 11.3.

2.3. Incorporation of Plasma Membrane Proteins into Gel and In-Gel Digestion

1. 0.5 M Tris–HCl, pH 6.8: Dissolve 60.57 g Tris in 800 mL water. Adjust pH with 1 M HCl to 6.8, and add water to a final volume of 1 L. Store at room temperature.
2. 1.5 M Tris–HCl, pH 8.8: Dissolve 181.71 g Tris in 800 mL water. Adjust pH with 1 M HCl to 8.8, and add water to a final volume of 1 L. Store at room temperature.
3. 30% (w/v) acrylamide:bisacrylamide (29:1). Store at 4°C.

   Neurotoxic if unpolymerized, use gloves!
4. 10% (w/v) SDS: Dissolve 25 g sodium dodecyl sulfate (SDS) in 250 mL ddH_2O (see Note 3). Store at RT.
5. TEMED (*N,N,N,N*-tetramethyl-ethylendiamine). Store at 4°C. Avoid light.
6. Fresh 10% (w/v) ammonium persulfate solution (APS): Dissolve 100 mg ammonium peroxodisulfate in 1 mL ddH_2O.
7. Extraction buffer: 4% (w/v) SDS, 10 mM DTT in 0.1 M Tris–HCl pH 7.6.
8. 25 mM Ammonium bicarbonate: 0.04 g in 20 mL water.
9. Reduction buffer 1: 0.1 M Na_2CO_3, 10 mM DTT, pH 11.3.
10. Reduction buffer 2: 10 mM DTT in 25 mM ammonium bicarbonate. Dissolve 0.031 g in 20 mL in 25 mM ammonium bicarbonate solution.

11. Alkylation solution: 55 mM iodoacetamide, 0.2 g in 20 mL ammonium bicarbonate solution.
12. Proteinase trypsin (proteomics sequencing grade, Sigma Aldrich, St. Louis, MO, USA).

2.4. Mass Spectrometry

1. Acetonitrile (chromatography grade, Hunan Fine Chemistry Institute, Hunan, China).
2. Formic acid (98% purity, Merck KGaA, Darmstadt, Germany).
3. Solvent B: 0.1% formic acid in acetonitrile.
4. Solvent A: 0.1% formic acid in water.
5. Dionex C18 PepMap column 150 mm × 0.18 mm (DIONEX Corp., USA.) or 100 mm × 0.15 mm C18 column (Column Technology Inc.).
6. Liquid chromatography system, such as Agilent 1200 LC system (Agilent Technologies Inc., Santa Clara, CA, USA).
7. LTQ mass spectrometer (Thermo Electron, Bremen, Germany).

3. Methods

Rat liver plasma membranes separates were first isolated by a combination of sucrose density gradient centrifugation differential centrifugation and two-phase aqueous polymer partition. Sucrose density gradient centrifugation separates membranes and organelles according to their differences in buoyant density. Two-phase aqueous polymer partition separates plasma membranes according to their different affinity for two immiscible aqueous polymer phases. The use of two-dimensional purification method results in the higher purity of plasma membranes than the sucrose density gradient centrifugation or two-phase aqueous polymer partition alone. In the described method, the plasma membrane enrichment is determined by Western blotting for flotillin and Na^+/K^+ ATPase, two commonly used plasma membrane markers (Fig. 1).

Fig. 1. Western blot analysis of proteins from rat liver subfractions using Na^+/K^+ ATPase, *CM* crude plasma membrane, *WCL* whole tissue lystate, *PM* purified plasma membrane.

After isolation, plasma membrane proteins were solubilized in SDS, subsequently incorporated into a polyacrylamide gel matrix. After the gel is formed, all the detergents and salts that interfere with the subsequent enzymatic digestion and LC-MS/MS analysis are washed away. The isolation/purification method described here with respect to rat liver plasma membrane. However, we have found the isolation/purification method to be equally applicable to other samples, such as rat brain, cell culture sample, and clinical tissue sample.

3.1. Preparation of Crude Membranes

1. After starving for 18–24 h, rats were sacrificed and the livers were excised. Remove the gall bladder and blood vessels, and slice the liver into 1–3 mm pieces in homogenization buffer (see Note 4).
2. Homogenize 5 g of live tissue in 20 mL of homogenization buffer with a Tissue-Tearor (IKA products, T8 ULTRA-TURRAX, Staufen, Germany) at 20,000 rpm five times for 30 s (see Note 5).
3. Transfer the homogenate into 50-mL conical tubes and centrifuge at 600 × *g* for 10 min at 4°C in order to sediment nuclei and tissue debris.
4. Re-extract the pellet with Tissue-Tearor T8 in the same volume of homogenization buffer, as used for initial homogenization (see Note 6).
5. Repeat step 3.
6. Combine all two supernatants, transfer the supernatant into two clean tubes, and centrifuge for 30 min at 100,000 × *g* (4°C) by SW28 rotor or T70 rotor (Beckman Coulter).
7. Discard supernatant and resuspend the pellets with 60% (w/v) sucrose (see Note 7).
8. Transfer the resuspended membrane pellet to SW-28 tubes and prepare a discontinuous sucrose step gradient by carefully overlaying 50, 44, and 42.3% (w/v) sucrose gradient solutions on top of each other. Avoid perturbation/mixing the layers by pipetting slowly and continuously down the side of the tube.
9. Carefully place the sample tube and a balance tube into an ultracentrifuge and centrifuge at 100,000 × *g* for 2.5 h by Beckman SW28 rotor or SW41 rotor, not including spin-up and spin-down times, at 4°C.
10. After centrifugation, a grigio chiaro band will be visible at the top of 42.3% sucrose layer representing crude membranes. Collect the crude plasma membrane fraction from the top of the gradient immediately (carefully collect 1 mL fractions from the top of the gradient using a pipette).

11. Wash the crude plasma membranes with 1 mM sodium bicarbonate solution and pellet it at 100,000 × *g* for 40 min in Beckman Ti70 or Ti80 rotors. Repeat the wash to remove all other salts, sucrose, etc.
12. Discard the supernatant; the resulting pellet represents the crude plasma membranes that are further purified by aqueous two-phase partitioning. The pellet can be stored at –80°C.

3.2. Aqueous Two-Phase Partitioning

The basic principle of two-phase partitioning is the hydrophilic polymer Dextran T-500 and PEG 3350 (or PEG 4000) cannot be fully fused in a certain range of concentration, and according to the microsomes surface charge and hydrophilic difference in the two-phase system, plasma membranes tended to partition in the PEG-rich upper phase. Other membranes distribute in the bottom phase or middle interface. Due to the strong influence of ions on membrane partitioning in the two-phase systems, there are some factors that have strong influence on membrane partitioning in two-phase partition system, such as polymers or salt concentration, top-to-bottom-phase volume ratio, ion concentration, and amount of added microsomes. The greatest impact factors are both polymer and salt concentrations. For different tissue or cell other than liver, optimization of the concentration of the polymer and salt is preferred.

All following steps of the two-phase partitioning protocol have to be performed at 4°C. Working at room temperature prevents phase separation.

1. Prepare 16 g of the two-phase system (6.4% PEG3350/6.4% Dextran T-500 w/w, in 5 mM phosphate buffer pH 7.2, containing 250 mM sucrose, and deionized water) in two tubes with the compositions indicated in Table 1 (see Note 8).
2. Resuspend the crude plasma membrane pellets in resuspended buffer (1 mM sodium bicarbonate).
3. Add the resuspended sample 2 mL (about 1 g crude plasma membrane pellets) to the tube 1 (two-phase system 1), and add water to 16.0 g.

Table 1
Composition of two-phase systems

20% (w/v) Dextran	5.12 g
40% (w/v) PEG3350	2.56 g
0.2 M Phosphate buffer, pH 7.2	0.4 mL
1 M Sucrose	1.6 mL
Water	4.32 mL (Bring weight to 14 g)

4. Add the 2 mL resuspended buffer (1 mM sodium bicarbonate) to the tube 1 (two-phase system 2); make sure that the weight of system 2 is 16 g.
5. Thoroughly mix (invertations 30–40 times) tube 1 and tube 2, and accelerate phase separation by centrifugation for 5 min at 750 × *g*.
6. Remove the upper phase of tube 1 and tube 2 and store it at 4°C until further use.
7. Transfer the upper phase of tube 1 to tube 2 containing fresh lower phase, and transfer the upper phase from tube 2 to the lower phase of tube 1. Mix the tubes thoroughly and centrifuge again for 5 min at 750 × *g*.
8. Collect and combine the upper phase of tube 1 and tube 2 that contains enriched plasma membrane fraction.
9. Dilute the upper phase with a tenfold volume of 1 mM $NaHCO_3$ solution and pellet the plasma membranes by centrifugation at 100,000 × *g* for 2 h in an SW 28 or a Ti70 rotor.
10. Resuspend the final pellets obtained in the upper phase in 0.1 M sodium carbonate, containing 10 mM DTT, pH 11.3 and agitate for 1 h on ice in order to release noncovalently attached proteins. Pellet the plasma membranes by centrifugation for 1 h at 100,000 × *g*. The pellets representing purified plasma membranes (PPMs) shall be stored at −80°C.
11. Extract membrane proteins from PPM by use of the extraction buffer. Estimate the protein amounts with the RC DC Protein Assay (Bio-Rad, Hercules, CA, USA) (see Note 9).

3.3. Incorporation of Plasma Membrane Proteins into Gel and In-Gel Digestion

Make 42 μL of 10% polyacrylamide gel as follows. For lower or higher amounts of protein, adjust the volume of polyacrylamide gel accordingly.

1. Mix 16 μL (about 150 μg protein) protein solution with 14 μL 30% acrylamide stock (30%, 29:1) solution, 10 μL separating acrylamide gel buffer, 2 μL 1% (w/v) ammonium persulfate (AP), and 0.4 μL 10% (v/v) TEMED in a small glass tube or a 500 μL size Eppendorf lobind tube. Close the tubes.
2. Incubate at room temperature for 1 h. The gel should be formed in the glass tube.
3. Fix the gel for 30 min with 50%v/v methanol and 7%v/v acetic acid.
4. Cut the gel into small pieces and wash twice with 200 μL 25 mM NH_4HCO_3 and 50% (v/v) acetonitrile. Discard supernatant.

5. Wash in 100 μL 100% acetonitrile for 10 min. Remove supernatant (gel pieces should dehydrate and go white).
6. Dry the gel in SpeedVac vacuum concentrator (Thermo Fisher Scientific, San Jose, CA, USA) for 10 min (all acetonitrile should have evaporated).
7. Add 50 μL 10 mM DTT for 30 min at 37°C and discard the supernatant.
8. Wash the gel pieces twice with 25 mM ammonium bicarbonate solution.
9. Wash in acetonitrile until gel pieces are white. Discard the supernatant.
10. Add 50 μL 55 mM iodoacetamide for 60 min (in the dark).
11. Wash the gel pieces twice with 100 μL 25 mM ammonium bicarbonate for 10 min.
12. Wash the gel pieces with 100 μL 100% acetonitrile until they are white. Discard the supernatant.
13. Dry the gel pieces in SpeedVac for 10 min.
14. Cover the gel with 12.5 ng/μL trypsin solution in 25 mM NH_4HCO_3 containing 0.1% (w/v) *n*-octyl glucoside, 0.1% (w/v) sodium deoxycholate (SDC), or 0.1% Rapigest (W/V) (see Note 10).
15. Digest overnight at 37°C.
16. Collect the supernatant in a new Eppendorf tube. Extract the gel pieces with 100 μL of 60% (v/v) ACN containing 0.1% (v/v) formic acid for 10 min with ultrasonication twice.
17. Pool the supernatant and dry it quickly in a SpeedVac to about 10 μL for mass spectrometric analysis.

3.4. LC-MS/MS Analysis

We have much experience in two different types of LC-MS/MS system, Q-Tof (from Bruker Daltonics and Waters Corporation) and Ion trap mass spectrometer (Bruker Daltonics and Thermo Scientific). Based on our experience, the LTQ mass spectrometer (Thermo Scientific) is robust and reliable for high-throughput proteome analysis. The LTQ system plus Orbit analyzer gains more and more attention recently. The technical details and experimental conditions of LC-MS/MS system are given below:

1. Digested samples are introduced via an LC system, such as Agilent 1200 LC system (Agilent Technologies Inc., Santa Clara, CA, USA) or thermo Surveyor LC (Thermo Electron, San Jose, CA).
2. Gradient elution is performed as follows: (1) 5% solvent B from 0 to 10 min, (2) linear slope from 5 to 40% solvent B during 180 min, (3) linear slope to 80% B from 180 to

200 min, (4) regeneration of the column at constant 80% B from 200 to 210 min, (5) linear slope from 80% B to 5% B from 210 to 220 min, and (6) equilibration with 5% B from 220 to 240 min.

3. The electrospray voltage was set to 3.1 kV. The capillary temperature was set to 160°C. Normalized collision energy was at 35.0%. Use an automated gain control function to manage the number of ions injected into the ion trap. The mass spectrometer full MS scan range was 400–1,800 *m/z*. Number of MS/MS components was set to 10. Dynamic exclusion was set at repeat count 2, repeat duration 30 s, and exclusion duration 90 s. System control and data collection were performed with Xcalibur software version 1.4 or higher version (Thermo).

3.5. Database Search

Use SEQUEST to perform the database search with the following parameters:

1. Search data against the Rat International Protein Index protein sequence database (ftp://ftp.ebi.ac.uk/pub/databases/IPI/current/) or latest Swiss-Prot protein database.
2. Use 2.5 Da mass tolerance for the peptide mass and 0.5 Da for the fragment ion tolerance.
3. Allow two missed cleavage sites.
4. Use carbamidomethyl-Cys and methionine oxidation as fixed and variable modifications, respectively.
5. Use sequence-reversed database searching to provide a >99% overall confidence level for the entire set of unique peptide identifications (<1% false positive rate). For example, accept peptides with a ΔCn score above 0.13 (regardless of charge state). Accept peptides with a +1 charge state if they were fully tryptic digested and had a cross correlation (Xcorr) of at least 1.55. Accept peptides with a +2 charge state if they had an Xcorr 2.35. Accept peptides with a +3 charge state if they had an Xcorr 2.75.
6. Elucidate the subcellular location of the identified proteins by the gene ontology (GO) (Table 2).

4. Notes

1. 1 M Phenylmethanesulfonyl fluoride (PMSF) in acetone (Sigma L7626). PMSF is toxic if swallowed and contact with acid liberates very toxic gas. Use only in a chemical fume hood.

Table 2
Statistical analysis of identified proteins (protein groups)

Number of identified proteins (protein groups)	883
GO annotation	
Total number of annotated proteins	658
With subcellular location	490
Membrane	294
Plasma membrane	96
Integral to plasma membrane	54
Mitochondrial inner/outer membrane proteins	37
Endoplasmic reticulum membrane	29
With 1 or more predicted transmembrane helices	333
With more than 1 predicted transmembrane helices	175

2. Dextran can contain up to 10% water and has to be freeze-dried. For freeze-drying, take some amount of dextran in a 50 mL tube and freeze it at −80°C under vacuum. Store the freeze-dried dextran in closed plastic tubes sealed tightly with parafilm at −20°C. Let chilled dextran to be at room temperature before opening in order to protect it from humidity. Dissolve 5 g of freeze-dried Dextran T-500 in 25 mL of Mili-Q water in a glass beaker. It is difficult to dissolve Dextran T-500 in water in a short time, so prepare it before the day you use.
3. SDS is harmful by inhalation and if swallowed, irritating to eyes, respiratory system, and skin. Use only in a chemical fume hood.
4. All the procedure is performed on ice or in cold room.
5. The ration of homogenization buffer to tissue is 4–5 mL/g. The liver tissue should be enough homogenized; otherwise, no band will appear at the top of sucrose solution. For smaller amounts of liver tissue, adjust the volume of homogenization buffer accordingly.
6. Make sure that the homogenization is enough. Re-extract the pellet with Tissue-Tearor T8 would be very helpful for enhancing the productivity of membranes.
7. The final concentration of the sucrose should be above 50% (v/v).
8. Prepare two-phase systems one day prior to use. Mix them by 30–40 invertations, and vortex, followed by another 30–40 invertations. Store the mixtures at 4°C overnight. Two-phase systems with the top phase enriched in PEG and the bottom phase enriched in dextran will form overnight. For 8 g of the two-phase system, reduce all of the volume/weight in half.

9. High concentration of SDS and urea will affect the gel mix form in the incorporation steps.
10. According to our test, 0.1% SDC shall be added in order to optimize the in-gel digestion of membrane proteins.

Acknowledgements

This work was supported by National Natural Science Foundation of China (31000375 and 30770437) and grant from National 973 Project of China (2007CB914203).

References

1. He F (2005) Human liver proteome project: plan, progress, and perspectives. Mol Cell Proteomics 4:1841–1848
2. Fagerberg L, Jonasson K, von Heijne G, Uhlen M, Berglund L (2010) Prediction of the human membrane proteome. Proteomics 10(6): 1141–1149
3. Hopkins AL, Groom CR (2002) The druggable genome. Nat Rev Drug Discov 1:727–730
4. Morre DJ, Morre DM (1989) Preparation of mammalian plasma membranes by aqueous two-phase partition. Biotechniques 7:946–948, 950–944, 956–948
5. Speers AE, Wu CC (2007) Proteomics of integral membrane proteins—theory and application. Chem Rev 107:3687–3714
6. Cao R, Li X, Liu Z, Peng X, Hu W, Wang X, Chen P, Xie J, Liang S (2006) Integration of a two-phase partition method into proteomics research on rat liver plasma membrane proteins. J Proteome Res 5:634–642
7. Cao R, He Q, Zhou J, Liu Z, Wang X, Chen P, Xie J, Liang S (2008) High-throughput analysis of rat liver plasma membrane proteome by a nonelectrophoretic in-gel tryptic digestion coupled with mass spectrometry identification. J Proteome Res 7:535–545
8. Lu X, Zhu H (2005) Tube-gel digestion: a novel proteomic approach for high throughput analysis of membrane proteins. Mol Cell Proteomics 4:1948–1958

Chapter 9

Quantitative Analysis of Liver Golgi Proteome in the Cell Cycle

Xuequn Chen, Philip C. Andrews, and Yanzhuang Wang

Abstract

During mitosis, the Golgi membranes in mammalian cells undergo a continuous disassembly process and generate mitotic fragments that are distributed into the daughter cells and reassembled into new Golgi after mitosis. This disassembly and reassembly process is critical for Golgi biogenesis during cell division, but the underlying molecular mechanism is poorly understood. In this study, we have recapitulated this process using an in vitro assay and analyzed the proteins that are associated with interphase and mitotic Golgi membranes using quantitative proteomics that combines the isobaric tags for relative and absolute quantification approach with OFFGEL isoelectric focusing separation and LC-MALDI-MS/MS. A total of 1,193 Golgi-associated proteins were identified and quantified. These included broad functional categories: Golgi structural proteins, Golgi resident enzymes, SNAREs, Rab GTPases, and secretory and cytoskeletal proteins. More importantly, the combination of the quantitative proteomic approach with Western blot analysis allowed us to unveil 86 proteins with significant changes in abundance under the mitotic condition compared to the interphase condition. Altogether, this systematic quantitative proteomic study revealed candidate proteins of the molecular machinery that controls the Golgi disassembly and reassembly processes in the cell cycle.

Key words: Liver Golgi, Cell cycle, Cell-free assay, Quantitative proteomics, iTRAQ, LC-MALDI-MS/MS

1. Introduction

The Golgi is the central organelle in the secretory pathway, essential for posttranslational modifications, sorting, and trafficking of secretory and membrane proteins and lipids in all eukaryotic cells. The unique feature of the Golgi in almost all eukaryotic cells, with only a few rare exceptions (1), is the densely packed stacks of flattened cisternal membranes. Processing enzymes in the Golgi,

Djuro Josic and Douglas C. Hixson (eds.), *Liver Proteomics: Methods and Protocols*, Methods in Molecular Biology, vol. 909, DOI 10.1007/978-1-61779-959-4_9, © Springer Science+Business Media New York 2012

including those involved in modifying bound oligosaccharides, are arranged across the stack in the *cis* to *trans* order in which they function, which ensures the accuracy of glycosylation. Despite its complicated morphology and function, the Golgi is dynamic, capable of rapid disassembly and reassembly during mitosis or upon drug treatment. At the onset of mitosis, the characteristic stacked organization of the Golgi undergoes extensive disassembly. The resulting mitotic Golgi fragments are subsequently distributed to the daughter cells, where they are reassembled into new Golgi stacks during cytokinesis (2, 3).

The unique morphology of the Golgi apparatus and its high dynamic have intrigued a large number of cell biologists, and many studies have aimed to understand how the stacked Golgi structure is formed and how the Golgi disassembly and reassembly process during the cell cycle is regulated (2, 3). It is generally believed that Golgi structure formation is regulated by cytosolic and membrane proteins, and many studies have been performed in efforts to identify these proteins using biochemical and cell biology approaches. Early morphological studies showed that there are proteinaceous cross-bridges linking adjacent Golgi cisternae (4). Later on, a protein complex that was resistant to detergent and salt was isolated and named as the "Golgi matrix" proteins (5), which were identified in further studies as a group of coiled-coil proteins (i.e., golgins) and Golgi Reassembly Stacking Proteins (GRASPs). These proteins play essential roles in vesicle tethering, cisternal stacking, and ribbon formation as well as cell cycle regulation (6, 7). The function of these structural proteins is regulated by cytosolic factors, which results in the disassembly and reassembly of the Golgi during the cell cycle (7–9). However, the identities of these cytosolic proteins are still far from clear. Recently, several labs have attempted to identify Golgi membrane proteins using organellar proteomics, a fundamental and fast-expanding research technology in proteomics and cell biology that combines biochemical fractionation and comprehensive protein identification (10–15). However, quantitative studies documenting the comprehensive protein changes on the Golgi membranes during the cell cycle have not been reported thus far. Analysis of the dynamics of proteins in the Golgi membrane during disassembly has the advantage of identifying peripheral membrane proteins that exhibit changes, which are more likely to be specific regulators for the morphological change rather than contaminating components.

Biochemical reconstitution experiments have provided powerful tools with which we can dissect biological processes. One widely used method to study the Golgi disassembly and reassembly processes during the cell cycle involves purified Golgi membranes to which mitotic or interphase cytosol is added (16–20). After incubation, the membranes are separated from the cytosol by centrifugation through a sucrose cushion and then processed

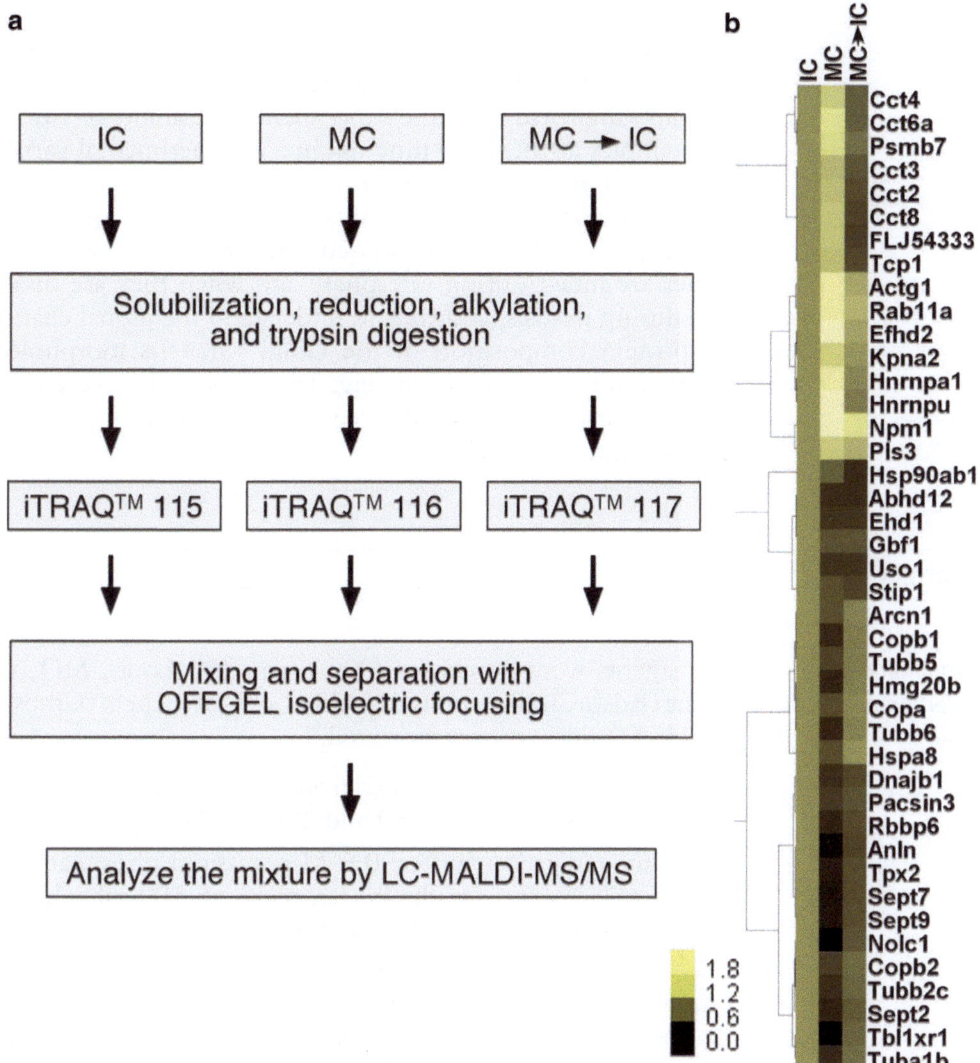

Fig. 1. Flowchart of the experimental procedures and hierarchical clustering of identified Golgi proteins whose binding to the membranes undergoes cell cycle regulation. (**a**) Experiment flowchart. Purified rat liver Golgi membranes were incubated with indicated cytosols and subjected to the subsequent treatments to identify and quantify Golgi-associated proteins. (**b**) Hierarchical clustering of Golgi proteins under indicated conditions. The protein amount after treatment with interphase cytosol was normalized to 1 and the relative amounts under other conditions were shown.

for biochemical and morphological analyses. When combined with modern quantitative proteomics approaches, this in vitro reconstitution system is expected to provide a powerful tool to further dissect the molecular mechanism that regulates Golgi membrane dynamics during the cell cycle (for strategy see Fig. 1a). In this study, we applied isobaric tags for relative and absolute quantification (iTRAQ) and LC-MALDI-MS/MS analysis to quantify protein changes on the Golgi membranes of different phases of the cell cycle (21). iTRAQ is a chemical labeling reagent that is widely used in proteomics studies in the recent years. Its advantage is to simultaneously identify and quantify proteins in up to eight samples in the same experiment. The ability to run multiplex samples at the same time minimizes experimental variations, resulting in the ability to detect relatively small changes in protein levels. This provides us an unbiased approach to quantitatively analyze proteins that are associated with the Golgi when the membranes are intact during interphase and when they are disassembled during mitosis, and thus to understand the global changes in the protein composition of the Golgi when its morphology is changed during cell division (Fig. 1b). This study revealed candidate proteins involved in the regulation of Golgi morphological changes during the cell cycle.

2. Materials

2.1. Purification of Golgi Membranes from Rat Liver

All reagents were purchased from Sigma (St. Louis, MO, USA), Roche (Roche, Indianapolis, IN, USA), or Calbiochem (Gibbstown, NJ, USA), unless otherwise stated.

1. Typically, livers are obtained from six Sprague Dawley rats with body weight between 150 and 200 g (22).
2. Homogenization buffer: 0.5 M potassium phosphate buffer, pH 6.7. To prepare the buffer, make up 500 ml solutions of 0.5 M anhydrous K_2HPO_4 and 0.5 M anhydrous KH_2PO_4. To 400 ml of the latter, gradually add the former until the pH reaches 6.7.
3. Gradient buffers: Make up sucrose buffers A–E in 100 mM phosphate buffer, pH 6.7, and 5 mM $MgCl_2$ from the indicated stock solutions and ice-cold water as shown in Table 1. 50 ml of each buffer A, B, and E and 100 ml of buffers C and D are needed. EDTA-free protease inhibitor tablet (Roche, Indianapolis, IN, USA) and pepstatin A are added to buffers C and D. Water should be precooled to 4°C overnight to ensure that all the buffers are ice cold. It is important to be as accurate as possible when mixing various components and to check the refractive index of each buffer using a refractometer (Reichert, Depew, NY, USA). The final refractive index should be adjusted using water or 2 M sucrose to within ±0.5% sucrose (about 0.001 in refractive index) for buffers C and D in particular. Leave the buffers on ice all the time during the experiment.
4. Beckman L8-70M ultracentrifuge, SW-41 rotor, and centrifuge tubes (Beckman, Cat. #: 344059, Brea, CA, USA), or equivalents.

Table 1
Sucrose solutions for Golgi purification

Buffer	A	B	C	D	E
Sucrose concentration (M)	0	0.25	0.5	0.86	1.3
0.5 M Phosphate buffer, pH 6.7 (ml)	10	10	20	20	10
2 M Sucrose (ml)	0	6.25	25	43	32.5
2 M $MgCl_2$ (ml)	0.125	0.125	0.25	0.25	0.125
Water (ml)	39.9	33.6	54.8	36.8	7.4
Total volume (ml)	50	50	100	100	50
Sucrose % (w/v)	0	8.6	16.0	26.4	38.6
Refractive index	1.3330	1.3456	1.3574	1.3747	1.3973

2.2. In Vitro Reconstitution of Mitotic Golgi Disassembly and Postmitotic Golgi Reassembly Using a Cell-Free System

2.2.1. Preparation of Cytosol

1. The interphase (IC) and mitotic (MC) cytosols are prepared from HeLa S3 cells (16, 18, 20).
2. EBS buffer: 0. 1 M sucrose, 80 mM β-glycerophosphate, 20 mM EGTA, 15 mM $MgCl_2$, 2 mM ATP, 1 mM glutathione, EDTA-free protease inhibitor cocktail (Roche, 1 in 50 ml), 5 mM pepstatin A. Adjust to pH 7.2 with KOH. Freshly made before the experiment and leave on ice.
3. Cytosols are changed into related buffers using Bio-spin 6-Tris columns (Bio-Rad, Hercules, CA, USA) following the manufacturer's instructions.
4. Beckman Max-XP ultracentrifuge, TLA 100.3 rotor, and centrifuge tubes (Beckman, Brea, CA, USA), or equivalents.

2.2.2. The Golgi Disassembly and Reassembly Assay

1. MEB buffer: 50 mM Tris–HCl, pH 7.4 (adjusted with HCl), 0.2 M sucrose, 50 mM KCl, 20 mM β-glycerophosphate, 15 mM EGTA, 10 mM $MgCl_2$, 2 mM ATP, 1 mM GTP, 1 mM glutathione, and EDTA-free protease inhibitors. Freshly made before the experiment and leave on ice.
2. ATP regeneration system (10×): 100 mM creatine phosphate (200 mM stock, in distilled water), 1 mM ATP (10 mM stock, in 250 mM Hepes-KOH, pH 7.3), 0.2 mg/ml cytochalasin B (2 mg/ml stock, in DMSO), 0.2 mg/ml creatine kinase (2 mg/ml stock, in 50% glycerol). Freshly mix the stock solutions with distilled water before use. Dilute 1:10 into the reactions.
3. Beckman Max-XP ultracentrifuge, TLA55 rotor, and centrifuge tubes, or equivalents.

4. Eppendorf 5417R microcentrifuge (Eppendorf, Hauppauge, NY, USA, or a similar centrifuge without cooling in a 4°C cold room) with a swing bucket rotor.
5. KHM buffer: 20 mM HEPES, pH 7.4, 0.2 M sucrose, 60 mM KCl, 5 mM $Mg(OAc)_2$, 2 mM ATP, 1 mM GTP, 1 mM glutathione, protease inhibitors (1 EDTA-free tablet for 50 ml buffer).
6. Electron microscopy (EM) fixation buffer: 2% glutaraldehyde in KHM buffer. Dilute glutaraldehyde from a 25% or 50% stock into KHM buffer freshly, warm to room temperature (18–22°C) before use.
7. EM post-fixative: 1% (w/v) OsO_4, 1.5% $K_3(Fe(CN)_6)$, 0.1 M sodium cacodylate, pH 7.4.
8. Epon-mix: Before use, mix 14.5 g Embed 812, 10.5 g NMA, and 5.0 g DDSA (Electron Microscopy Sciences, Hatfield, PA, USA). Mix well until no streaks are left; add 0.54 g DMP-30 and mix. Prepare freshly before use.
9. Uranyl acetate solution: Dissolve 2% uranyl acetate in H_2O. Push the solution through a 0.22 μm syringe filter before use.
10. Lead citrate solution: Weigh 1.33 g lead nitrate and 1.76 g sodium citrate, add 30 ml H_2O, and shake for 2 min at room temperature. Add 8 ml 1 M NaOH, adjust volume to 50 ml with H_2O. Push through a 0.22 μm syringe filter before use.

2.3. In-Solution Digestion and iTRAQ™ Labeling

1. Sequencing grade modified trypsin (Promega, Madison, WI, USA).
2. iTRAQ™ reagents (Applied Biosystems, Foster City, CA, USA).
3. Resuspension buffer: 0.5 M triethyl ammonium bicarbonate (TEAB), 8 M urea, and 0.4% SDS.
4. Bradford assay kit (Bio-Rad, Hercules, CA, USA).

2.4. OFFGEL Isoelectric Focusing and LC-MALDI-MS/MS

1. A 3100 OFFGEL Fractionator (Agilent Technologies, Santa Clara, CA, USA).
2. IPG strip pH 3–10 (GE healthcare, Pittsburgh, PA, USA).
3. C18 spin columns (Thermo Scientific, Rockford, IL, USA).
4. Reversed-phase C18 trap columns (Zorbax C18; 5 mm by 0.3 mm i.d.; 5 μm particles) (Agilent Technologies, Santa Clara, CA, USA).
5. Reversed-phase C18 analytical columns (Zorbax 300 SB C18 column, 75 μm×150 mm, 3.5 μm particles) (Agilent Technologies, Santa Clara, CA, USA).
6. Agilent 1100 HPLC system with micro-collection/spotting system.

7. Solvent A: 0.1% TFA in Optima™ H_2O (LC/MS grade) (Fisher Scientific, Pittsburgh, PA, USA) and solvent B: 90% acetonitrile and 0.1% TFA in Optima™ H_2O.
8. MALDI matrix: α-cyano-4-hydroxycinnamic acid (2 mg/ml) (Agilent Technologies, Santa Clara, CA, USA).
9. Infusion pump (Harvard Apparatus, Holliston, MA, USA).
10. MALDI-TOF/TOF instrument: Applied Biosystems 4800 Proteomics Analyzer (AB Sciex, Foster City, CA, USA).
11. Peptide Mass Standards Kit (AB Sciex, Foster City, CA, USA).

2.5. Database Searching, Quantification, and Categorization of the Identified Proteins

1. ProteinPilot™ v3.0 (AB Sciex, Foster City, CA, USA).
2. IPI rat (version 3.49) and IPI human databases (version 3.53).
3. Hierarchical clustering using Cluster 3.0 and viewed with the TreeView program.
4. "Compute pI/Mw" tool accessible on the ExPASY Web site.

2.6. Confirmation of the Proteomic Analysis by Western Blotting

The antibodies used are monoclonal antibodies against Bet1, Gos28, GM130, and HSP90 from BD Transduction Laboratories (Rockville, MD, USA); monoclonal antibodies against α-actin (St. Louis, MO, USA), ARF1 (1D9, Abcam, Cambridge, MA, USA), β-COP (M3A5, gift of Dr. I Mellman, Genentech, South San Francisco, CA, USA), GRASP65 (23), PP2A C subunit (Millipore, Billerica, MA, USA), and β-tubulin (Dr. K. Gull, University of Oxford, UK); polyclonal antibodies against ARF1 (Dr. D. Sheff, University of Iowa, IA, USA), cdc2 (Millipore, Billerica, MA, USA), β-COP (EAGE, Dr. I Mellman, Genentech, South San Francisco, CA, USA), ERK2 (Millipore, Billerica, MA, USA), GM130 (MLO7, M. Lowe, University of Manchester, UK), GRASP55 (Dr. J. Seemann, University of Texas, Dallas, USA) (24), GRASP65 (23, 25), HSP70 (Synaptic Systems, Goettingen, Germany), α-Mannosidase (Man) I (Dr. G. Warren, Max F. Perutz Laboratories, Vienna, Austria) and II (Dr. K. Moremen, University of Georgia, GA, USA), golgin-84 (Dr. A. Satoh, Okayama University, Okayama, Japan), NSF (Dr. G. Warren), p115 (Dr. G. Warren), Rab1A (Santa Cruz, Santa Cruz, CA, USA), Rab6 (Santa Cruz), Rab11 (Santa Cruz), rat serum albumin (Dr. G. Warren), sec31 (Dr. F. Gorelick, Yale University, New Haven, CT, USA), syntaxin 5 (Dr. G. Warren), SNAP29 (Synaptic Systems), and TGN38 (Dr. G. Warren). Antibodies to β′-, γ-, δ-, and ε-COPs were kindly provided by Dr. D. Sheff. The monoclonal antibody for phosphorylated GRASP65 was raised in mouse with recombinant GST-GRASP65 (aa 202–446) treated with mitotic HeLa cell cytosol. Secondary antibodies for immunofluorescence and for Western blotting were from Molecular Probes (Carlsbad, CA, USA) and Jackson Immunoresearch Laboratories (West Grove, PA, USA), respectively.

3. Methods

3.1. Purification of Golgi Membranes from Rat Liver

The Golgi membranes are purified from rat liver homogenate by two sequential sucrose gradient centrifugations as previously described (22). This method uses rat liver as the source since liver cells have abundant Golgi membranes due to the high activity in protein and lipid secretion. Liver tissue is homogenized by passing through a 150 μm mesh sieve. This method is relatively mild and is essential for the preservation of the stacked structure of the Golgi membranes. The first gradient separates the Golgi membranes from unbroken cells, other cellular organelles (e.g., nuclei and mitochondria), and the majority of cytosol and free lipids. The second sucrose gradient further separates the Golgi membranes from cytosolic contaminants and concentrates the Golgi membranes. A typical preparation can yield 5–6 mg Golgi membranes that are purified about 100-fold over the homogenate, and 60–70% in stacks analyzed by electron microscopy.

1. Starve six female Sprague–Dawley rats (body weight 150–200 g) for 24 h.
2. Sacrifice the rats using high doses of CO_2 followed by cervical dislocation. Rapidly remove the livers into a large volume of ice-cold buffer C (protease inhibitors can be omitted in the wash buffer) to wash off the blood. Transfer the livers into ice-cold buffer C with EDTA-free protease inhibitors and pepstatin A. Mince the liver into small pieces with a pair of scissors.
3. Weigh livers, which are normally 40–45 g from six rats.
4. Homogenize the tissue by pressing through a 150 mm mesh stainless-steel sieve with the bottom of a 200 ml conical flask in a rolling action. Adding a small amount of buffer C (with protease inhibitor) to the sieve will help the liver homogenate press through more easily. The final volume of the homogenate should be about 50 ml.
5. Make the gradients: Place 6 ml of buffer D (Table 1) in each of the 12 SW-41 Ultraclear tubes, overlay 4.5 ml of homogenates, and 1.5 ml of buffer B on the top of the gradient, and balance the tubes.
6. Centrifuge at 103,800 × *g* (29,000 rpm) in an SW-41 rotor for 60 min at 4°C.
7. Aspirate the lipid at the top and the cytosol indicated by red color of hemoglobin, and collect Golgi membranes that accumulate at the 0.5/0.86 M interfaces with a plastic Pasteur pipette.
8. Pool the collected fractions. Measure the sucrose concentration of the sample by a refractometer, and adjust the sucrose

concentration to 0.25 M (refractive index 1.3456) using buffer A (no sucrose).

9. Pool the second gradient with 1 ml 1.3 M sucrose (buffer E) and 2 ml 0.5 M sucrose (buffer C) in each tube, and overlay 9 ml diluted Golgi sample.
10. Centrifuge at 7,900×*g* (8,000 rpm) in an SW-41 rotor for 30 min at 4°C.
11. Discard the supernatant and collect the membranes at the 0.5 M/1.3 M sucrose interface. Adjust the sucrose concentration to 0.5 M (refractive index 1.3574) using buffer A.
12. Check protein concentration using Bio-Rad Bradford assay. Check morphology and purity of the purified Golgi membranes by EM.
13. Aliquot and freeze samples in liquid nitrogen and store at −80°C (see Note 1).

3.2. In Vitro Reconstitution of Mitotic Golgi Disassembly and Postmitotic Golgi Reassembly Using a Cell-Free System

1. The interphase and mitotic cytosols were prepared from HeLa S3 cells as previously described (16, 18, 20). Cytosols are centrifuged for 30 min at full speed in a tabletop centrifuge to remove protein precipitates and changed into indicated buffers using Bio-Spin 6 columns before used to treat Golgi membranes.
2. The Golgi disassembly assay is performed as described previously (19, 20). In brief, purified Golgi membranes (200 μg) are mixed with 10 mg of mitotic cytosol (see Note 2), 1 mM GTP, and an ATP-regenerating system in MEB buffer with a final volume of 1 ml.
3. After incubation for 60 min at 37°C, mitotic Golgi fragments (MGFs) are isolated and soluble proteins are removed by centrifugation (55,000 rpm or 136,000×*g*, for 30 min in a TLA55 rotor) through a 0.4 M sucrose cushion in KHM buffer onto a 6 μl 2 M sucrose cushion (see Note 3).
4. The membranes are resuspended in KHM buffer, and aliquots are succeeded to fixation and processing for EM, to reassembly reactions described below, or kept frozen at −80°C until further use for the proteomics analysis.
5. Incubation of Golgi membrane with interphase cytosol in KHM buffer is used as control.
6. For Golgi reassembly, 200 μg of MGFs are resuspended in 4 mg IC in KHM buffer in the presence of an ATP regeneration system in a final volume of 300 μl and incubated at 37°C for 60 min. Gently mix the samples every 20 min by tapping the Eppendorf tube.
7. Membranes are pelleted by centrifugation as described above, processed for EM, or kept at −80°C for the proteomic analysis or Western blotting. All results are confirmed by three or more independent experiments.

8. For EM analysis, the percentage of membranes in cisternae or in vesicles is determined by the intersection method (16, 18, 19, 23, 26) (see Note 4).

3.3. In-Solution Digestion and iTRAQ™ Labeling

1. For quantitative proteomics analysis, membranes are collected under three experimental conditions: (1) incubation of membranes with interphase cytosol alone; (2) with mitotic cytosol alone; and (3) sequential incubation with mitotic cytosol and interphase cytosol (21).
2. After incubation, Golgi membrane is separated from cytosol by centrifugation through a 0.4 M sucrose cushion and the Golgi membrane pellets are stored at –80°C until use.
3. For in-solution digestion, Golgi membrane pellets are solubilized on ice in 50 μl resuspension buffer for 30 min.
4. Protein concentrations are determined with the Bio-Rad Bradford assay kit.
5. Golgi membrane proteins (30 μg) from each of the previously mentioned three groups are reduced with 5 mM tris(2-carboxyethyl)phosphine (TCEP) for 1 h at 37°C. Cysteines are then blocked with 10 mM methyl methanethiosulfonate (MMTS) for 20 min at room temperature.
6. The protein solution is diluted four times with 0.5 M TEAB containing 3 μg trypsin (1:10 w/w) and incubated at 37°C overnight.
7. To quantitatively compare Golgi membrane treated with mitotic cytosol to that treated with interphase cytosol, multiplexed isobaric tags (iTRAQ™ reagents) are used to label tryptic peptides from different conditions: 115 for interphase cytosol treatment, 116 for mitotic cytosol treatment, and 117 for sequential mitotic and interphase cytosol treatments. The labeling procedure is performed according to the manufacturer's instructions, with more details described below.
8. Dissolve the iTRAQ™ reagents by adding 70 μl of ethanol, vortex thoroughly for a minute, and then spin down briefly in a minicentrifuge.
9. The tryptic peptides from three different treatments are then mixed with the 115, 116, and 117 iTRAQ™ reagents, respectively, at room temperature for an hour.
10. The labeling reactions are terminated by diluting the mixture with 10 volumes of H_2O.

3.4. OFFGEL Isoelectric Focusing and LC-MALDI-MS/MS

1. The combined peptide mixture is first fractionated based on isoelectric points (pIs) on a 3100 OFFGEL Fractionator (Agilent Technologies) using pH 3–10 IPG strip with a 12-well manifold according to the manufacturer's instruction.

2. The iTRAQ-labeled peptides are diluted in the peptide-focusing buffer to a final volume of 1.8 ml, 150 μl of which is loaded into each well.
3. The samples are focused with a maximum current of 50 μA until 50 kV h is achieved.
4. Peptide fractions are harvested and desalted using Pierce C18 spin columns.
5. The resulting elute is dried using a vacuum centrifuge and resuspended in 40 μl 0.1% trifluoroacetic acid (TFA) and 5% acetonitrile.
6. The resuspended peptide sample from each OFFGEL fraction is concentrated on a reversed-phase C18 cartridge (Agilent Technologies) and then separated by reversed-phase C18 analytical column (Agilent Technologies) using an Agilent 1100 HPLC system with the following binary gradient with a flow rate of 300 nl/min: 0 min, 6.5% B; 9 min, 6.5% B; 12 min, 15% B; 92 min, 45% B; 97 min, 60% B; 102 min, 100% B; 104 min, 100% B; 105 min, 6.5% B; and 115 min, 6.5% B. Solvent A is 0.1% TFA and solvent B is 90% acetonitrile and 0.1% TFA.
7. The column effluent is mixed with MALDI matrix (2 mg/ml α-cyano-4-hydroxycinnamic acid) (Agilent Technologies) through a 25 nl mixing tee and spotted on 1536 well OptiTOF™ MALDI target plates which are later analyzed by tandem mass spectrometry (see Note 5).
8. The MS and MS/MS spectra are acquired on an Applied Biosystems 4800 Proteomics Analyzer (TOF/TOF) in positive ion reflection mode with a 200 Hz Nd:YAG laser operating at 355 nm. The accelerating voltage is 20 kV with a 400-ns delay.
9. For MS/MS spectra, the collision energy is 2 keV and the collision gas is air.
10. Each 1536 well OptiTOF™ MALDI plate is first calibrated on nine calibration wells using standards from Applied Biosystems with a 20 ppm mass accuracy in the MS mode.
11. Both MS and MS/MS data are then acquired in the sample wells using the instrument default calibration. Typical MS spectra are obtained with the minimum possible laser energy in order to maintain the best resolution. Single-stage MS spectra for the entire samples are first collected and in each sample well MS/MS spectra are acquired from the 12 most intense peaks above the signal-to-noise ratio threshold of 30.

3.5. Database Searching, Quantification, and Categorizing the Identified Proteins

1. Database searching is performed using ABI's ProteinPilot™ v3.0 using the Paragon algorithm (27). This software interacts directly with the Oracle database in which the mass spectrometer stores its data, and submits monoisotopic peak lists in batch to a local instance of the search engine for protein identities. No additional peak list filtering is specified.
2. Peak lists are generated by the mass spectrometer during data acquisition based on a specified signal-to-noise threshold (30 in this case).
3. ProteinPilot™ also performs protein grouping to remove redundant hits and comparative quantifications using iTRAQ ratios.
4. In an effort to comprehensively identify both rat (originated from the Golgi membranes) and human proteins (recruited from HeLa cell cytosol), a combined human and rat database is created by manually concatenating IPI rat (version 3.49, with 40,131 of *Rattus norvegicus* proteins) and IPI human databases (version 3.53, with 73,748 of *Homo sapiens* proteins). All data sets are then searched against this database using ProteinPilot™.
5. All reported proteins are identified with 90% or greater confidence as determined by ProteinPilot™ Unused scores (≥1.0) with the corresponding false positive discovery rate below 1%. The iTRAQ peak area data were normalized for loading error by biased corrections calculated using ProteinPilot™. The default ion intensity threshold in ProteinPilot™ for calculating the peptide ratios is 40 counts.
6. Hierarchical clustering is performed based on Pearson correlation using Cluster 3.0 program and clusters are viewed with the TreeView program.
7. The theoretical pIs of the peptides in each OFFGEL fraction are calculated using "Compute pI/Mw" tool accessible on the ExPASY Web site. The identified peptides with a confidence greater than 90% are included in the peptide list for the calculation.

3.6. Confirmation of the Proteomic Analysis by Western Blotting

1. To validate the iTRAQ-based results, Western blot analyses are conducted on a large number of proteins, including many in the major clusters.
2. The protein levels in the interphase and mitotic cytosols are first compared to confirm that the change of Golgi-associated proteins is not due to the availabilities of the proteins in either cytosol.
3. Then the proteins associated with either interphase or mitotic Golgi membranes are compared. As expected, Golgi resident proteins such as TGN38 and enzymes such as α-Mannosidase II have no change across all conditions. Consistent with our

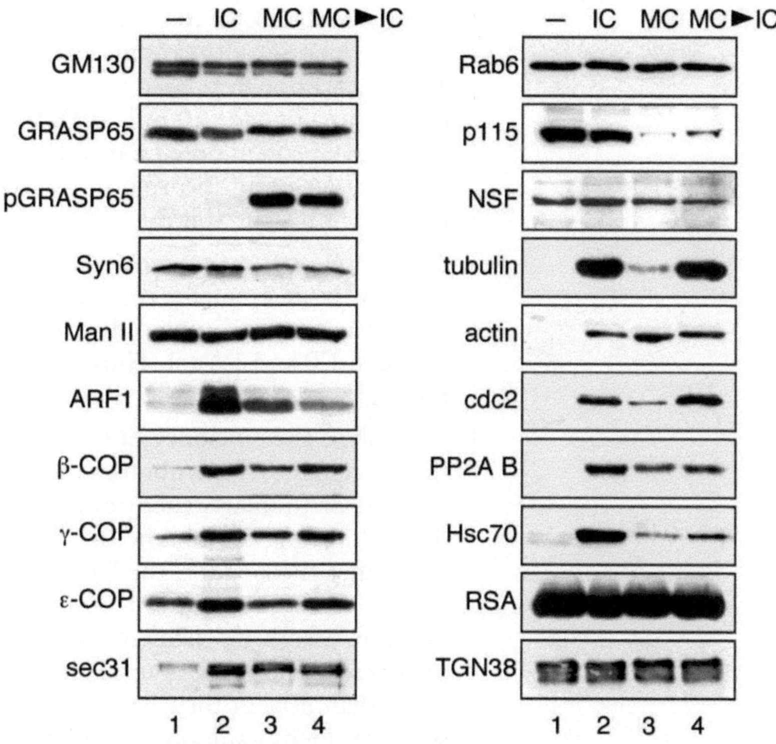

Fig. 2. Western blot analysis of major Golgi-associated proteins. Purified rat liver Golgi membranes were treated with buffer (*lane 1*), interphase cytosol (IC, *lane 2*), mitotic cytosol (MC, *lane 3*), or with subsequent treatments with interphase and mitotic cytosol (MC→IC, *lane 4*) followed by isolation of the total membranes by centrifugation. Equal fractions were analyzed by Western blot for indicated proteins; representative images from three independent experiments were shown. pGRASP65, phospho-specific GRASP65; NSF, N-ethylmaleimide-sensitive fusion protein; RSA, rat serum albumin. Modified from Chen et al. (21) with permission from the journal.

iTRAQ results, the amount of membrane proteins such as GM130 and GRASP65 does not change upon different treatments (Fig. 2, also see Note 6).

4. The unique patterns of changes of COPI coatomer subunits are confirmed by Western blot with antibodies against five out of all six subunits identified by iTRAQ. The different patterns of the heat-shock proteins and the cytoskeleton proteins actin and tubulin are also confirmed (Fig. 2, see Note 7).

4. Notes

1. Golgi membrane pellets can be used immediately after preparation or stored in the centrifuge tubes at –80°C with the caps sealed with parafilm.

2. The protein ratio of mitotic cytosol:Golgi in the assay is based on the quantification results of these two components in rat liver cells. Prior to use, spin the mitotic cytosol in a benchtop microcentrifuge for 20 min at 18,000 × g (14,000 rpm) at 4°C to remove protein precipitates. Desalt the mitotic cytosol into MEB buffer using Bio-spin 6-Tris column following the manufacturer's instruction.
3. This step is critical to reduce excess cytosolic proteins. If the reaction is in a smaller volume, increase the volume of the 0.4 M sucrose cushion so that the total volume in the tube is between 1.0 and 1.4 ml. The centrifuge tube may collapse after the high-speed spin if the volume is too small.
4. To analyze the efficiency of Golgi disassembly and reassembly, the percentage of membranes in Golgi stacks, single cisternae, vesicles, or tubular structures can be quantified. Cisternae are defined as long membrane profiles with a length greater than four times their width, the latter being not more than 60 nm. Normal cisterna ranges 20–30 nm in width and is longer than 200 nm. Stacks are defined as two or more cisternae that were separated by no more than 15 nm and overlapped in parallel by more than 50% of their length. Vesicles are spherical or nearly spherical membrane profiles with 50–200 nm diameter, often can be seen with the protein coat. The vesicles in our study are approximately 70 nm in diameter. Tubular network are single tubules with a length more than 1.5 times but less than four times their maximal width. The width is often uneven, sometimes more than 60 nm, and sometimes with branches.
5. The MALDI matrix is delivered to the mixing tee by external infusion pump at 800 nl/min. A reduced concentration of α-cyano-4-hydroxycinnamic acid is used to prevent matrix crystals from clogging the capillary in the nanoflow pump. In addition, the capillaries are washed with matrix solvent using the infusion pump after each experiment to prevent matrix crystals' formation.
6. Western blot analysis for GRASP65 is performed using two antibodies, one recognizes all forms of GRASP65, which shows that the amount of GRASP65 is not changed upon different treatments of the membranes although the protein migrates slower on the SDS-PAGE when treated with mitotic cytosol; the other recognizes only the phosphorylated form, pGRASP65.
7. Altogether, the comprehensive Western analysis results (Fig. 2) are consistent with the iTRAQ-based quantitative proteomics results (Fig. 1b).

Acknowledgements

We gratefully acknowledge Drs. F. Gorelick, K. Gull, T. Kreis, M. Lowe, K. Moremen, A. Price, A. Satoh, J. Seemann, D. Sheff, D. Shields, and G. Warren for generously providing antibodies. We thank J. Williams and D. Tang for suggestions and reagents. We thank Sarah Volk for her assistance in OFFGEL electrophoresis. This work was supported by National Institute of Health grant P41 RR018627 to P. Andrews, and was partially supported by the National Institutes of Health (GM087364) and the American Cancer Society (RGS-09-278-01-CSM) to Y.W.

References

1. Mowbrey K, Dacks JB (2009) Evolution and diversity of the Golgi body. FEBS Lett 583(23):3738–3745
2. Wang Y (2008) Golgi apparatus inheritance. In: Mironov A, Pavelka M, Luini A (eds) The Golgi apparatus State of the art 110 years after Camillo Golgi's discovery. Springer, New York, pp 580–607
3. Shorter J, Warren G (2002) Golgi architecture and inheritance. Annu Rev Cell Dev Biol 18: 379–420
4. Cluett EB, Brown WJ (1992) Adhesion of Golgi cisternae by proteinaceous interactions: intercisternal bridges as putative adhesive structures. J Cell Sci 103:773–784
5. Slusarewicz P, Nilsson T, Hui N, Watson R, Warren G (1994) Isolation of a matrix that binds medial Golgi enzymes. J Cell Biol 124(4): 405–413
6. Barr FA, Short B (2003) Golgins in the structure and dynamics of the Golgi apparatus. Curr Opin Cell Biol 15(4):405–413
7. Ramirez IB, Lowe M (2009) Golgins and GRASPs: holding the Golgi together. Semin Cell Dev Biol 20(7):770–779
8. Lupashin V, Sztul E (2005) Golgi tethering factors. Biochim Biophys Acta 1744(3):325–339
9. Short B, Haas A, Barr FA (2005) Golgins and GTPases, giving identity and structure to the Golgi apparatus. Biochim Biophys Acta 1744(3):383–395
10. Bell AW, Ward MA, Blackstock WP, Freeman HN, Choudhary JS, Lewis AP, Chotai D, Fazel A, Gushue JN, Paiement J, Palcy S, Chevet E, Lafreniere-Roula M, Solari R, Thomas DY et al (2001) Proteomics characterization of abundant Golgi membrane proteins. J Biol Chem 276(7):5152–5165
11. Wu CC, MacCoss MJ, Mardones G, Finnigan C, Mogelsvang S, Yates JR 3rd, Howell KE (2004) Organellar proteomics reveals Golgi arginine dimethylation. Mol Biol Cell 15(6): 2907–2919
12. Wu CC, Taylor RS, Lane DR, Ladinsky MS, Weisz JA, Howell KE (2000) GMx33: a novel family of trans-Golgi proteins identified by proteomics. Traffic 1(12):963–975
13. Wu CC, Yates JR 3rd, Neville MC, Howell KE (2000) Proteomic analysis of two functional states of the Golgi complex in mammary epithelial cells. Traffic 1(10):769–782
14. Taylor RS, Wu CC, Hays LG, Eng JK, Yates JR 3rd, Howell KE (2000) Proteomics of rat liver Golgi complex: minor proteins are identified through sequential fractionation. Electrophoresis 21(16):3441–3459
15. Mogelsvang S, Howell KE (2006) Global approaches to study Golgi function. Curr Opin Cell Biol 18(4):438–443
16. Misteli T, Warren G (1994) COP-coated vesicles are involved in the mitotic fragmentation of Golgi stacks in a cell-free system. J Cell Biol 125(2):269–282
17. Rabouille C, Kondo H, Newman R, Hui N, Freemont P, Warren G (1998) Syntaxin 5 is a common component of the NSF- and p97-mediated reassembly pathways of Golgi cisternae from mitotic Golgi fragments in vitro. Cell 92(5):603–610
18. Rabouille C, Misteli T, Watson R, Warren G (1995) Reassembly of Golgi stacks from mitotic Golgi fragments in a cell-free system. J Cell Biol 129(3):605–618
19. Tang D, Mar K, Warren G, Wang Y (2008) Molecular mechanism of mitotic Golgi disassembly and reassembly revealed by a defined

reconstitution assay. J Biol Chem 283(10): 6085–6094

20. Tang D, Xiang Y, Wang Y (2010) Reconstitution of the cell cycle regulated Golgi disassembly and reassembly in a cell free system. Nat Protoc 5(4):758–772
21. Chen X, Simon ES, Xiang Y, Kachman M, Andrews PC, Wang Y (2010) Quantitative proteomics analysis of cell cycle regulated Golgi disassembly and reassembly. J Biol Chem 285(10):7197–7207
22. Wang Y, Taguchi T, Warren G (2006) Purification of rat liver golgi stacks. In: Celis J (ed) Cell biology: a laboratory handbook, 3rd edn. Elsevier Science, San Diego, pp 33–39
23. Wang Y, Seemann J, Pypaert M, Shorter J, Warren G (2003) A direct role for GRASP65 as a mitotically regulated Golgi stacking factor. EMBO J 22(13):3279–3290
24. Xiang Y, Wang Y (2010) GRASP55 and GRASP65 play complementary and essential roles in Golgi cisternal stacking. J Cell Biol 188(2):237–251
25. Wang Y, Satoh A, Warren G (2005) Mapping the functional domains of the Golgi stacking factor GRASP65. J Biol Chem 280(6):4921–4928
26. Xiang Y, Seemann J, Bisel B, Punthambaker S, Wang Y (2007) Active ADP-ribosylation factor-1 (ARF1) is required for mitotic Golgi fragmentation. J Biol Chem 282(30):21829–21837
27. Shilov IV, Seymour SL, Patel AA, Loboda A, Tang WH, Keating SP, Hunter CL, Nuwaysir LM, Schaeffer DA (2007) The Paragon Algorithm, a next generation search engine that uses sequence temperature values and feature probabilities to identify peptides from tandem mass spectra. Mol Cell Proteomics 6(9): 1638–1655

Chapter 10

Glycoproteins and Glycosylation: Apolipoprotein C3 Glycoforms by Top-Down MALDI-TOF Mass Spectrometry

Yan Zhang, Alan R. Sinaiko, and Gary L. Nelsestuen

Abstract

Apolipoprotein C3 (ApoC3) is synthesized in liver so that levels or isoform distributions may constitute indicators of liver pathogenesis. The glycoforms of intact protein ApoC3 in serum or plasma can be readily analyzed by matrix-assisted laser desorption/ionization-time of flight mass spectrometry after a one-step extraction using a C4 reverse-phase ZipTip. Glycoform distributions were sensitive to severe systemic diseases such as sepsis or liver diseases such as chronic hepatitis C and alcoholic liver cirrhosis. Glycoisoform distributions were also altered in persons with elevated body mass index and were corrected to normal distribution by metformin, a common drug used for diabetes therapy. This simple method may offer an approach to analysis of health and the mechanism of drug therapies.

Key words: Glycoforms, Glycosylation, Apolipoprotein C3, Plasma biomarker, Mass spectrometry, MALDI-TOF, Protein profiling, Liver diseases

1. Introduction

Glycosylation is a common posttranslational modification of secreted animal proteins. The two most common types of glycosylation consist of N-linked glycosylation of asparagine and O-linked glycosylation of serine or threonine. Heterogeneity of glycosylation is well known, resulting in a distribution of glycoprotein variants or glycoisoforms. Alteration of isoform distribution has been shown to have value in diagnosis, pathogenesis, and progression of breast cancer (1), prostate cancer (2), colorectal cancer (3), liver diseases (4, 5), and other medical conditions (6, 7).

Large-scale analysis of the structures and functions of glycans (i.e., glycomics) based on state-of-the-art technologies, such as

Djuro Josic and Douglas C. Hixson (eds.), *Liver Proteomics: Methods and Protocols*, Methods in Molecular Biology, vol. 909, DOI 10.1007/978-1-61779-959-4_10, © Springer Science+Business Media New York 2012

arrays, HPLC, and mass spectrometry (MS), has become an attractive approach for biomarker research (8). However, many challenges exist for the study of protein glycosylation, especially for intact proteins (9). Our laboratory developed a simple and quick method for detection of a specific glycoisoform distribution that utilizes rapid extraction of plasma or serum samples by reverse-phase (C4 ZipTip) followed by matrix-assisted laser desorption/ionization-time of flight (MALDI-TOF) MS analysis (10). Briefly, 0.5 μL of plasma or serum is diluted in acidic buffer containing 5% acetonitrile (ACN) and subjected to C4 ZipTip extraction. The bound proteins were eluted, mixed with 85% saturated sinapinic acid matrix solution, and applied to a MALDI target. The mass spectrometer was operated in linear mode and 500 laser shots collected for each sample. The spectra were processed by the manufacturer's software and quantified by relative internal peak intensity ratios.

Plasma protein profiles contained 15 common components that included apolipoprotein C1, C2, C3, transthyretin (TTR), and their isoforms (see Fig. 1). Of special interest was the unusual accessibility to glycoisoforms of Apolipoprotein C3 (ApoC3) (Fig. 2). These consisted of ApoC3 without glycosylation (ApoC3-0, *m/z*=8,765) or with a GalNAc-Gal-Sialic Acid glycan chain containing either one (ApoC3-1, *m/z*=9,422) or two (ApoC3-2,

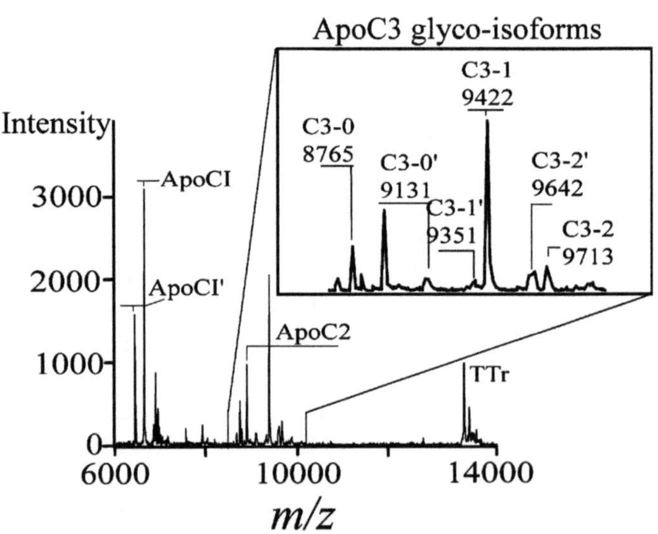

Fig. 1. A typical MALDI-TOF spectrum of normal human plasma. The profile was obtained from 0.5 μL of human plasma that was extracted with a C4 ZipTip and subjected to MALDI-TOF MS analysis by the method described in this document. A total of 15 components were always observed that included ApoC1, ApoC2, ApoC3, and transthyretin (TTr). The expanded region highlights the glycoforms of ApoC3.

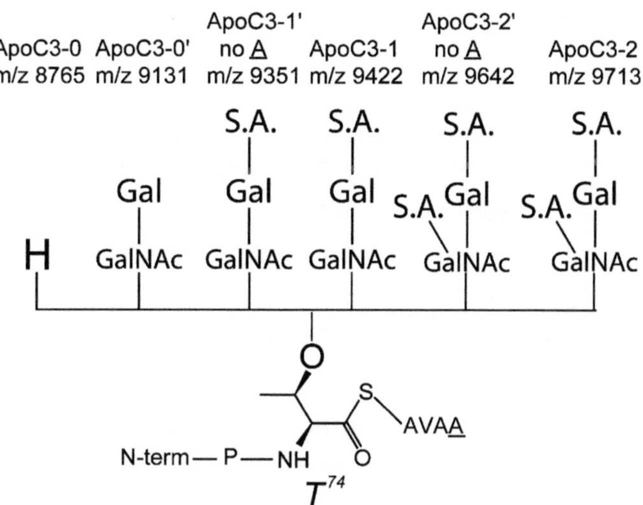

Fig. 2. The structures of ApoC3 glycoforms. The O-linked glycan structure Acetylgalactosamine-galactose (GalNAc-Gal*N*) is attached to Threonine 74 of ApoC3. ApoC3 glycoforms included ApoC3 without glycosylation (ApoC3-0, m/z=8,765). ApoC3 with glycan containing zero (ApoC3-0', m/z=9,131), one (ApoC3-1, m/z=9,422), or two (ApoC3-2, m/z=9,713) sialic acid (S.A.) groups. Proteolytic removal of C-terminal Alanine resulted in two addition components of the profile (ApoC3-1', m/z=9,351, and ApoC3-2', m/z=9,642).

m/z=9,713) sialic acids. C-terminal proteolysis provided additional isoforms consisting of removal of C-terminal Alanine (ApoC3-1', m/z=9,351, ApoC3-2', m/z=9,642) (Fig. 2). The protein profiles were analyzed using internal relative peak intensity ratios, such as the homologous ratio of ApoC3-2/ApoC3-1 or the heterologous ratio of ApoC3-1/ApoC1. Reproducibility was best for homologous peak ratios that gave 4–10% coefficient of variation for replicate measurements (10).

ApoC3 is synthesized in liver, so it was not surprising that the glycoform distribution was sensitive to liver diseases. Analysis of subjects with Hepatitis C demonstrated a reduced total level of ApoC3 as detected by ELISA, but an increased proportion of ApoC3-2 as detected by MALDI-TOF MS (5). Furthermore, significant correlation was observed between the ratio of ApoC3-2/ApoC3-1 and fibrosis score (R=0.55, P=0.002). The statistical correlation was superior to that of current enzyme assays, i.e., alanine amino transferase and aspartate amino transferase (5).

This method was also applied to conditions ranging from obesity and the effect of drug therapy and/or bariatric surgery to graft vs. host disease and severe sepsis (5). Other variations of the general method have been shown to be valuable in analyzing protein profiles in urine for kidney diseases (11, 12) and bronchoalveolar lavage fluid (BALF) for lung transplant rejection (13, 14).

2. Materials

1. Serum or plasma samples are obtained by standard methods and stored at −80°C until use (see Note 1).
2. ACN HPLC grade.
3. Trifluoroacetic acid (TFA) HPLC grade.
4. Sinapinic acid.
5. Water at resistivity of 18 MΩ cm.
6. C4 ZipTip.
7. Sample reconstitution buffer: ACN:water:TFA (5:95:0.1, v/v). Add 50 μL ACN to 940 μL water plus 10 μL 10% TFA (see Note 2).
8. ZipTip activation buffer: ACN:water:TFA = 50:50:0.1, v/v. Mix 500 μL ACN, 490 μL water, and 10 μL of 10% TFA. Make fresh daily, and cap the tube when not in use to prevent evaporation. Store at room temperature.
9. ZipTip washing buffer: 0.1% TFA in water. Add 10 μL 10% TFA to 990 μL water. Make fresh daily and store at room temperature.
10. ZipTip elution buffer: ACN:water:TFA (75:25:0.1, v/v). Mix 750 μL ACN with 240 μL water and add 10 μL of 10% TFA. For each sample, pipette 1.6 μL of elution buffer into a 0.6 mL Eppendorf tube at room temperature just before elution of proteins from the ZipTip. Do not pipette ahead of time or store the elution buffer at low temperature (see Note 3). Make the reagent fresh daily and store at room temperature. Cap the tube when not in use to prevent evaporation.
11. Sinapinic acid solution (see Note 4): Place a small amount of sinapinic acid powder in 100 μL of ACN:water:TFA (50:50:0.1, v/v) and vortex vigorously for 1 min. Add more sinapinic acid as needed to provide undissolved solid at the end of mixing. Centrifuge the solution at 1,000 × *g* for 10 min. Remove 85 μL of supernatant and mix with 15 μL ACN:water:TFA (50:50:0.1) to produce a solution that is 85% saturated with respect to sinapinic acid. Make the matrix fresh daily and store at room temperature.
12. Stainless steel MALDI-TOF MS target appropriate for the instrument.
13. An appropriate mass spectrometer such as the Bruker Biflex III MALDI-TOF MS (see Note 5).

3. Methods

Plasma or serum samples can be analyzed after direct application of unextracted plasma or serum onto the target. However, the resulting profiles are of low intensity and do not provide the consistency of extracted samples. For high intensity and reproducibility, the samples were extracted by reverse phase to remove salts and other components that may interfere with analysis. The C4 ZipTip was optimum for processing intact proteins. In order to avoid saturation of the ZipTip, a small amount of plasma or serum (0.5 μL) was added to 15 μL of reconstitution buffer for extraction. The samples were then incubated at room temperature for 1 h. This incubation resulted in greater consistency of the profile and may serve to dissociate some of the components from the lipid. ZipTip extraction requires about 3 min per sample. Therefore an efficient method consisted of dilution of one sample at each 3-min interval for 1 h (see Note 6). This was followed at 1 h by extraction of each sample at corresponding 3-min intervals so that every sample was incubated for exactly 1 h. A total of about 2 h was required to prepare 20 samples. Naturally, it is important to avoid interruptions during this part of the analysis.

After extraction and elution of samples from the C4 ZipTip, each sample was applied to a MALDI target and analyzed by MALDI-TOF MS in linear mode (review Note 5) using a mass range of 1,000–20,000. For reasons outlined in Note 5, the reflectron mode was also avoided due to problems with post-source decay. The MALDI-TOF spectrum could be analyzed by peak area or peak height. We found that relative peak height ratios within the same spectrum provided the most reproducible values, especially for homologous components such as glycoforms of the same protein. The coefficient of variation for multiple analyses of the same sample varied from 4 to 10%. This high reproducibility was illustrated by the consistency of individual subjects over time. For example, the consistency of ApoC3 glycoisoforms (ApoC3-1/ApoC3-2) was shown by analysis of individuals at ages 13 and 19 (Fig. 3). Each line represented a different individual. The difference between subjects and the horizontal curve shape illustrated the individual consistency of isoform ratio. Heavy adolescents (body mass index (BMI) > 25) showed significantly higher ApoC3-1/ApoC3-2 ratios than thin subjects of the same age (Fig. 3). In fact, heavy and thin adolescents gave ratios similar to their heavy and thin adult counterparts. Prior studies found that the ApoC3-1/ApoC3-2 ratio returned to values typical of thin individuals following either bariatric surgery or metformin therapy (5). Thus, glycoisoform ratios showed that the effect of elevated BMI was evident at early age and that a person's individual isoform ratio may extend throughout that individual's lifetime.

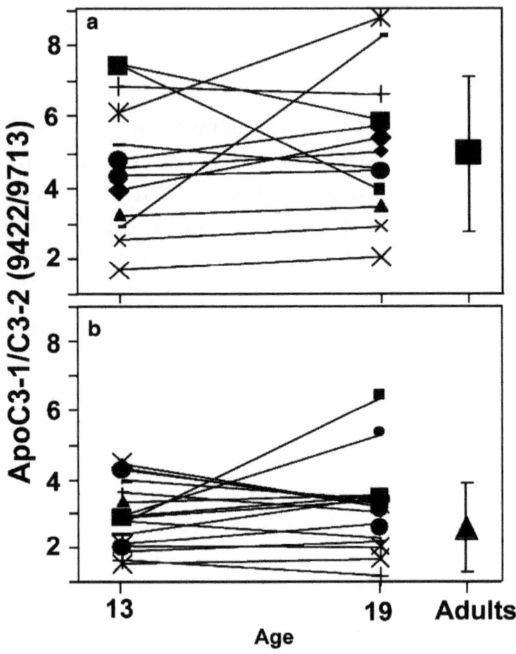

Fig. 3. Effect of BMI and the long-term stability of protein profiles of thin and heavy adolescents. (**a**) ApoC3-1/ApoC3-2 ratios from MALDI-TOF profiles of plasma from heavy individuals at ages 13 and 19 (n = 13, average BMI at age 19 = 37.6 ± 2.6). Each line represents a different individual. The average and standard deviation for this ratio among adults prior to bariatric surgery is shown for reference (*solid square* at "Adults," BMI = 46.6 ± 8.0 kg/m^2, age = 49.7 ± 9.8). (**b**) Stability of ApoC3-1/ApoC3-2 ratios at ages 13 and 19 among thin individuals (BMI < 25 at age 19, average BMI at age 19 = 21.9 ± 2.1, n = 18). The average and standard deviation for a group of thin adults (average BMI = 22.8 ± 1.0, *solid triangle*, average age 43.9 ± 9.7 years) is shown for comparison.

3.1. Sample Preparation and C4 ZipTip Extraction

1. A maximum of 20 serum or plasma samples are processed in each batch. Prepare 1 mL of reconstitution solution, 2 mL washing solution, 1 mL elution solution, and 100 μL 85% saturated sinapinic acid solution, and provide an empty Eppendorf tube for waste. All solutions are made fresh daily (review Notes 2 and 3).
2. Label two 0.6 mL Eppendorf tubes for each of the samples to be analyzed. The first tube should contain 15 μL of reconstitution buffer that is capped to avoid evaporation. The second tube should be empty but available for addition of 1.6 μL of elution buffer just prior to elution of the ZipTip.
3. Add serum or plasma (0.5 μL) to 15 μL of reconstitution buffer (ACN:water:TFA = 5:95:0.1) at 3-min intervals and mix until all samples have been prepared. After mixing each sample, check the pH of the solution (1.0 μL) with pH paper to make sure that it is acidic. It is important to realize that pH paper will not function properly at high ACN concentrations (e.g., 50% or 95%).

4. Allow the samples to each stand at room temperature for 1 h (see Note 7).
5. Apply a C4 ZipTip to an air displacement pipettor such as the Eppendorf that is set at a volume of 10 μL (see Note 8). Activate the C4 ZipTip by pipetting two 10 μL volumes of activation buffer (ACN:water:TFA = 50:50:0.1). Wash the tip with two 10 μL volumes of washing buffer (0.1% TFA). Discard the solutions into the waste tube. Discard is best accomplished by ejecting the liquid from the tip followed by touching the extruded droplet to the surface of a waste tube, taking care not to touch the tip itself to the waste tube.
6. Aspirate 10 μL of the sample in reconstitution buffer (20 μL total) and expel the sample back into the tube slowly, avoiding introduction of air bubbles into the reverse-phase gel. Continue for 1 min or about 15 repeated aspiration and expulsion steps (review Note 8).
7. Expel the reconstitution buffer to the surface of the reverse-phase gel and wash the ZipTip eight times with 10uL of 0.1% TFA. Discard each wash into the waste tube. Leave the last wash solution in the ZipTip until the elution solution is prepared (step 8). Then expel that solution as usual.
8. With a separate pipette, place 1.6 μL of elution solution into a clean Eppendorf tube making certain that all volume is deposited into a single droplet at the bottom of the tube. Expel and discard the final wash solution from the ZipTip (step 7). Draw and expel the elution solution into the ZipTip at least eight times, assuring that no air gets into the reverse-phase gel. Note that, although the Pipettor is set at 10 μL, you draw only 1.6 μL of solution.
9. Immediately following elution of the sample from the ZipTip, mix 0.75 μL of the elution solution with 0.75 μL of the 85% saturated sinapinic acid solution on the MALDI-TOF target. Use clean tips for each solution. Mix by drawing and expulsion of 0.75 μL several times, taking care to avoid formation of bubbles. Agitate the mixture on the target by rubbing the target surface with the pipette tip until the droplet is cloudy due to extensive crystallization of the sinapinic acid matrix (see Note 9).

3.2. MALDI-TOF MS Analysis

1. Analyze the samples on the target on Bruker Biflex III MALDI-TOF mass spectrometer in linear mode (review Note 5).
2. The mass spectrometer can be calibrated externally with the +1 and +2 charge state of cytochrome C. Alternatively, serum and plasma samples are so consistent that the instrument can be calibrated internally using the ApoC1 isoform peak at $m/z = 6{,}434$ and either the ApoC3-1 peak at $m/z = 9{,}423$ or the TTR peak at $m/z = 13{,}760$.

3. During sample analysis, the laser power is set at the highest attenuation that still provides good signal (see Note 10). Typically, 500 laser shots at ten or more different sites on the target are collected. A satisfactory profile should have a base peak (either ApoC1 or ApoC3-1) of at least 2,000 counts.

3.3. MALDI-TOF MS Protein Profile Analysis

1. Software provided by the manufacturer is used to process the raw spectra as follows:
   (a) The profile is smoothed at 15 points by the Golay–Savitzky formula.
   (b) Baseline is subtracted.
   (c) Peaks are labeled.
   (d) The peak intensity list is exported as a text file (.txt).
2. Open the peak list in Microsoft Excel and calculate peak intensity ratios for each sample. These include homologous peak ratios such as the glycoisoforms of ApoC3, ApoC3-1/ApoC3-2.
3. Compare peak intensity ratios of subjects with and without disease or before and after treatment (e.g., bariatric surgery or metformin therapy (5)) or longitudinally after time (e.g., Fig. 3).
4. Compare peak intensity ratios among different subject groups using appropriate statistical analysis to detect population differences and trends (5).

4. Notes

1. All samples described herein were obtained with informed consent and under approval of the Institutional Review Board of the University of Minnesota. Repeated freeze and thaw processes (up to ten times) had little effect on the MALDI-TOF profile pattern. Nonetheless, it is recommended to minimize freeze/thaw when possible. Most important is that storage temperature should be −80°C. Storage at −20°C resulted in unusual levels of protein oxidation that are detected by peaks at +16 and/or +32 relative to those of the profile in Fig. 1. Oxidation was minimal for samples that were repeatedly freeze–thawed as long as they are replaced at −80°C. This issue came to light in a most dramatic fashion during a study of samples from control subjects vs. those with liver disease (http://www.ncbi.nlm.nih.gov/pubmed/17709793). Although all mass spectrometry and other laboratory findings of the study were correct as presented, the publication was withdrawn before print when continued studies discovered that the disease

samples had been stored at –20°C while the controls were at –80°C. Continued research has revealed that oxidation can occur within days under improper storage conditions. Interpretation of oxidized peaks should be made with caution. Lipid-associated proteins appear to be uniquely susceptible to oxidation when stored at –20°C.

2. Make a stock solution of 10% TFA and store in a glass container at room temperature. Pure TFA (100%) must be handled with glass syringes, while solutions of <10% TFA were processed by regular pipette tips.
3. The elution buffer should be added to the Eppendorf at room temperature immediately prior to elution of the ZipTip. On several occasions, new operators attempted to streamline the process by preparing several tubes containing 1.6 μL of elution buffer with storage on ice. This was always detrimental and destroyed consistency of the method.
4. Use of sinapinic acid matrix is critical. Of matrices that were tested, only sinapinic acid gave the consistency and profiles shown in Fig. 1. Other matrices showed additional peaks such as multiple-charged states of serum albumin. The basis for the sinapinic acid-specific profile is not known but provided the extremely reproducible profiles of these specific components.
5. It is critical that the mass spectrometer have a strictly linear flight path when operated in the "linear" mode. An example of an appropriate instrument is the Bruker Biflex III mass spectrometer. MALDI-TOF instruments from some manufacturers involve deflection of the ion beam, even when operated in the linear mode. This results in loss of components that undergo post-source decay. Sialic acid is quite labile and highly subject to loss from post-source decay. As a result, glycoproteins or glycopeptides that contain sialic acid are poorly detected in these instruments. They are either not observed at all or are detected in a very inconsistent manner. In short, the linear mode for the mass spectrometer must involve an uninterrupted linear flight path from source to detector.
6. An experienced operator can process one sample every 3 min to generate 20 samples per batch. However, an inexperienced operator might use 5-min intervals and 12 samples per batch.
7. The 1-h incubation after mixing with reconstitution buffer has a modest impact on the profile that resulted in greater consistency.
8. Throughout all pipette manipulations, it is critical to avoid introduction of air into the reverse-phase gel of the ZipTip. Air bubbles cause channeling and uneven solvent flow. All wash and elution steps should involve extrusion of liquid only to the surface of the reverse-phase gel in the ZipTip. It is also necessary

to avoid aspiration of air while drawing solution into the pipette.

9. Sinapinic acid is not a popular matrix for several reasons. Under normal conditions, it forms large crystals that result in extreme differences across the sample with hot and cold spots on the target. Agitation of the sample in the manner described results in formation of small crystals that are uniformly distributed on the target and produce a homogeneous spot with equal intensity and higher intensity across the entire spot.

10. The optimum laser power has to be determined for each instrument. Excess laser power produces in-source decay that can be detected by loss of sialic acid. The resulting ApoC3-0' is observed at $m/z=9{,}131$ (Fig. 1). This peak is entirely the result of in-source decay and should be an inconsequential component of the profile, as illustrated by the data in Fig. 1.

References

1. Kirmiz C, Li B, An HJ, Clowers BH, Chew HK, Lam KS, Ferrige A, Alecio R, Borowsky AD, Sulaimon S, Lebrilla CB, Miyamoto S (2007) A serum glycomics approach to breast cancer biomarkers. Mol Cell Proteomics 6:43–55
2. Kyselova Z, Mechref Y, Al Bataineh MM, Dobrolecki LE, Hickey RJ, Vinson J, Sweeney CJ, Novotny MV (2007) Alterations in the serum glycome due to metastatic prostate cancer. J Proteome Res 6:1822–1832
3. Qiu Y, Patwa TH, Xu L, Shedden K, Misek DE, Tuck M, Jin G, Ruffin MT, Turgeon DK, Synal S, Bresalier R, Marcon N, Brenner DE, Lubman DM (2008) Plasma glycoprotein profiling for colorectal cancer biomarker identification by lectin glycoarray and lectin blot. J Proteome Res 7:1693–1703
4. Blomme B, Van Steenkiste C, Callewaert N, Van Vlierberghe H (2009) Alteration of protein glycosylation in liver diseases. J Hepatol 50:592–603
5. Harvey SB, Zhang Y, Wilson-Grady J, Monkkonen T, Nelsestuen GL, Kasthuri RS, Verneris MR, Lund TC, Ely EW, Bernard GR, Zeisler H, Homoncik M, Jilma B, Swan T, Kellogg TA (2009) O-glycoside biomarker of apolipoprotein C3: responsiveness to obesity, bariatric surgery, and therapy with metformin, to chronic or severe liver disease and to mortality in severe sepsis and graft vs host disease. J Proteome Res 8:603–612
6. Ceciliani F, Pocacqua V (2007) The acute phase protein alpha1-acid glycoprotein: a model for altered glycosylation during diseases. Curr Protein Pept Sci 8:91–108
7. Durand G, Seta N (2000) Protein glycosylation and diseases: blood and urinary oligosaccharides as markers for diagnosis and therapeutic monitoring. Clin Chem 46:795–805
8. Turnbull JE, Field RA (2007) Emerging glycomics technologies. Nat Chem Biol 3: 74–77
9. Siuti N, Kelleher NL (2007) Decoding protein modifications using top-down mass spectrometry. Nat Methods 4:817–821
10. Nelsestuen GL, Zhang Y, Martinez MB, Key NS, Jilma B, Verneris M, Sinaiko A, Kasthuri RS (2005) Plasma protein profiling: unique and stable features of individuals. Proteomics 5:4012–4024
11. Zhang Y, Oetting WS, Harvey SB, Stone MD, Monkkonen T, Matas AJ, Cosio FG, Nelsestuen GL (2009) Urinary Peptide patterns in native kidneys and kidney allografts. Transplantation 87:1807–1813
12. Akkina SK, Zhang Y, Nelsestuen GL, Oetting WS, Ibrahlm HN (2009) Temporal stability of the urinary proteome after kidney transplant: more sensitive than protein composition? J Proteome Res 8:94–103
13. Nelsestuen GL, Martinez MB, Hertz MI, Savik K, Wendt CH (2005) Proteomic identification of human neutrophil alpha-defensins in chronic lung allograft rejection. Proteomics 5:1705–1713
14. Zhang Y, Wroblewski M, Hertz MI, Wendt CH, Cervenka TM, Nelsestuen GL (2006) Analysis of chronic lung transplant rejection by MALDI-TOF profiles of bronchoalveolar lavage fluid. Proteomics 6:1001–1010

Chapter 11

Phosphoproteomic Analysis of Liver Homogenates

Gokhan Demirkan, Arthur R. Salomon, and Philip A. Gruppuso

Abstract

Regulation of protein function via reversible phosphorylation is an essential component of cell signaling. Our ability to understand complex phosphorylation networks in the physiological context of a whole organism or tissue remains limited. This is largely due to the technical challenge of isolating serine/threonine phosphorylated peptides from a tissue sample. In the present study, we developed a phosphoproteomic strategy to purify and identify phosphopeptides from a tissue sample by employing protein gel filtration, protein strong anion exchange and strong cation exchange (SCX) chromatography, peptide SCX chromatography, and TiO_2 affinity purification. By applying this strategy to the mass spectrometry-based analysis of rat liver homogenates, we were able to identify with high confidence and quantify over 4,000 unique phosphopeptides. Finally, the reproducibility of our methodology was demonstrated by its application to analysis of the mammalian Target of Rapamycin (mTOR) signaling pathways in liver samples obtained from rats in which hepatic mTOR was activated by refeeding following a period of fasting.

Key words: Signal transduction, Liver, Mass spectrometry, Phosphoproteins, Ion exchange chromatography, Metal oxide chromatography

1. Introduction

A primary goal of systems biology is to understand the control of complex signaling pathways in the physiological context of an organism. Our studies have utilized the laboratory rat and, more specifically, analysis of signaling in liver. The liver is the largest vital organ in vertebrates and it has been widely used as a model for the study of regeneration, development, drug metabolism, and signal transduction pathways. Only five previously published studies have aimed to characterize the phosphoproteome in liver tissue and none of these examined drug-sensitive phosphorylation events

Djuro Josic and Douglas C. Hixson (eds.), *Liver Proteomics: Methods and Protocols*, Methods in Molecular Biology, vol. 909, DOI 10.1007/978-1-61779-959-4_11, © Springer Science+Business Media New York 2012

(1–5). Of these studies, only one explored the reproducibility of their findings (2).

Recent advances in the technology of mass spectrometry (MS) have made possible the quantitative analysis of dynamic phosphorylation events. However, these newly developed techniques have some limitations, particularly in the analysis of global serine/threonine phosphorylation. In the last few years, several MS studies have been conducted on tyrosine-phosphorylation changes following various stimuli to cell signaling (6–8). In these cases, the enrichment of tyrosine phosphorylated peptides has been carried out by peptide immunoprecipitation with antibodies specific to tyrosine phosphorylation sites. The lack of equivalent antibodies against phosphoserine and phosphothreonine precludes this approach for studying these far more prevalent phosphorylation events.

To meet the goal of profiling the liver phosphoproteome, we had to refine and adapt the available methods and ascertain the reproducibility of our analyses. In effect, we combined several standard methods for protein and peptide fractionation. This allowed us to reduce sample complexity, thereby improving the sensitivity of LC/MS analysis (4, 9). In addition to traditional methods, a newer strategy, the enrichment of phosphopeptides by using metal oxide chromatography (MOC), was employed. In the MOC approach, phosphorylated peptides are enriched based on their affinity for metal oxides, including zirconium dioxide (10), titanium dioxide (11), and aluminum hydroxide (12). For our experiments, we used the more recently developed methodology of titanium dioxide (TiO_2) chromatography in the presence of lactic acid.

Most phosphoproteomic studies focusing on signal transduction have been conducted using cell lines. In addition, identification of the greatest number of phosphorylation changes has been given primary importance over reproducibility of the analyses. "Label-free" quantitation has been substituted by some investigators as an alternative to quantitative methods that employ isotope labeling (13). However, data on the reproducibility of label-free quantitation for tissue analysis are lacking. We therefore used the approach of analyzing multiple technical and biological replicate samples using a phosphoproteomic platform that employs automated desalting and reversed-phase separation of peptides in a highly consistent and reproducible manner. All loading steps and column elutions are accurately replicated with computer controls (14).

The use of this experimental approach has enabled us to identify more than 4,000 unique phosphopeptides from rat liver with high reproducibility on a linear trap quadrupole (LTQ)-Fourier-transform ion cyclotron resonance (FTICR) mass spectrometer. To assess the applicability of our proteome-wide analysis of cell-signaling pathways and drug treatment effects in vivo, we extended our studies to the analysis of a particular signaling pathway that involves the mammalian Target of Rapamycin (mTOR).

2. Materials

2.1. Animal Studies, Preparation of Liver Homogenates, and Gel Filtration Chromatography

1. Adult male Sprague–Dawley rats (Charles River Laboratories, Wilmington, MA) that were of age 35–40 days.
2. Scalpels, scissors, forceps.
3. Liquid nitrogen.
4. Buffer A: 16 mM β-glycerophosphate, pH 7.4, 150 mM NaCl, 100 μM sodium orthovanadate, 10 mM NaF, 1 mM EGTA, and 5 mM EDTA. Store at 4°C.
5. Glass-Teflon homogenizer (Wheaton Science Products, Millville, NJ).
6. Bradford Protein Assay (Bio-Rad, Hercules, CA).
7. Buffer B: 10 mM Tris, pH 8.6, 100 μM sodium orthovanadate, 10 mM NaF, 0.5 mM EGTA, 2 mM EDTA.
8. Sephadex G-25 Coarse (GE Healthcare Bio-Sciences Corp., Piscataway, NJ).
9. 2.5 × 20 cm Econo-column chromatography column (Bio-Rad).

2.2. Ion Exchange Separation of Proteins

1. Q Sepharose Fast Flow beads (GE Healthcare Bio-Sciences Corp.).
2. SP Sepharose Fast Flow beads (GE Healthcare Bio-Sciences Corp.).
3. 1.5 × 5 cm Econo-column chromatography columns (Bio-Rad).
4. Buffer C: 10 mM Tris, pH 7.3, 100 μM sodium orthovanadate, 10 mM NaF, 0.5 mM EGTA, 2 mM EDTA.

2.3. Protein Reduction, Alkylation, and Digestion

1. Urea, electrophoresis grade (Fisher Scientific, Pittsburgh, PA).
2. 1 M Dithiothreitol.
3. 550 mM Iodoacetemide.
4. 10 mM Tris, pH 8.0.
5. Sequencing grade modified trypsin (Promega, Madison, WI).
6. Trifluoroacetic acid (TFA).
7. Oasis HLB extraction cartridges (Waters Corp., Milford, MA).
8. SpeedVac Concentrator.

2.4. Strong Cation Exchange Fractionations of Peptides

1. Ammonium formate, 99% (Acros Organics, Belgium).
2. 0.1% formic acid (Acros Organics), 30% acetonitrile HPLC grade (Fisher Scientific).
3. Buffer D: 5 mM ammonium formate, pH 2.65, 30% acetonitrile (pH adjusted with formic acid).
4. Buffer E: 1.2 M ammonium formate, pH 2.65, 30% acetonitrile (pH adjusted with formic acid).
5. 1 ml Resource S column (GE Healthcare Bio-Sciences Corp.).

2.5. Titanium Dioxide Chromatography

1. Titansphere® Phos-TiO Kit (GL Sciences, Torrance, CA).
2. 1% ammonium hydroxide (Sigma) in water.
3. 1% ammonium hydroxide in 40% acetonitrile.
4. 5% pyrrolidine in water.

2.6. Automated MS Analysis

1. 0.1% Acetic acid in water.
2. 5 μm Monitor C18 particles.
3. Solvent A (aqueous buffer): 0.1 M acetic acid in distilled water.
4. Solvent B (organic buffer): 0.1 M acetic acid in acetonitrile.
5. 0–70% 0.1 M acetic acid in acetonitrile.
6. LTQ-Fourier Transform [FT] (Thermo Fisher Scientific, Waltham, MA).
7. C18 analytical column (360 μm outer diameter × 75 μm inner diameter); fused silica with 12 cm of 5 μm Monitor C18 particles and an integrated 4 μm ESI emitter tip fritted as described previously (15).
8. Xcalibur for mass spectrometer instrument data analysis (Thermo Fisher Scientific).
9. Bioworks software using the SEQUEST algorithm (Thermo Fisher Scientific).
10. High-Throughput Autonomous Proteomic Pipeline (HTAPP) (14).
11. PeptideDepot software, available from http://peptidedepot.com (16).

3. Methods

The approach used for phosphoproteomic profiling of liver homogenates involved the development of methods for homogenate preparation, protein fractionation, and peptide fractionation

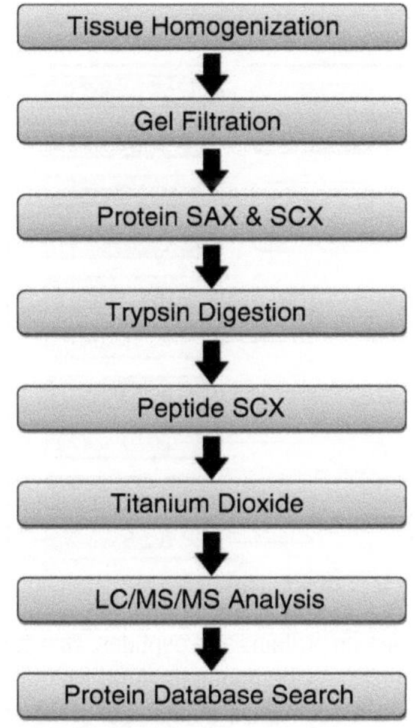

Fig. 1. Experimental approach to the phosphoproteomic profiling of liver homogenates.

(Fig. 1). The first step in our methodology focuses on the preparation of liver homogenates containing soluble liver proteins (see Note 1) and removing phospholipids from these preparations. This is achieved by passing liver homogenates through a Sephadex G-25 gel filtration column. In order to fractionate the final MS sample to achieve greater sensitivity in phosphopeptide detection (see Note 2), we followed gel filtration chromatography with a series of protein ion exchange chromatography steps (Fig. 2). First, strong anion exchange (SAX) gradient chromatography is used to fractionate the crude homogenate into three fractions. Strong cation exchange (SCX) chromatography is used to fractionate the SAX flow through into two additional fractions. After trypsin digestion of the five liver homogenate ion exchange fractions (see Note 3), the resultant peptides are desalted by solid-phase extraction and separated by peptide SCX chromatography (Fig. 3). Finally, phosphopeptides recovered from SCX fractions by TiO_2 chromatography in the presence of lactic acid are identified by LC/MS/MS.

3.1. Animal Studies, Preparation of Liver Homogenates, and Gel Filtration Chromatography

1. Liver tissue should be obtained under appropriate conditions that minimize stress to the rat. Harvested liver tissue should be immediately flash frozen in liquid nitrogen and stored at –70°C until further processing (see Note 4).
2. Bring tissue to balance on dry ice and weigh frozen tissue, being careful not to let it thaw (see Note 5).
3. Add 10 ml of ice-cold Buffer A (see Note 6) for every gram of liver and mince tissue in beaker with scissors.

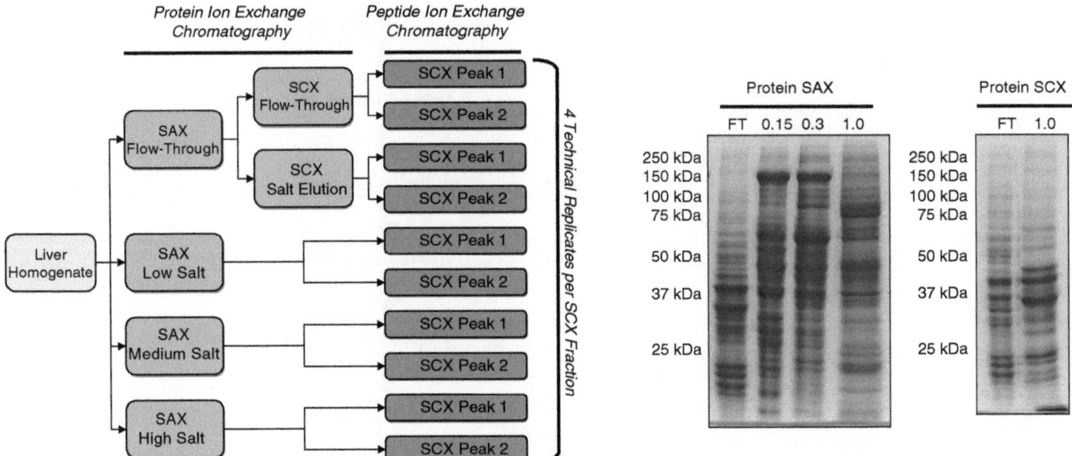

Fig. 2. Patterns of protein expression in five protein ion exchange fractions. The panel on the *left* shows the strategy for fractionation of liver homogenate proteins and peptides. This strategy, including the analysis of four technical replicates per fraction, resulted in 40 samples per homogenate. A rat liver homogenate was subjected to SAX and SCX chromatography. The panel on the *right* shows a Coomassie-stained gel of the resulting fractions. Those derived from SAX chromatography (flow through [FT] and 0.15, 0.3, and 1 M NaCl eluates) and SCX chromatography (flow through and 1 M NaCl eluate) are shown following the lanes containing molecular weight markers (M) and 50 μg of the unfractionated liver homogenate (H). The numbers to the *left* of the stained gel represent molecular weight in kilodaltons.

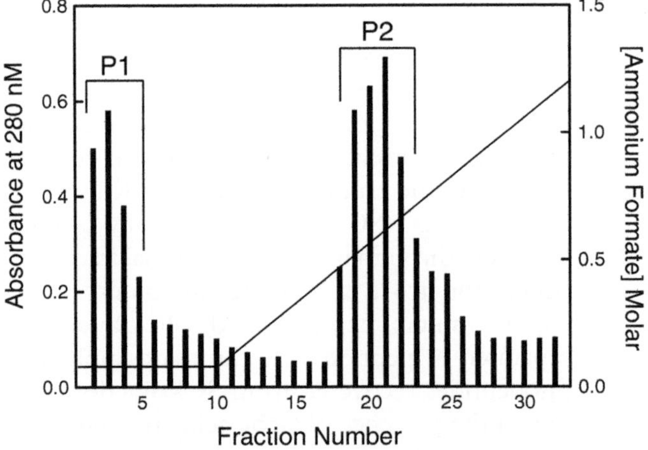

Fig. 3. Peptide SCX chromatography separation at pH 2.65 of a liver homogenate fraction. The graph shows an analysis of a single 300 mM NaCl SAX fraction subjected to peptide SCX chromatography. For this representative sample fractionation and all other analyses, digested peptides (10 mg) were applied to a Resource S SCX column and eluted with isocratic ammonium formate (5 mM) followed by a 5 mM to 1.5 M ammonium formate gradient. Peptide content of each fraction was estimated by measuring the absorbance at 280 nm. Two fraction pools containing the highest peptide amounts (designated as Peaks 1 and 2) were recovered from each column run.

4. Homogenize tissue at 700 rpm for 20 s on ice using a glass homogenizer fitted with a Teflon plunger. Repeat this process seven times.
5. Centrifuge the resulting homogenate at 1,086 × *g* for 15 min at 4°C.
6. Transfer the supernatants to new tubes and centrifuge at 100,000 × *g* for 1 h at 4°C.
7. Discard the resulting pellets and retain the supernatants. Perform a protein assay to determine protein concentration and use the remaining homogenate for gel filtration, or store at −70°C.
8. Dilute liver homogenate 1:3 in ice-cold Buffer B.
9. At 4°C, load 25 ml of diluted homogenate to the 80 ml Sephadex G-25 Coarse column equilibrated in Buffer B (see Note 7).
10. Elute the column with 80 ml Buffer B by gravity flow. Discard the first 30 ml that comes off the column and collect the next 30 ml flow through in a clean tube. Freeze the flow through at −70°C.

3.2. Ion Exchange Separation of Proteins

1. Set up 3 ml Q-Sepharose and SP-Sepharose columns at 4°C.
2. Apply 50 mg (approximately 30 ml) of homogenate protein to the Q-Sepharose Fast Flow column at 0.8 ml/min.
3. Collect flow through and wash the column with 5 ml of Buffer C at 0.8 ml/min. Collect Buffer C wash, and combine with the flow through. Designate this as the Q-Sepharose flow through fraction.
4. Elute the column at 0.8 ml/min with 5 ml of Buffer C containing 150 mM NaCl. Collect this eluate as a single fraction in a clean tube.
5. Elute the column at 0.8 ml/min with 5 ml of Buffer C containing 300 mM NaCl. Collect this eluate as a single fraction in a clean tube.
6. Elute the column at 0.8 ml/min with 7.5 ml of Buffer C containing 1 M NaCl. Collect this eluate as a single fraction in a clean tube.
7. Equilibrate the SP-Sepharose column with Buffer C.
8. Apply the Q-Sepharose flow through/wash (pH adjusted to 7.3, if necessary, using HCl) to the SP-Sepharose column.
9. Collect flow through and wash column with 5 ml of Buffer C at 0.8 ml/min. Collect the Buffer C wash, and combine with the flow through as the SP-Sepharose flow-through fraction.
10. Elute proteins from the SP-Sepharose column at 0.8 ml/min with 10 ml of Buffer C containing 1 M NaCl. Collect this eluate as a single fraction in a clean tube.

11. Measure protein concentrations in five liver homogenate fractions using Bradford reagent (see Note 8).

3.3. Protein Reduction, Alkylation, and Digestion

1. Denature proteins in the ion exchange fractions by adding urea to each to a final concentration of 8 M.
2. Reduce the sample for 1 h at 56°C by addition of 1 M DTT to a final concentration of 1 mM DTT (see Note 9).
3. Alkylate the samples by adding iodoacetemide to a final concentration of 5 mM and then incubating for 1 h at room temperature in the dark (see Note 9).
4. Dilute the resulting reduced and alkylated samples fourfold with 10 mM Tris, pH 8.0, and incubate overnight at 37°C with sequencing grade modified trypsin at a 1:100 protease:protein ratio.
5. Acidify the resulting tryptic peptides with TFA to pH 2.7. Clarify the peptide mixture by centrifugation at 1,000 × *g* for 5 min.
6. Desalt the clarified peptide supernatants using Oasis HLB extraction cartridges according to the manufacturer's directions. Dry the desalted samples using a SpeedVac Concentrator or equivalent (see Note 10).

3.4. Strong Cation Exchange Fractionation of Peptides

1. Reconstitute the dried samples in 1 ml of Buffer D and centrifuge them at 12,000 × *g* for 8 min at 4°C.
2. Equilibrate a 1 ml Resource S SCX column with Buffer D.
3. Using an HPLC system, apply the peptide mixture to the Resource S column and separate the peptides using the following protocol: flow rate, 0.5 ml/min; Buffer D for 20 min; and linear ammonium formate gradient, using Buffers D and E, from 5 mM to 1.2 M over 44 min. Entire process yields 32 fractions, 1 ml each.
4. Estimate peptide content of each fraction by measuring the absorbance at 280 nm (see Note 11).
5. Pool fractions to obtain two pools of approximately four 1 ml fractions each (see Note 12). Split each combined fraction pool into four microfuge tubes to speed the drying process. Dry the fraction pools in SpeedVac Concentrator and store at –70°C.

3.5. Titanium Dioxide Chromatography

1. Dissolve the peptides in each SCX pool in 400 μl of 0.1% formic acid, and 30% acetonitrile. Separate each of these samples into four 100 μl samples for the performance of technical replicates.
2. Enrich phosphopeptides using the Titansphere Phos-TiO Kit following the manufacturer's protocol up to the elution step. In this protocol, lactic acid is included in the loading buffer to

reduce the interaction between acidic non-phosphopeptides and titanium dioxide (see Note 13).

3. Elute phosphopeptides from the Phos-TiO tips by first applying 25 μl of 1% ammonium hydroxide in water. Then treat the tips with 25 μl of 1% ammonium hydroxide in 40% acetonitrile. Finally, elute with 50 μl of 5% pyrrolidine in water.
4. Dry all samples in a SpeedVac Concentrator and store at −70°C.

3.6. Automated MS Analysis

1. Dissolve dry phosphopeptide-enriched mixtures in 0.1% acetic acid in water and elute from the nano HPLC reverse-phase column into the mass spectrometer with a gradient of 0–70% 0.1 M acetic acid in acetonitrile developed over 30 min at a flow rate of 1.8 μl/min. Perform static peak parking as previously described by flow rate reduction from 200 nl/min to 20 nl/min at the elution time previously determined from a BSA peptide scouting run (17). Apply electrospray voltage of 2.0 kV in a split flow configuration and collect spectra in positive ion mode, essentially as described (18). Perform one full MS scan in the Fourier Transform (from m/z 400 to 1,800) followed by nine data-dependent MS/MS spectra in the linear ion trap from the nine most abundant ions. Dynamically exclude selected ions for 30 s and screen for charge-states of +1, +2, and +3. Use previously described settings for FTMS and ion trap mass spectrometry scans (19).
2. MS/MS spectra were automatically searched against separate mouse, hamster, and rat National Center for Biotechnology Information (NCBI) nonredundant protein databases using the Sequest algorithm provided with Bioworks 3.2. Quantitative data were automatically calculated by PeptideDepot for every assigned peptide in all LC/MS runs based on exact precursor mass and retention time (see Notes 14–18).

4. Notes

1. The homogenization buffer described here results only in the extraction of soluble proteins. Since the presence of detergents in samples complicates MS analysis, extension of our methods to lipid-soluble proteins will require the application of additional methods for removal of detergent. One such method has been described by Lu et al. (20). As an alternative approach, livers can be homogenized in a buffer containing 8 M urea to extract both cytoplasmic and membrane-associated proteins (9). Furthermore, tryptic peptides derived from liver

homogenates can be fractionated using reverse-phase chromatography at intermediate pH in place of ion exchange chromatography (21).

2. Very-low-abundance proteins or proteins with a low stoichiometry of phosphorylation could presumably go undetected in our MS analysis due to sensitivity limits of the LTQ-FTICR mass spectrometer and the suppression of ionization efficiency from the addition of negatively charged phosphorylation groups detected in positive ion mode. With advances in technology, we anticipate that the sensitivity of peptide detection will increase substantially.
3. Identification of some phosphorylation sites requires treatment with proteases other than trypsin due to a suboptimal density of tryptic sites in a given protein. More specifically, proteins with too few or too many tryptic sites to yield phosphopeptides of optimal length will require the use of alternative proteases.
4. If two samples are to be compared, the experimental design should be appropriate for paired analyses. That is, control and experimental samples should be obtained, prepared, processed, and analyzed in parallel. This approach allows for the statistical analyses of the data to also be done in a paired manner.
5. It is essential that all homogenates and ion exchange fractions be maintained at 4°C to prevent proteolysis and protein dephosphorylation.
6. Prepare all buffers and solutions in deionized water and store at 4°C unless otherwise stated.
7. Protein recovery during the gel filtration step was determined by protein quantification and gel electrophoresis to be greater than 90%.
8. As expected, gel electrophoresis of the five protein ion exchange fractions showed different patterns of protein expression but similar protein content (Fig. 3).
9. Dithiothreitol and iodoacetemide solutions should be prepared fresh on the day of use.
10. Sarstedt Eppendorf tubes should be used to minimize sample loss during sample drying in SpeedVac.
11. Amino acids with aromatic rings absorb UV light at 280 nm. Although absorbance reading at 280 nm does not yield a precise measure of peptide content in solution, it is a convenient way to determine relative peptide abundance.
12. Peptide SCX peak 1 was shown by MS analysis to be highly enriched with phosphopeptides and successfully separated from the more complex, positively charged multivalent peptides that eluted later in the peptide SCX peak 2.

13. The lactic acid-treated Titansphere Phos-TiO kit was shown by MS analysis to provide an improved recovery of phosphopeptides when compared to Fe^{3+} immobilized metal affinity chromatography (IMAC) or loose TiO_2 beads in the presence of 2,5-dihydroxybenzoic acid.

14. The applicability of our method to analysis of drug treatment effects and cell signaling was determined by fasting followed by refeeding, a well-characterized rat model for mTOR activation (22). This model had been previously employed in our laboratory to study liver growth in the absence of hepatocyte proliferation (23). During a period of food deprivation lasting 48 h, liver mass and protein content decrease by approximately one-quarter and one-third, respectively. Within 1 h of refeeding, marked activation of mTOR Complex 1 (mTORC1) occurs, as assessed by phosphorylation of ribosomal protein S6 and S6 kinase. Prior to refeeding, injection of rapamycin, a specific mTORC1 inhibitor, prevents this activation while injection of vehicle (DMSO) has no effect. (23). Thus, this model is suitable for studies on mTORC1 signaling in liver tissue.

15. To assess the reproducibility of our phosphopeptide enrichment strategy in the context of in vivo studies, we compared LC/MS runs from three control biological replicates as well as four technical replicates obtained from each biological replicate. LC/MS retention time alignments of multiple datasets were carried out to correct for chromatographic shifts between experiments. Eighty-two percent of the phosphopeptides were identified in all three control biological replicates, while only 13 and 5% of the phosphopeptides were identified in two and one of the control biological replicates, respectively. This high degree of reproducibility, and the detection of over 4,000 phosphopeptides, was dependent on the performance of the technical replicates.

16. The final dataset derived from three biological control replicates contained 3,231 nonredundant phosphorylation sites representing 1,400 proteins and 4,238 phosphopeptides at a 1% false discovery rate. Eighty-five percent of all phosphorylation sites represented serine phosphorylation. Thirteen percent represented phosphorylated threonine residues, and 2% were phosphorylated tyrosine residues.

17. Identification of 5 known rapamycin-sensitive phosphorylation sites supported the reliability of our analysis. In addition, we identified 62 rapamycin-sensitive candidate phosphorylation events. A high number of these candidate phosphorylation sites were found on proteins known to be involved in transcription, translation, or cell growth, or known to interact with mTOR itself.

18. We observed similar proportions of enhanced and reduced phosphorylation in response to rapamycin. Gene ontology analysis showed overrepresentation of mTOR pathway-related proteins among rapamycin-sensitive phosphopeptide candidates.

References

1. Jin WH, Dai J, Zhou H, Xia QC, Zou HF, Zeng R (2004) Phosphoproteome analysis of mouse liver using immobilized metal affinity purification and linear ion trap mass spectrometry. Rapid Commun Mass Spectrom 18: 2169–2176
2. Moser K, White FM (2006) Phosphoproteomic analysis of rat liver by high capacity IMAC and LC-MS/MS. J Proteome Res 5:98–104
3. Villen J, Beausoleil SA, Gerber SA, Gygi SP (2007) Large-scale phosphorylation analysis of mouse liver. Proc Natl Acad Sci USA 104: 1488–1493
4. Han G, Ye M, Zhou H, Jiang X, Feng S, Jiang X, Tian R, Wan D, Zou H, Gu J (2008) Large-scale phosphoproteome analysis of human liver tissue by enrichment and fractionation of phosphopeptides with strong anion exchange chromatography. Proteomics 8:1346–1361
5. Han G, Ye M, Liu H, Song C, Sun D, Wu Y, Jiang X, Chen R, Wang C, Wang L, Zou H (2010) Phosphoproteome analysis of human liver tissue by long-gradient nanoflow LC coupled with multiple stage MS analysis. Electrophoresis 31:1080–1089
6. Cao L, Yu K, Salomon AR (2006) Phosphoproteomic analysis of lymphocyte signaling. Adv Exp Med Biol 584:277–288
7. Salomon AR, Ficarro SB, Brill LM, Brinker A, Phung QT, Ericson C, Sauer K, Brock A, Horn DM, Schultz PG, Peters EC (2003) Profiling of tyrosine phosphorylation pathways in human cells using mass spectrometry. Proc Natl Acad Sci U S A 100:443–448
8. Zhang Y, Wolf-Yadlin A, Ross PL, Pappin DJ, Rush J, Lauffenburger DA, White FM (2005) Time-resolved mass spectrometry of tyrosine phosphorylation sites in the epidermal growth factor receptor signaling network reveals dynamic modules. Mol Cell Proteomics 4: 1240–1250
9. Villen J, Gygi SP (2008) The SCX/IMAC enrichment approach for global phosphorylation analysis by mass spectrometry. Nat Protoc 3:1630–1638
10. Kweon HK, Hakansson K (2006) Selective zirconium dioxide-based enrichment of phosphorylated peptides for mass spectrometric analysis. Anal Chem 78:1743–1749
11. Larsen MR, Thingholm TE, Jensen ON, Roepstorff P, Jorgensen TJ (2005) Highly selective enrichment of phosphorylated peptides from peptide mixtures using titanium dioxide microcolumns. Mol Cell Proteomics 4:873–886
12. Wolschin F, Wienkoop S, Weckwerth W (2005) Enrichment of phosphorylated proteins and peptides from complex mixtures using metal oxide/hydroxide affinity chromatography (MOAC). Proteomics 5:4389–4397
13. Huber A, Bodenmiller B, Uotila A, Stahl M, Wanka S, Gerrits B, Aebersold R, Loewith R (2009) Characterization of the rapamycin-sensitive phosphoproteome reveals that Sch9 is a central coordinator of protein synthesis. Genes Dev 23:1929–1943
14. Yu K, Salomon AR (2010) HTAPP: high-throughput autonomous proteomic pipeline. Proteomics 10(11):2113–2122
15. Ficarro SB, Zhang Y, Lu Y, Moghimi AR, Askenazi M, Hyatt E, Smith ED, Boyer L, Schlaeger TM, Luckey CJ, Marto JA (2009) Improved electrospray ionization efficiency compensates for diminished chromatographic resolution and enables proteomics analysis of tyrosine signaling in embryonic stem cells. Anal Chem 81:3440–3447
16. Yu K, Salomon AR (2009) PeptideDepot: flexible relational database for visual analysis of quantitative proteomic data and integration of existing protein information. Proteomics 9:5350–5358
17. Ficarro SB, Salomon AR, Brill LM, Mason DE, Stettler-Gill M, Brock A, Peters EC (2005) Automated immobilized metal affinity chromatography/nano-liquid chromatography/electrospray ionization mass spectrometry platform for profiling protein phosphorylation sites. Rapid Commun Mass Spectrom 19:57–71
18. Cao L, Yu K, Banh C, Nguyen V, Ritz A, Raphael BJ, Kawakami Y, Kawakami T, Salomon AR (2007) Quantitative time-resolved phosphoproteomic analysis of mast cell signaling. J Immunol 179:5864–5876

19. Nguyen V, Cao L, Lin JT, Hung N, Ritz A, Yu K, Jianu R, Ulin SP, Raphael BJ, Laidlaw DH, Brossay L, Salomon AR (2009) A new approach for quantitative phosphoproteomic dissection of signaling pathways applied to T cell receptor activation. Mol Cell Proteomics 8:2418–2431
20. Lu X, Zhu H (2005) Tube-gel digestion: a novel proteomic approach for high throughput analysis of membrane proteins. Mol Cell Proteomics 4:1948–1958
21. Song C, Ye M, Han G, Jiang X, Wang F, Yu Z, Chen R, Zou H (2010) Reversed-phase-reversed-phase liquid chromatography approach with high orthogonality for multidimensional separation of phosphopeptides. Anal Chem 82:53–56
22. Anand P, Gruppuso PA (2005) The regulation of hepatic protein synthesis during fasting in the rat. J Biol Chem 280:16427–16436
23. Anand P, Gruppuso PA (2006) Rapamycin inhibits liver growth during refeeding in rats via control of ribosomal protein translation but not cap-dependent translation initiation. J Nutr 136:27–33

Chapter 12

A Combination of Affinity Chromatography, 2D DIGE, and Mass Spectrometry to Analyze the Phosphoproteome of Liver Progenitor Cells

Enrique Santamaría, Virginia Sánchez-Quiles, Joaquín Fernández-Irigoyen, and Fernando J. Corrales

Abstract

Reversible protein phosphorylation is a ubiquitous posttranslational modification that regulates cellular signaling pathways in multiple biological processes. A comprehensive analysis of protein phosphorylation patterns can only be achieved by employing different complementary experimental strategies all aiming at selective enrichment of phosphorylated proteins/peptides. In this chapter, we describe a method that utilizes a phosphoprotein affinity chromatography (Qiagen) to isolate intact phosphoproteins. These are subsequently detected by difference in two-dimensional gel electrophoresis and identified by mass spectrometry techniques. Additional experiments using a specific stain for phosphoproteins demonstrated that phosphoprotein affinity column was an effective method for enriching phosphate-containing proteins. Further validating the method, this workflow was applied to probe changes in the activation patterns of intermediates involved in different signaling pathways, such as NDRG1 and stathmin, in liver progenitor cells (MLP-29) upon proteasome inhibition.

Key words: Liver progenitor cells, Mass spectrometry, Phosphorylation

1. Introduction

Proteasome inhibitors appear to induce apoptosis in proliferating cells and are protective in quiescent or terminally differentiated cells (1, 2). The ability of proteasome inhibitors to inhibit cell proliferation and induce apoptosis in hepatoma and hepatic stellate cells, together with their ability to inhibit angiogenesis (2, 3), makes these agents attractive candidates for the treatment of liver fibrosis and hepatocellular carcinomas (4–6). Transcriptional profiling studies have been conducted to elucidate the molecular mechanisms

Djuro Josic and Douglas C. Hixson (eds.), *Liver Proteomics: Methods and Protocols*, Methods in Molecular Biology, vol. 909, DOI 10.1007/978-1-61779-959-4_12, © Springer Science+Business Media New York 2012

triggered by proteasome inhibition in different cell lines (7–9). However, in addition to transcriptomics, the understanding of protein abundance, localization, and regulation is central to fully define the mechanisms underlying the cellular response to these drugs. The reversible phosphorylation of proteins plays a major role in many vital cellular processes by modulating protein function and transmitting signals within cellular pathways and networks. Since phosphorylation is dynamic, the sites of phosphorylation cannot be predicted by an organism's genome, and therefore, proteomic measurements are required to identify the targets and the specific sites where the phosphate is incorporated (10). However, analysis of phosphoproteins is not straightforward for four main reasons. First, the stoichiometry of phosphorylation is in general relatively low, meaning that only a small fraction of the available pool of a protein is phosphorylated at any given time. Second, phosphorylated residues of a protein may vary in response to adequate stimuli, implying that a particular protein may exist as different phosphorylated isoforms. Third, many of the signaling molecules are present at low abundance in cells and, in these cases, enrichment is a prerequisite before analysis. Finally, phosphatases could dephosphorylate residues unless precautions are taken to inhibit their activity during preparation and purification steps of cell lysates (11).

Many strategies based on phosphopeptide enrichment such as Immobilized Metal Affinity Chromatography (IMAC), titanium dioxide enrichment, and strong cation exchange (SCX) chromatography have been developed in combination with stable isotope labeling by amino acids in culture (SILAC) or chemical labeling by tandem mass tags (known as iTRAQ™) to perform global, quantitative, and site-specific phosphoproteomics in cell signaling studies (12). This chapter is focused on a phosphoprotein enrichment method based on affinity chromatography (Qiagen) that takes advantage of the negative nature of the phosphoryl group. The method enables an optimal separation of the phosphorylated from the unphosphorylated cellular protein fraction. The affinity chromatography procedure, in which phosphorylated proteins are bound to a column while unphosphorylated proteins are recovered in the flow-through fraction, reduces complexity and greatly facilitates phosphorylation-profiling studies. Both fractions retain full biological activity and can be further purified if desired. This phosphoprotein enrichment method has been successfully used for analysis of phosphoproteins in plants (13), yeast (14), rat liver (15), as well as in myocytes, embryonic stem cells, and keratinocytes (16–18). It covers different applications like global phosphoproteome analysis (19), targeted-phosphosite analysis (20), and can also be applied as a prefractionation step in combination with phosphopeptide enrichment techniques such as IMAC (21), and titanium dioxide chromatography (22). To validate this method, we have used Pro-Q Diamond phosphoprotein dye technology (Molecular Probes) suitable for the fluorescent detection of phosphoserine-, phosphothreonine-, and phosphotyrosine-containing proteins directly in sodium dodecyl sulfate (SDS)-polyacrylamide gels (23, 24). As protein phosphorylation is a highly dynamic process that often is only present for a limited period of time, it is convenient to perform not only quantitative but also time-dependent analysis. To study the temporal-phosphoproteome alterations induced by proteasome inhibitors in MLP-29 cells, we have used two-dimensional difference gel electrophoresis (2D-DIGE), one of the most common methods for quantitative proteomics (25). The study of phosphorylation dynamics revealed stathmin and NDRG1 protein as potential downstream effectors upon impaired proteasomal function, providing new insights into the molecular mechanisms involved during the apoptotic process induced by proteasome inhibitors (19).

2. Materials

2.1. Cell Culture and Lysis

1. MLP-29 is an epithelial homogeneous cellular clone obtained by limiting dilution from a mouse embryonic liver cell line (26).
2. Dulbecco's modified Eagle's medium (DMEM) (Invitrogen, Auckland, New Zealand) supplemented with 10% (v/v) fetal bovine serum (FBS) (Invitrogen) and 1% penicillin–streptomycin–glutamine (Invitrogen).
3. Carbobenzoxy-L-leucyl-L-leucyl-L-leucinal, Z-LLL-CHO (MG132; purity > 98% by HPLC) and epoxomicin (purity > 95% by HPLC) (Calbiochem, San Diego, CA, USA) are dissolved in dimethyl sulfoxide (DMSO), stored in aliquots at –20°C, and then added to cell culture dishes as required.
4. Wash buffer: Saline solution (0.9 g sodium chloride per 100 ml; pH = 4.5–7.0) (Grifols Laboratories, Barcelona, Spain).
5. Cell scrapers 25 cm (Sarstedt, Newton, NC, USA).

2.2. Phosphoprotein Affinity Chromatography

1. Phosphoprotein Purification Kit (ref: 37101, Qiagen GmbH, Hilden, Germany) contains phosphoprotein purification columns, buffers, protease inhibitor tablets, benzonase, and Nanosep® ultrafiltration columns (10 kDa molecular weight cutoff).
2. Bradford protein assay (Bio-Rad, Hercules, CA) for determination of protein concentration in all eluate fractions.

2.3. SDS-PAGE and Specific Phosphoprotein Gel Stain

1. Resolving buffer: 1.5 M Tris–HCl, pH 8.8, 0.4% (w/v) SDS. Store at room temperature.
2. Stacking buffer: 0.5 M Tris–HCl, pH 6.8, 0.4% SDS. Store at room temperature.

3. Forty percent (w/v) acrylamide/bis solution (37.5:1 with 2.6% C) (this is a neurotoxin when unpolymerized and so care should be taken not to receive exposure) (Bio-Rad) and *N*,*N*,*N*,*N'*-Tetramethyl-ethylenediamine (TEMED, Bio-Rad).
4. Ammonium persulfate: Prepare 10%(w/v) solution in water and immediately freeze in single-use (200 μl) aliquots at −20°C.
5. Loading buffer: XT Sample buffer 4× (Bio-Rad).
6. Running buffer: Tris/glycine/SDS buffer (Bio-Rad). 1× solution is 25 mM Tris, 192 mM glycine, and 0.1% (w/v) SDS, pH 8.3. Store at room temperature.
7. Fixing solution: 50% (v/v) methanol, 10% (v/v) acetic acid. Prepare in a chemical extractor hood.
8. Washing solution: Ultrapure water (18 MΩ-cm or equivalent).
9. Staining solution: Pro Q-Diamond solution (Molecular Probes, Invitrogen, Eugene, OR, USA).
10. Destaining solution: 20% (v/v) acetonitrile, 50 mM sodium acetate, pH 4.
11. Rotary shaker and a scanner with a visible-light laser-based (excitation at 532–560 nm) (i.e.: Typhoon Trio from GE Healthcare, Uppsala, Sweden).

2.4. Multiplexing Fluorescent Two-Dimensional Difference Gel Electrophoresis

2.4.1. Materials

1. ReadyPrep™ 2-D Cleanup kit (Bio-Rad). Alternative sources are 2D-clean up kit (GE Healthcare) or 2-D Sample Prep kit (Pierce Biotechnology, Rockford, IL, USA).
2. Sample buffer: 30 mM Tris–HCl, 2 M thiourea, 7 M urea, and 4% CHAPS, pH 8.5. Ensure that all the solutions containing urea are prepared freshly and do not heat above 37°C to prevent protein carbamylation and subsequent formation of charge trains on the 2D gel.
3. Bradford protein assay (Bio-Rad).
4. CyDye DIGE Fluor minimal dyes 2, 3, and 5 (GE Healthcare).
5. >99.5% Pure dimethylformamide (DMF, Sigma-Aldrich, St. Louis, MO, USA).
6. 10 mM lysine (Sigma-Aldrich).
7. Rehydration buffer: 2 M thiourea, 7 M urea, 2% dithiothreitol (DTT), 4% CHAPS, and 2% Pharmalyte (GE Healthcare).
8. Immobiline DryStrips 24 cm, pH 3–11 NL (GE Healthcare). Alternative sources are Bio-Rad, Sigma-Aldrich, and Isogen Life Sciences (Badalona, Spain).
9. Reducing buffer: 6 M urea, 1.5 M Tris–HCl, pH 8.8, 30% (v/v) glycerol, 2% (w/v) SDS, and 2% (w/v) DTT.

10. Alkylating buffer: 6 M urea, 1.5 M Tris–HCl, pH 8.8, 30% (v/v) glycerol, 2% (w/v) SDS, and 2.5% (w/v) iodoacetamide.
11. 12.5% Polyacrylamide gels (21 cm × 24 cm).
12. Fixing solution: 30% (v/v) methanol and 7.5% (v/v) acetic acid. Prepare in a chemical extractor hood.
13. Sypro Ruby protein gel stain (Bio-Rad or Molecular Probes).
14. Destaining solution: 10% (v/v) methanol and 6% (v/v) acetic acid. Prepare in a chemical extractor hood.

2.4.2. Equipment

1. Ettan™ IPGphor™ apparatus (GE Healthcare). Alternative equipment is PROTEAN IEF system (Bio-Rad), Multiphor II Horizontal electrophoresis unit (PerkinElmer, Waltham, MA), or Uniphor Horizontal Electrophoresis unit (Sigma-Aldrich).
2. Ettan Dalt system (GE Healthcare). Alternative equipment is PROTEAN Plus Dodeca™ Cell (Bio-Rad).
3. Typhoon Trio scanner (GE Healthcare). Alternative equipment is Typhoon 9400/9410, Typhoon FLA 9000 (GE Healthcare), or FLA 5100 Imaging System (FujiFilm, Tokyo, Japan).
4. Decyder™ 6.5 software (GE Healthcare). Alternative software are Progenesis SameSpots 3.0 (Nonlinear Dynamics, Newcastle/uT, UK) or Dymension 3 (Syngene, Cambridge, UK).

2.5. In-Gel Digestion

1. Equipment: MassPrep Station (Waters, Milford, MA, USA).
2. Destaining solution: 50 mM ammonium bicarbonate/50% (vol/vol) acetonitrile.
3. Reduction solution: 10 mM DTT in 100 mM ammonium bicarbonate.
4. Alkylation solution: 55 mM iodoacetamide in 100 mM ammonium bicarbonate.
5. Trypsin (12,5 ng/µl per protein band) (Promega, Fitchburg, WI, USA).

2.6. Mass Spectrometry

1. Microcapillary reversed phase LC (CapLC™; Atlantis, C_{18}, 3 µm, 75 µm × 10 cm Nano Ease™ fused silica capillary column) (Waters).
2. Equilibration buffer: 5% (v/v) acetonitrile, 0.2% (v/v) formic acid.
3. Elution: Linear gradient of 5–50% acetonitrile (v/v) (30 min; flow rate of 0.2 µl/min).
4. Tandem mass spectrometer. In our case, we have used a Q-TOF Micro mass spectrometer (Waters).

2.7. Computational Analysis

1. MassLynx 4.0 software (Waters).
2. Phenyx 2.2 software (GeneBio, Geneva, Switzerland).

3. Database: Uniprot knowledegebase (UniprotKB/Swiss-Prot and UniprotKB/TrEMBL) (http://www.uniprot.org).

2.8. Western Blotting

1. Transfer buffer: 25 mM Tris, 190 mM glycine, 20% (v/v) methanol. Store in the transfer apparatus at room temperature with cooling during use.
2. Supported nitrocellulose transfer membrane and 3MM Chromatography paper from Whatman (Dassel, Germany).
3. Tris-buffered saline with Tween (TBS-T): Prepare 10× stock with 1.37 M NaCl, 27 mM KCl, 250 mM Tris–HCl, pH 7.4, 1% Tween-20. Dilute 100 ml with 900 ml water for use.
4. Blocking buffer: 5% (w/v) nonfat dry milk in TBS-T. For detection of specific phosphorylated residues, 5% (w/v) bovine serum albumin (BSA, Sigma-Aldrich) in TBS-T.
5. Primary antibody dilution buffer: TBS-T supplemented with 5% (w/v) BSA or 5% (w/v) nonfat dry milk.
6. Primary antibodies: Anti-ERK1/2, anti-pERK1/2 (Thr202/Tyr204), anti-pCDK2 (Thr160), anti-pCaMKII (Thr286), anti-phosphothreonine antibody (Cell Signaling, Danvers, MA, USA), anti-CDK2 (Santa Cruz Biotechnology, Heidelberg, Germany), anti-pStathmin (Ser16) antibody (a generous gift from Dr. Sobel, París, France), anti-pStathmin (Ser25 and Ser38) antibodies, anti-stathmin, anti-beta actin antibody (Abcam, Cambridge, MA), anti-phosphoserine antibody (Chemicon, Temecula, CA, USA), and anti-phosphotyrosine clone 4G10 antibody (Upstate Biotechnology, Waltham, MA, USA).
7. Secondary antibodies: Goat anti-rabbit IgG (H+L) conjugated to horseradish peroxidase (Bio-Rad) or goat anti-mouse IgG horseradish peroxidase conjugated (Santa Cruz Biotechnology).
8. Enhanced chemiluminiscent (ECL) reagents from Perkin-Elmer and Amersham Hyperfilm™ ECL from GE Healthcare.

3. Methods

The present method was used for MLP-29 cells and should be adaptable to other cell lines or tissues (15–17). Since the aim of the study was to understand the MLP-29 phosphoproteome response to proteasome inhibition, the samples were taken at different time points (3, 10, and 24 h) after MG132 addition (Fig. 1). Success and comprehensiveness of the method will largely depend on the amount of biological starting material. The methodology

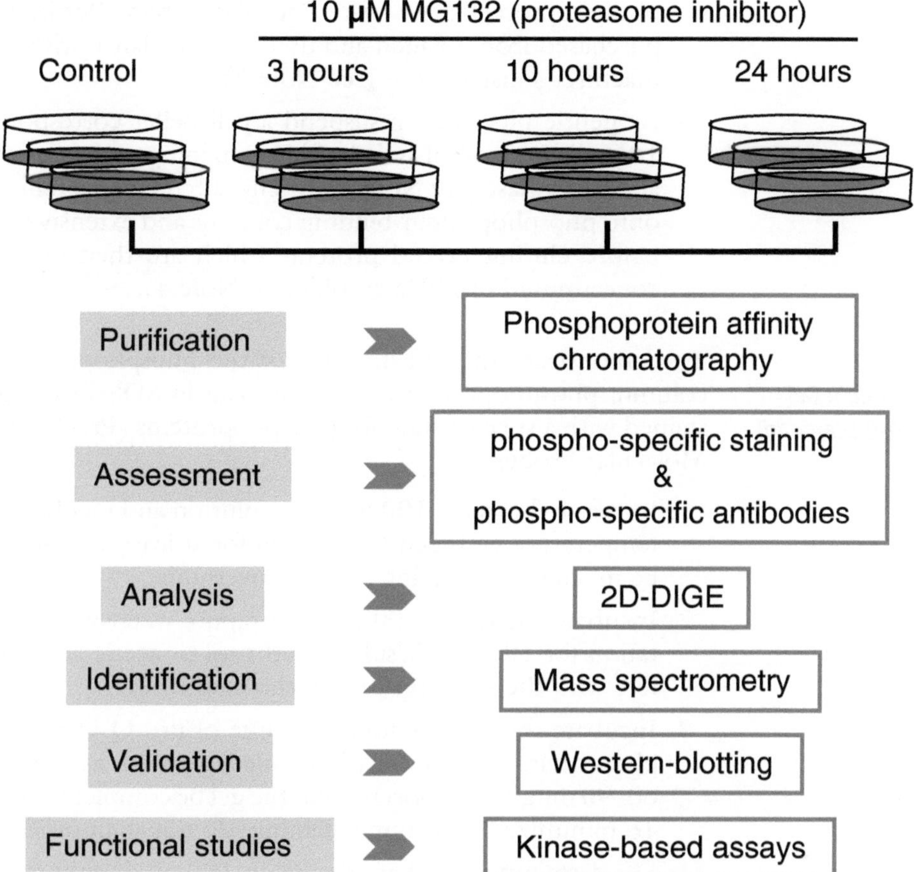

Fig. 1. Overview of methods used to monitor global phosphorylation changes in MLP-29 cells upon impaired proteasomal function. The strategy used by our group combines phosphoprotein affinity chromatography, 2D-DIGE, and nanoLC-MS/MS. The approach allowed us to check potential expression changes in 2,000–2,500 phosphorylated protein species upon proteasome inhibition revealing deregulated proteins involved in response to stress, RNA processing, and cell cycle control (19).

described here uses 2.5 mg of protein lysate in each condition (~10^7 cells). In MLP-29 cells, from 2.5 mg of protein lysate, the yield of phosphorylated proteins obtained using this method is approximately 125–175 μg, which allows performance of 2D-DIGE experiments (~100 μg of protein/condition). In order to reduce experimental variability, we recommend preparing all samples that need to be analyzed on the same day under identical conditions.

3.1. Cell Culture and Phosphoprotein Affinity Chromatography

1. MLP-29 cells are grown in DMEM supplemented with 10% FBS and 1% penicillin–streptomycin–glutamine and treated with 10 μM MG132 or 10 μM epoxomicin for the indicated periods of time (3, 10, and 24 h).
2. Wash cells before harvesting in HEPES-based or saline buffer (see Note 1). To obtain MLP-29 cell lysates enriched for

phosphoproteins, we use Phosphoprotein Purification Kit purchased from Qiagen and used in accordance with the manufacturer's instructions (see Note 2).

3. By gentle pipetting, resuspend a cell pellet corresponding to 10^7 cells in 5 ml of lysis buffer containing protease inhibitors and benzonase (see Note 3). Briefly, 2.5 mg of protein is loaded onto phosphoprotein-binding columns and extensively washed before eluting bound proteins which are then desalted and concentrated (10 kDa cutoff) (see Note 4).

3.2. Specific Phosphoprotein Gel Stain (Pro-Q Diamond)

To demonstrate the effectiveness of the phosphoprotein affinity column, phosphoprotein extracts are run in SDS-PAGE gels and stained with a specific stain for phosphoproteins (Pro-Q Diamond, Molecular Probes).

1. Immerse the gel in 100 ml of fix solution and incubate at room temperature with gentle agitation for at least 30 min. Gels can be maintained in fixing solution overnight.
2. Incubate the gel in 100 ml of ultrapure water with gentle agitation for 10 min. Residual methanol or acetic acid will interfere with the staining. Repeat this step twice.
3. Incubate the gel in enough volume of Pro Q-Diamond phosphoprotein gel stain with gentle agitation in the dark for 60–90 min. It is important that the gel be completely immersed. To minimize background staining, do not stain overnight.
4. For destaining, incubate the gel in 100 ml destaining solution with gentle agitation for 30 min at room temperature, protected from light. Repeat this procedure twice.
5. Wash twice with ultrapure water at room temperature for 5 min per wash. Pro-Q Diamond phosphoprotein stain has an excitation maximum at 555 nm and an emission maximum at 580 nm. Imaging instruments with light sources and filters that match the excitation and emission maxima result in the highest sensitivity. As shown in Fig. 2, the phosphoprotein affinity column is an effective method for enriching phosphoproteins (see Note 5).

3.3. Multiplexing Fluorescent Two-Dimensional Difference Gel Electrophoresis

1. All phosphoprotein extracts are cleaned and precipitated using ReadyPrep™ 2-D Cleanup kit (Bio-Rad) (see Note 6). Pellet the samples by centrifugation at 14,000 rpm (24,100 × g) for 5 min at 4°C and air-dry the pellet for 5 min. The pellet will not dissolve in sample buffer if overdried.
2. Resuspend the pellet by adding 50 μl of rehydration buffer and add additional 10 μl volumes until the pellet dissolves.
3. Determine the protein concentration using Bradford assay (Bio-Rad) or 2D-quant kit (GE Healthcare). Final protein

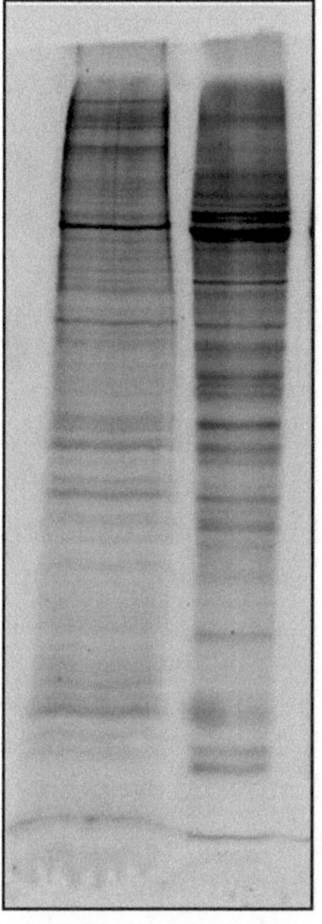

Fig. 2. Validation of phosphoprotein enrichment procedure. MLP-29 total protein extract and MLP-29 phosphoprotein fraction (15 μg each line) were separated in SDS-PAGE gel. Phosphoprotein-specific staining (Pro-Q Diamond) was used to visualize phosphorylated proteins. As shown in the figure, the affinity column (Qiagen) used here is an effective method for enriching phosphoproteins.

concentration must be in the range of 5–10 μg/μl and shall be adjusted by adequately concentrating or diluting the samples.

4. Prepare a normalized pool (standard) comprising equal amounts from all the samples present in all conditions to be compared (see Note 7). Make sure that the sample pH is between 8.0 and 9.0. The pH must be usually corrected by adding dilute base.
5. Reconstitute the CyDye Fluor minimal dye in 10 μl of DMF by centrifugating at 12,000 rpm for 30 s to make a stock of 1,000 pmol/μl. Wrap the tubes in aluminum foil and store at –20°C until needed (see Note 8).

6. Prepare a working solution of 200 pmol/μl of CyDye by adding 4 μl of DMF to 1 μl of stock solution. Label 50 μg of each sample with 1 μl of working solution of either Cy3 or Cy5 dye in a manner that Cy3 and Cy5 are swapped equally among the samples from the different conditions. At the same time, label an aliquot of 50 μg of internal standard with 1 μl of working solution of Cy2 label to be mixed with corresponding aliquots of Cy3- and Cy5-labeled samples for each gel.
7. Mix the sample and dye by vortexing and keep the sample on ice for 30 min in the dark.
8. Add 10 μl of 10 mM lysine to each sample to stop the labeling reaction.
9. Vortex, centrifuge at 14,000 rpm (24,100 × *g*) for 10 s, and keep on ice for 10 min. The labeled samples can be stored at −80°C for up to 3 months. At this phase, matched samples from conditions "control" and "disease/treatment" (labeled with Cy3 or Cy5) along with an aliquot of internal standard (Cy2) should be pooled. Each 2D-gel will have three labeled samples: the internal standard, and the two conditions to be compared.
10. Add adequate rehydration buffer to make the samples previously prepared up to 450 μl. The samples are ready for isoelectric focusing (IEF). Perform the IEF according to the setup indicated in the IEF apparatus (see Subheading 2.4.2). We usually use the 24 cm pH 3–11 NL Immobiline DryStrips (see Note 9). The processed IEF strips can be stored at −80°C for up to 3 months.
11. Equilibrate the IEF strips to reduce the disulfide bonding by gently rocking them in 10 ml/strip of reducing buffer for 10 min. Immediately after this, alkylate the –SH groups of proteins by gentle rocking in 10 ml/strip of alkylating buffer for 10 min.
12. Rinse the IEF strips in the SDS electrophoresis running buffer (see Note 10). Prepare a 0.6% (w/v) agarose overlay solution by heating. Slowly pipette the agarose solution, making sure that no bubbles are introduced between the glass plates of the gels for the SDS-PAGE. Place the IEF strip between the glass plates ensuring that the IEF strip rests on the SDS-PAGE. The second dimension is run at 1 W overnight until the bromophenol blue dye front reaches the bottom of the gel.
13. Scan the gels using a Typhoon scanner at 100 μm resolution with λex/λem of 488/520, 532/580, and 633/670 nm for Cy2, Cy3, and Cy5, respectively (see Note 11).
14. Analyze the image using the Decyder™ software. Detail explanation on how to perform 2D-DIGE technique and the comparative analysis can be found in Tannu et al. (25).

15. Prepare protein samples with 400–450 μg combined with 50 μg of Cy2-labeled internal standard for the preparative gel. Subsequently, perform the IEF and second dimension as mentioned above. This gel is necessary to pick the differential spots and to identify the corresponding proteins by mass spectrometry.
16. Fix the preparative gel in fixing solution followed by washing the gel twice with distilled water.
17. After fixation, stain the gel overnight with 500 ml of Sypro Ruby stain.
18. After removing Sypro Ruby solution, wash twice for 10 min with destaining solution.
19. Finally, scan the gel using λex/λem of 532/560 nm, localize the spots of interest, and excise them for further tryptic digestion (see Note 12).

3.4. Mass Spectrometry and Computational Analysis

Microcapillary reversed phase LC is performed with a CapLC™ (Waters) capillary system.

1. Reversed-phase separation of tryptic digests is performed with an Atlantis, C_{18}, 3 μm, 75 μm × 10 cm Nano Ease™ fused silica capillary column (Waters) equilibrated in 5% (v/v) acetonitrile, containing 0.2% (v/v) formic acid.
2. After injection of 6 μl of sample (peptide mixture obtained after digestion with trypsin corresponding to each spot), the column is washed for 5 min with the same buffer and the peptides are eluted using a linear gradient of 5–50% (v/v) acetonitrile in 30 min at a constant flow rate of 0.2 μl/min. The column is coupled online to a Q-TOF Micro mass spectrometer (Waters) using a PicoTip nanospray ionization source (Waters). The heated capillary temperature is 80°C and the spray voltage is 1.8–2.2 kV.
3. MS/MS data are collected in an automated data-dependent mode. The three most intense ions in each survey scan are sequentially fragmented by collision-induced dissociation (CID) using an isolation width of 2.5 and a relative collision energy of 35%. Spectra are processed with MassLynx 4.0 and database searching is done with Phenyx 2.2 (GeneBio, Geneva, Switzerland) against Uniprot knowledgebase.
4. The search is enzymatically constrained for trypsin and allowed for one missed cleavage site. Further search parameters are as follows: no restriction on molecular weight and isoelectric point; fixed modification, carbamidomethylation of cysteine; and variable modification, oxidation of methionine.

3.5. Western Blotting

1. To validate the proteomic data, equal amounts of protein from control and treated MLP-29 cells (15 μg) are resolved in 12.5% SDS-PAGE gels. A detailed description of how to perform the

Western blotting technique can be found in Mattingly et al. (27).

2. Briefly, protein samples are electrophoretically transferred onto nitrocellulose membranes for 45 min at 120 V.
3. To monitor the protein transfer, equal loading of the gel is assessed by Ponceau staining.
4. Membranes are probed with primary antibodies at 1:1,000 dilution in 5% (w/v) BSA or 5% (w/v) nonfat milk, depending on the primary antibody used.
5. After incubation with the appropriate horseradish peroxidase-conjugated secondary antibody (1:5,000), the immunoreactivity is visualized by enhanced chemiluminiscence (Perkin Elmer).
6. Finally, protein loading control is assessed by hybridization with a beta-actin-specific antibody.

An example of proteasome inhibition-induced changes in the phosphorylation profile of MLP-29 cells characterized by our group is shown in Fig. 3.

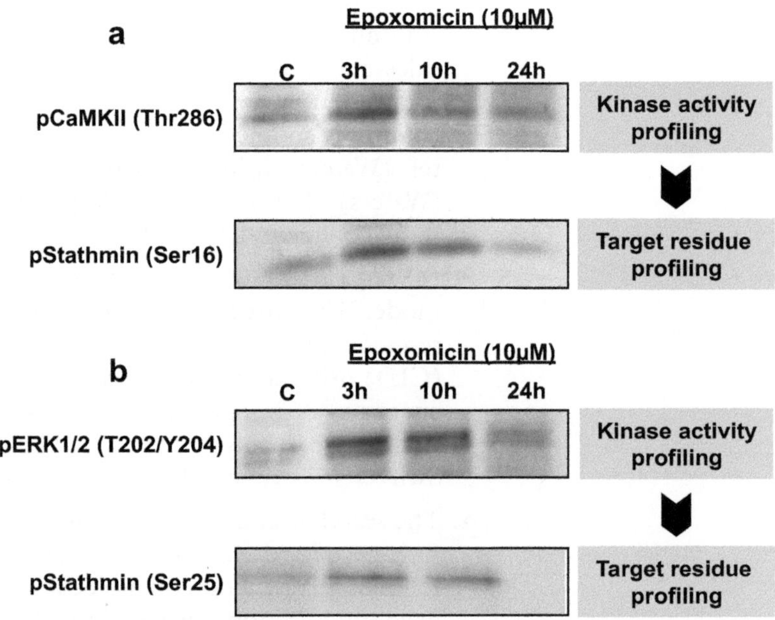

Fig. 3. Phosphorylation pattern of stathmin during epoxomicin-impaired proteasomal function. Activation of CaM kinase II and the corresponding transient increase in the phosphorylation of Ser16 at 3 and 10 h in epoxomicin-treated MLP-29 cells (**a**). Activation of ERK1/2 in parallel with increased phosphorylation of Ser25 during epoxomicin treatment (**b**).

4. Notes

1. Do not use phosphate buffer (i.e.: PBS) for washing the cells as this will interfere with binding of phosphorylated proteins to the column.
2. Although the most commonly used kit is Phosphoprotein Purification Kit from Qiagen at present, different phosphoprotein enrichment kits commercially available may also be used: Pro-Q Diamond Phosphoprotein Enrichment Kit (Invitrogen), BD™ Phosphoprotein Enrichment Kit (BD Biosciences), and Phosphoprotein Enrichment Kit (Merck, Darmstadt, Germany). All these kits are also based on affinity chromatography that enables an efficient isolation of phosphorylated proteins from complex cellular extracts.
3. Ensure that protein concentration in the lysate is adjusted to 0.1 mg/ml. This concentration adjustment is made to ensure that all phosphate groups are easily accessible during purification and are not hidden within protein complexes. Under the lysis buffer conditions used, no phosphatase activity is measurable, ensuring that the phosphorylated protein fraction obtained accurately reflects the in vivo phosphorylation profile.
4. In order to allow complete binding, flow through should not pass through the column at a flow rate greater than 0.5 ml/min (~50 min). Expected concentration of proteins in the eluate is around 0.2 μg/μl. It is important to note that the maximum binding capacity of the Qiagen column is approximately 500 μg of phosphorylated protein.
5. In order to validate the phosphoprotein enrichment, we also analyzed a peptide mixture (from 5 μg of phosphoprotein fraction obtained from MLP-29 cells) by nanoLC-MS/MS, and more than 125 proteins that are regulated by phosphorylation were identified. Additional approaches may be used to control the phosphoprotein purification method such as alkaline phosphatase treatment, immunodetection with antibodies to phosphothreonine, phosphoserine, and phosphotyrosine or other phosphoprotein staining kits commercially available such as GelCode Phosphoprotein Staining Kit (Thermo Fisher, Manchester, UK) or PhosDecor™ Fluorescent Phosphoprotein in Gel Detection Kit (Sigma-Aldrich).
6. Alternative sources are 2D-clean up kit (GE Healthcare) or 2-D Sample Prep kit (Pierce Biotechnology). The cleaning-up step facilitates preparation of low-conductivity samples suitable for IEF, reducing streaking, background staining, and other gel artifacts associated with substances contaminating 2D/IEF samples.

7. This method has the advantage of being able to quantify protein spots which are only present in one group due to the presence of internal standard so make sure that each sample is used in the exact same amount to make up a correct internal standard.
8. Ensure that DMF is water free as the reactive groups of the dyes are water sensitive. Once reconstituted, the dyes should be used within 3–4 weeks.
9. The length of the pH strip, its pH range, as well as the IEF setup depend on the samples to be analyzed and the scientific objective of the researcher.
10. Use fresh DTT and iodoacetamide in the reducing and alkylating buffers, respectively. Perform the reduction and alkylation steps as close as possible to running the second dimension.
11. It is essential that the maximum pixel intensities of all the three images do not differ from each other by more than 10,000–15,000. If an image is saturated (>100,000), the gel should be re-scanned using a lower photomultiplier tube (PTM) value to bring all the protein spots in the linear dynamic range. This is crucial to obtain any meaningful quantitative comparison between the gel images.
12. 2D-electrophoresis presents some inherent disadvantages such as the following: (a) many protein spots likely comprise multiple proteins; (b) poor spot resolution at higher p*I* values; and (c) difficulty in electrophoresing large and hydrophobic proteins in the first dimension separation. Alternative proteomics methods may be also used to perform quantitative analysis of phosphoproteome. SILAC is probably one of the most sensitive, accurate, and robust approaches in MS-based quantitative phosphoproteomics in combination with SCX and titanium dioxide (TiO_2) chromatography for phosphopeptide enrichment (12, 28). Chemical labeling strategies can be also employed when SILAC is not feasible. Chemical labeling by tandem mass tags (iTRAQ™, from Applied Biosystems or TMT reagents from Proteome Science plc.) or stable isotope dimethyl labeling (29) has been already used in phosphoproteomic studies (30, 31). Moreover, a label-free quantitative approach is also possible to apply (32, 33).

Acknowledgements

This laboratory is member of the National Institute of Proteomics Facilities, Proteored. Parts of the work described here were supported by the agreement between FIMA and the "UTE project

CIMA"; grants Plan Nacional I+D+I SAF2008-0154 from Ministerio de Ciencia e Innovación to FJC; ISCIII-RETIC RD06/0020 to MAA and FJC. We like to acknowledge technical contributions from Carmen Miqueo, María I. Mora, and Manuela Molina from Center for Applied Medical Research (CIMA, University of Navarra).

References

1. Drexler HC (1997) Activation of the cell death program by inhibition of proteasome function. Proc Natl Acad Sci USA 94:855–860
2. Drexler HC, Risau W, Konerding MA (2000) Inhibition of proteasome function induces programmed cell death in proliferating endothelial cells. FASEB J 14:65–77
3. Oikawa T, Sasaki T, Nakamura M, Shimamura M, Tanahashi N, Omura S, Tanaka K (1998) The proteasome is involved in angiogenesis. Biochem Biophys Res Commun 246:243–248
4. Ganten TM, Koschny R, Haas TL, Sykora J, Li-Weber M, Herzer K, Walczak H (2005) Proteasome inhibition sensitizes hepatocellular carcinoma cells, but not human hepatocytes, to TRAIL. Hepatology 42:588–597
5. Anan A, Baskin-Bey ES, Bronk SF, Werneburg NW, Shah VH, Gores GJ (2006) Proteasome inhibition induces hepatic stellate cell apoptosis. Hepatology 43:335–344
6. Anan A, Baskin-Bey ES, Isomoto H, Mott JL, Bronk SF, Albrecht JH, Gores GJ (2006) Proteasome inhibition attenuates hepatic injury in the bile duct-ligated mouse. Am J Physiol Gastrointest Liver Physiol 291:G709–G716
7. Landowski TH, Megli CJ, Nullmeyer KD, Lynch RM, Dorr RT (2005) Mitochondrial-mediated disregulation of Ca2+ is a critical determinant of Velcade (PS-341/bortezomib) cytotoxicity in myeloma cell lines. Cancer Res 65:3828–3836
8. Poulaki V, Mitsiades CS, Kotoula V, Negri J, McMillin D, Miller JW, Mitsiades N (2007) The proteasome inhibitor bortezomib induces apoptosis in human retinoblastoma cell lines in vitro. Invest Ophthalmol Vis Sci 48: 4706–4719
9. Yang W, Monroe J, Zhang Y, George D, Bremer E, Li H (2006) Proteasome inhibition induces both pro- and anti-cell death pathways in prostate cancer cells. Cancer Lett 243: 217–227
10. Goshe MB (2006) Characterizing phosphoproteins and phosphoproteomes using mass spectrometry. Brief Funct Genomic Proteomic 4:363–376
11. Mann M, Ong SE, Gronborg M, Steen H, Jensen ON, Pandey A (2002) Analysis of protein phosphorylation using mass spectrometry: deciphering the phosphoproteome. Trends Biotechnol 20:261–268
12. Macek B, Mann M, Olsen JV (2009) Global and site-specific quantitative phosphoproteomics: principles and applications. Annu Rev Pharmacol Toxicol 49:199–221
13. Laugesen S, Messinese E, Hem S, Pichereaux C, Grat S, Ranjeva R, Rossignol M, Bono JJ (2006) Phosphoproteins analysis in plants: a proteomic approach. Phytochemistry 67: 2208–2214
14. Makrantoni V, Antrobus R, Botting CH, Coote PJ (2005) Rapid enrichment and analysis of yeast phosphoproteins using affinity chromatography, 2D-PAGE and peptide mass fingerprinting. Yeast 22:401–414
15. Davis MA, Hinerfeld D, Joseph S, Hui YH, Huang NH, Leszyk J, Rutherford-Bethard J, Tam SW (2006) Proteomic analysis of rat liver phosphoproteins after treatment with protein kinase inhibitor H89 (N-(2-[p-bromocinnamylamino-]ethyl)-5-isoquinolinesulfonamide). J Pharmacol Exp Ther 318: 589–595
16. Puente LG, Carriere JF, Kelly JF, Megeney LA (2004) Comparative analysis of phosphoprotein-enriched myocyte proteomes reveals widespread alterations during differentiation. FEBS Lett 574:138–144
17. Puente LG, Borris DJ, Carriere JF, Kelly JF, Megeney LA (2006) Identification of candidate regulators of embryonic stem cell differentiation by comparative phosphoprotein affinity profiling. Mol Cell Proteomics 5:57–67
18. Herz C, Aumailley M, Schulte C, Schlotzer-Schrehardt U, Bruckner-Tuderman L, Has C (2006) Kindlin-1 is a phosphoprotein involved in regulation of polarity, proliferation, and motility of epidermal keratinocytes. J Biol Chem 281:36082–36090
19. Santamaria E, Mora MI, Munoz J, Sanchez-Quiles V, Fernandez-Irigoyen J, Prieto J, Corrales FJ (2009) Regulation of stathmin

phosphorylation in mouse liver progenitor-29 cells during proteasome inhibition. Proteomics 9:4495–4506

20. Li Z, Dong X, Wang Z, Liu W, Deng N, Ding Y, Tang L, Hla T, Zeng R, Li L, Wu D (2005) Regulation of PTEN by Rho small GTPases. Nat Cell Biol 7:399–404
21. Puente LG, Voisin S, Lee RE, Megeney LA (2006) Reconstructing the regulatory kinase pathways of myogenesis from phosphopeptide data. Mol Cell Proteomics 5:2244–2251
22. Klemm C, Otto S, Wolf C, Haseloff RF, Beyermann M, Krause E (2006) Evaluation of the titanium dioxide approach for MS analysis of phosphopeptides. J Mass Spectrom 41: 1623–1632
23. Steinberg TH, Agnew BJ, Gee KR, Leung WY, Goodman T, Schulenberg B, Hendrickson J, Beechem JM, Haugland RP, Patton WF (2003) Global quantitative phosphoprotein analysis using multiplexed proteomics technology. Proteomics 3:1128–1144
24. Schulenberg B, Aggeler R, Beechem JM, Capaldi RA, Patton WF (2003) Analysis of steady-state protein phosphorylation in mitochondria using a novel fluorescent phosphosensor dye. J Biol Chem 278:27251–27255
25. Tannu NS, Hemby SE (2006) Two-dimensional fluorescence difference gel electrophoresis for comparative proteomics profiling. Nat Protoc 1:1732–1742
26. Muller M, Morotti A, Ponzetto C (2002) Activation of NF-kappaB is essential for hepatocyte growth factor-mediated proliferation and tubulogenesis. Mol Cell Biol 22: 1060–1072
27. Mattingly RR (2003) Mitogen-activated protein kinase signaling in drug-resistant neuroblastoma cells. Methods Mol Biol 218:71–83
28. Van Hoof D, Munoz J, Braam SR, Pinkse MW, Linding R, Heck AJ, Mummery CL, Krijgsveld J (2009) Phosphorylation dynamics during early differentiation of human embryonic stem cells. Cell Stem Cell 5:214–226
29. Boersema PJ, Raijmakers R, Lemeer S, Mohammed S, Heck AJ (2009) Multiplex peptide stable isotope dimethyl labeling for quantitative proteomics. Nat Protoc 4:484–494
30. Leitner A, Lindner W (2009) Chemical tagging strategies for mass spectrometry-based phospho-proteomics. Methods Mol Biol 527:229–243, x
31. Boersema PJ, Foong LY, Ding VM, Lemeer S, van Breukelen B, Philp R, Boekhorst J, Snel B, den Hertog J, Choo AB, Heck AJ (2009) In-depth qualitative and quantitative profiling of tyrosine phosphorylation using a combination of phosphopeptide immunoaffinity purification and stable isotope dimethyl labeling. Mol Cell Proteomics 9:84–99
32. Palmisano G, Thingholm TE (2010) Strategies for quantitation of phosphoproteomic data. Expert Rev Proteomics 7:439–456
33. Xie X, Feng S, Vuong H, Liu Y, Goodison S, Lubman DMA (2010) Comparative phosphoproteomic analysis of a human tumor metastasis model using a label-free quantitative approach. Electrophoresis 31:1842–1852

Chapter 13

Identifying Acetylated Proteins in Mitosis

Carol Chuang and Li-yuan Yu-Lee

Abstract

Histone deacetylase (HDAC) inhibitors are currently used in anticancer therapy to perturb genomic targets involved in gene transcriptional responses. However, the role of HDAC inhibitors on the acetylation of proteins outside of the transcriptional network has not been thoroughly assessed. We recently discovered that one of the HDACs, HDAC3, is localized on the mitotic spindle and regulates proper mitotic progression (1). To determine potential HDAC targets, we undertook a proteomics approach to search for acetylated proteins in mitosis (2). First, we synchronized cells in mitosis and used a polyclonal anti-acetyl-Lysine antiserum to immunoprecipitate acetylated proteins, followed by their identification by LC-ESI-MS/MS. We then confirmed the acetylation status of several mitotic proteins by anti-acetyl-Lysine immunoprecipitation with a monoclonal antibody followed by Western blot analyses of the proteins of interest. We further confirmed by a reciprocal immunoprecipitation with protein-specific antibody followed by Western blot analysis with another monoclonal anti-acetyl-Lysine antibody. Interestingly, the acetylation of a subset of the mitotic proteins can be further enhanced by treatment with apicidin, a small molecule inhibitor with specificity for HDAC3, suggesting that their acetylation may be regulated by HDAC3 in mitosis. In this chapter, we describe the various techniques using NudC as an example of an acetylated protein that is sensitive to apicidin treatment in mitosis.

Key words: Acetylated proteins, Immunoprecipitation, Mass spectrometry, Mitosis, Thymidine block, Nocodazole enrichment, Apicidin, HDAC3, Reduced lysate

1. Introduction

Mitosis is a highly regulated process in which errors can lead to genomic instability, a hallmark of cancer. During mitosis, transcription is largely silent and RNA translation is globally inhibited. Thus, mitosis is largely driven by posttranslational modification of protein networks to activate, inhibit, localize, and/or degrade proteins. Phosphorylation of mitotic proteins has been shown to be one of the major mechanisms regulating cell cycle progression (3, 4).

Djuro Josic and Douglas C. Hixson (eds.), *Liver Proteomics: Methods and Protocols*, Methods in Molecular Biology, vol. 909, DOI 10.1007/978-1-61779-959-4_13,

A recent proteome-wide acetylation study suggested that the size of the acetylome approaches that of the phosphoproteome (5), thus implicating acetylation/deacetylation as a prominent regulatory mechanism for major cellular processes (5–7).

Currently, histone deacetylase (HDAC) inhibitors are used to treat various cancers aiming to perturb transcriptional responses. However, the role of HDAC inhibitors on the acetylation of proteins outside of the transcriptional network has not been thoroughly assessed. We showed that protein acetylation/deacetylation is prevalent in mitosis (1, 2). Immunoprecipitation (IP) followed by Western blot analyses confirmed that RNA processing proteins, eIF4G and RNA helicase A, and several cell cycle proteins, including APC1, anillin, and NudC, were acetylated in mitosis. Here, we use NudC as an example. We further show that acetylation of NudC was enhanced by treatment with apicidin, a small molecule inhibitor with specificity for HDAC3, suggesting that its acetylation may be regulated by HDAC3 in mitosis (2).

In this chapter, we present a general method of identifying and confirming that specific proteins are acetylated. First, identify acetylated proteins by IP with an anti-acetyl-lysine antibody followed by LC-ESI-MS/MS. Second, confirm LC-ESI-MS/MS analysis with IP using an alternate anti-acetyl-Lysine antibody from that used for the mass spectrometry identification, and then immunoblotting for specific proteins of interest. Third, perform the reciprocal experiment by IP for the specific proteins of interest and then immunoblotting using a third anti-acetyl-Lysine antibody. We note that the reciprocal IP yielded the best results when the cell lysates were first treated with a reducing agent to fully expose epitopes before protein-specific IP using anti-peptide antibody. Furthermore, treatment with apicidin for 3.5 h at the peak of mitosis prior to harvest allowed the observation of enhanced acetylation of proteins that are modulated by HDAC3.

2. Materials

2.1. Cell Culture

1. Dulbecco's modified Eagle's medium (D-MEM) (1×), liquid (high glucose) (Invitrogen/Gibco, Grand Island, NY, USA) supplemented with 10% fetal bovine serum (FBS, Invitrogen/Gibco).
2. Thymidine (Sigma-Aldrich, St. Louis, MO, USA) is dissolved in water (see Note 1) to 200 mM. Sterilize under the culture hood using a 0.2 μm syringe filter. Store in 1 mL aliquots at −20°C and add to culture media as needed.
3. Nocodazole (EMD Chemicals Inc./Calbiochem, Gibbstown, NJ, USA) is dissolved in sterile DMSO to 500 μg/mL under

the culture hood. Store in 1 mL aliquots at 4°C and add to culture media as needed.

4. Apicidin (Sigma-Aldrich) is dissolved in sterile ethanol to 500 μM under the culture hood. Store in 400 μL aliquots at –20°C and add to culture media as needed.
5. PBS (10×): 1.37 M NaCl, 26.8 mM KCl, 53.7 mM Na_2HPO_4, 17.6 mM KH_2PO_4, pH 7.26. Dilute to 1×, pH 7.4, and autoclave before use.
6. Mitotic shake off pipet: Beral® transfer pipets (Samco Scientific, San Fernando, CA, USA).

2.2. Antibodies

2.2.1. Antibodies for Immunoprecipitation

1. Antibody for IP followed by LC-ESI-MS/MS analysis: Anti-acetyl-Lysine, Cat. # 06-933 (rabbit antiserum) (Millipore/Upstate, Temecula, CA, USA).
2. Antibody for IP followed by Western blot analysis: Anti-acetyl-Lysine, clone 4G12 (mouse monoclonal antibody) (Millipore/Upstate).
3. Antibody for reduced lysate IP followed by Western blot analysis: NudC G1 antibody (affinity-purified goat anti-NudC C terminus 15 aa peptide antibody) (generated through Bethyl Laboratories, Inc., Montgomery, TX, USA; under submission).

2.2.2. Antibodies for Western Blot Analysis

1. Antibody for Western blot analysis following IP: Anti-acetyl-Lysine (rabbit polyclonal antibody) (Millipore/Upstate).
2. Antibody for Western blot analysis following reduced lysate IP: Anti-acetyl-Lysine, clone Ac-K-103 (mouse monoclonal antibody) (Cell Signaling Technology, Inc., Danvers, MA, USA).
3. Antibody for Western blot analysis following IP and reduced lysate IP: NudC R2 antibody (affinity-purified rabbit anti-NudC C terminus 15 aa peptide antibody, 1:3,000) (8).

2.3. General Inhibitors for Cell Lysis and Immunoprecipitation

1. Mammalian protease inhibitor (100×): Protease Inhibitor Cocktail for use with mammalian cell and tissue extracts, DMSO (Sigma-Aldrich). Aliquot and store at –20°C.
2. Serine–threonine phosphatase inhibitor (100×): Phosphatase Inhibitor Cocktail 3, DMSO (Sigma-Aldrich). Aliquot and store at 4°C.
3. Tyrosine phosphatase inhibitor (100×): Phosphatase Inhibitor Cocktail 2, aqueous (Sigma-Aldrich). Aliquot and store at 4°C.
4. Sodium orthovanadate (Na_3VO_4, Sigma-Aldrich) is dissolved in water at 500 mM and stored in aliquots at –20°C.
5. Sodium fluoride (NaF, Sigma-Aldrich) is dissolved in water at 500 mM and stored in aliquots at 4°C.
6. Phenylmethanesulfonylfluoride (PMSF, Sigma-Aldrich) is dissolved in anhydrous isopropanol and stored in aliquots at –20°C.

7. Sodium Butyrate (NaB, Sigma-Aldrich) is dissolved in water at 1 M and stored at 4°C.
8. DNase I (Roche Applied Science, Mannheim, Germany) is diluted to 15 U/μL in water and stored in aliquots at −20°C.

2.4. Immunoprecipitation

2.4.1. Anti-Acetyl-Lysine Immunoprecipitation for LC-ESI-MS/MS

1. IP beads: Protein G Sepharose™ 4 Fast Flow (GE Healthcare, Uppsala, Sweden).
2. IP Lysis Buffer: 150 mM NaCl, 50 mM Tris, 5 mM EGTA, 1.5 mM EDTA, pH 8.0, 0.5% (v/v) Triton X-100, 5% (v/v) Glycerol. Store at 4°C. Supplement with 1 mM PMSF, mammalian protease-inhibitor cocktail, 5 mM Na_3VO_4, 5 mM NaF, serine–threonine and tyrosine phosphatase inhibitor cocktails, and 10 mM NaB before use.
3. Anti-acetyl-Lysine, Cat. # 06-933 (rabbit antiserum) (Millipore/Upstate).

2.4.2. Anti-Acetyl-Lysine Immunoprecipitation for Western Blot Analysis

1. IP beads: Protein G Sepharose™ 4 Fast Flow (GE Healthcare).
2. IP Lysis Buffer: 150 mM NaCl, 50 mM Tris, 5 mM EGTA, 1.5 mM EDTA, pH 8.0, 0.5% (v/v) Triton X-100, 5% (v/v) Glycerol. Store at 4°C. Supplement with 1 mM PMSF, mammalian protease-inhibitor cocktail, 5 mM Na_3VO_4, 5 mM NaF, serine–threonine and tyrosine phosphatase inhibitor cocktails, and 10 mM NaB before use.
3. Anti-acetyl-Lysine, clone 4G12 (mouse monoclonal antibody) (Millipore/Upstate).

2.4.3. Immunoprecipitation Using Reduced Cell Lysate for Western Blot Analysis

1. IP beads: Protein G Sepharose™ 4 Fast Flow (GE Healthcare).
2. Reducing Buffer: 1% (w/v) SDS, 5 mM EDTA, 10 mM DTT. Make fresh (see Note 2).
3. RIPA Buffer: 150 mM NaCl, 25 mM Tris, 1 mM EDTA, 0.5% (w/v) deoxycholic acid (sodium salt), pH 7.4 (see Note 3), 1% (v/v) NP40. Store at 4°C. Supplement with 1 mM PMSF, mammalian protease-inhibitor cocktail, 5 mM Na_3VO_4, 5 mM NaF, serine–threonine and tyrosine phosphatase inhibitor cocktails, 1 mM NaB, and 15 U/mL DNase I before use.
4. NudC G1 antibody (affinity-purified goat anti-NudC C terminus 15 aa peptide antibody) (generated through Bethyl Laboratories, Inc., Montgomery, TX, USA; under submission).

2.5. SDS-Polyacrylamide Gel Electrophoresis

1. Precast gradient SDS-PAGE gel: NuPAGE® 4–12% Bis-Tris Gel 1.0 mm × 10 well (Invitrogen, Carlsbad, CA, USA).
2. Running buffer (20×): NuPAGE® MOPS SDS Running Buffer (Invitrogen).

3. Loading Buffer (3×): NuPAGE® LDS Sample Buffer (4×) (Invitrogen) supplemented with 5% β-mercaptoethanol and diluted with water to 3×.
4. Pre-stained molecular weight marker: Precision Plus Protein™ Standards (BIO-RAD, Hercules, CA, USA).

2.6. Coomassie Blue Stain and In-Gel Digestion for LC-ESI-MS/MS

2.6.1. Coomassie Blue Staining

1. Fixing Solution: 50% methanol and 7% acetic acid in water. Make fresh since methanol easily evaporates.
2. Coomassie Blue Stain: GelCode™ Blue Stain Reagent (Thermo Scientific, Rockford, IL, USA).

2.6.2. In-Gel Tryptic Digestion for LC-ESI-MS/MS

1. Ammonium bicarbonate (NH_4HCO_3; Sigma-Aldrich) is dissolved in water at 0.1 M. Store at room temperature. Dilute as needed.
2. Acetonitrile (ACN; Sigma-Aldrich) is stored at room temperature. Highly flammable and volatile. Use under chemical hood with proper protective gear.
3. Trypsin: 50 mM NH_4HCO_3, 5 mM $CaCl_2$, and 12.5 mg/mL trypsin (porcine, sequence grade, Sigma-Aldrich).
4. Dithiothreitol (DTT; Thermo Fischer Scientific, Pittsburgh, PA, USA) is dissolved in water at 0.1 M. Store at –20°C. Dilute as necessary.
5. Iodoacetamide (Sigma-Aldrich) is stored at room temperature. Highly toxic. Use under chemical hood with proper protective gear.
6. Formic acid (Thermo Fischer Scientific) is stored at room temperature. Highly flammable and corrosive. Use under chemical hood with proper protective gear.
7. Trifluoroacetic acid (Sigma-Aldrich) is stored at room temperature. Highly corrosive and toxic. Use under chemical hood with proper protective gear.

2.7. Western Blotting for Acetylated Proteins

1. Transfer Buffer (10×): 0.37% SDS, 480 mM Tris, 390 mM Glycine. Store at room temperature. Dilute to 1× supplemented with 20% (v/v) methanol before use.
2. TBS-T: 0.8% (w/v) NaCl, 20 mM Tris pH 7.4, 0.2% (v/v) Tween 20.
3. Antibody diluting solution: 5% (w/v) bovine serum albumin (BSA, Sigma-Aldrich) in TBS-T. Store at 4°C.
4. Anti-acetyl-Lysine (rabbit polyclonal antibody) (Millipore/ Upstate) or anti-acetyl-Lysine, clone Ac-K-103 (mouse monoclonal antibody) (Cell Signaling Technology, Inc.).

5. Secondary antibody: Peroxide labeled anti-mouse IgG (H+L) (Vector Laboratories, Inc., Burlingame, CA, USA).
6. Enhanced chemiluminescent (ECL) reagents: SuperSignal® West Pico Chemiluminescent Substrate (Thermo Scientific).

2.8. Sodium Azide Treatment and Blotting for Specific Proteins

1. Sodium Azide (NaN_3, Mallinkrodt Baker Inc., Paris, KY) is dissolved in water at 10% (w/v). Store at 4°C and use at 200×.
2. Blocking buffer and antibody diluting solution: 5% (w/v) ReliaBLOT® Block (Bethyl Laboratories, Inc., Montgomery, TX, USA) in TBS-Tween 20 (TBS-T) (Bethyl Laboratories, Inc.).
3. Primary antibody: NudC R2 antibody (affinity-purified rabbit anti-NudC C terminus 15 aa peptide antibody, 1:3,000) (8).
4. Secondary antibody: ReliaBLOT® HRP Conjugate (Bethyl Laboratories, Inc.).

3. Methods

We employed a well-established double thymidine block and release synchronization protocol to enrich for mitotic cells. The synchronized prometaphase-like cells were used for immunoprecipitation of acetylated proteins using a commercially available anti-acetyl-Lysine antibody, followed by their identification by LC-ESI-MS/MS analysis (2). After SDS-PAGE resolution of proteins, instead of electrophoretic transfer of the samples to a nitrocellulose filter, the protein bands are stained with GelCode™ Blue Stain Reagent (Thermo Scientific), digested in-gel by trypsin, and processed for LC-ESI-MS/MS analysis (2).

Two approaches are used to confirm protein acetylation: (1) IP with another anti-acetyl-Lysine antibody followed by Western blotting for the protein of interest and (2) reciprocal protein-specific IP in which the cell lysate was reduced to maximally expose peptide epitopes, diluted tenfold in RIPA Buffer, and IP with an anti-peptide protein-specific antibody followed by Western blot analysis for acetyl-Lysine reactivity. As protein acetylation is a dynamic process that is positively regulated by histone acetyltransferases (HATs) and negatively regulated by HDACs, it is important to supplement all wash buffers and sample buffers with sodium butyrate (NaB), a general HDAC inhibitor, to prevent spontaneous protein deacetylation during sample collection, preparation, and analysis.

3.1. Enrichment for Mitotic HeLa Cells (See Note 4)

[Day 1] Seed culture

1. Late afternoon—Subculture HeLa cells in pre-warmed DMEM supplemented with 10% (v/v) FBS (see Note 5). Split cells 1:8 to 1:10 from 80 to 90% confluent culture in a 10 cm dish.

[Day 2] First thymidine block

2. 6:00 p.m.—Replace medium with 37°C (see Note 6) pre-warmed media supplemented with 2 mM thymidine (see Note 7). Incubate cells for 15 h (can incubate for 14–18 h; in this and subsequent steps that require timed incubations, choose one timing for consistency).
3. For the next day's work, incubate just enough PBS and medium in the 37°C water bath overnight (see Note 8).

[Day 3] Release/Second thymidine block

4. 9:00 a.m.—Wash cells (see Note 9) twice with 10 mL of 37°C pre-warmed PBS and culture them in 37°C pre-warmed regular medium for 10.5 h (can release for 10.5–11 h).
5. 7:30 p.m.—Replace medium with 37°C pre-warmed medium supplemented with 2 mM thymidine and incubate cells for 13.5 h (can incubate for 13–15 h).
6. For the next day's work, incubate just enough PBS and medium in the 37°C water bath overnight.

[Day 4] Release/Nocodazole enrich/Harvest

7. 9:00 a.m.—Wash cells twice with 10 mL of 37°C pre-warmed PBS and culture in pre-warmed regular medium for 5 h (can release for 5–5.5 h).
8. 2:00 p.m.—Replace medium with pre-warmed media supplemented with 50 ng/mL Nocodazole (see Note 10) with or without 100 nM or 500 nM of Apicidin (see Note 11). Incubate cells for 3.5 h (can incubate for 3–5 h).
9. 5:30 p.m.—Harvest mitotic cells (see Note 12) by "shake off," which is gently pipetting the rounded mitotic cells with a Beral pipet. An example image of randomly cycling cells (most are interphase, "flat" cells) versus DTB synchronized cells (most are mitotic, "round" cells) is shown in Fig. 1.
10. Wash cells twice with 4°C cold PBS (see Note 13). Remove PBS completely and either process the cell pellets for immediate analysis, or store the cell pellets (see Note 14) at either –20°C for use within a month or –80°C for long-term storage.

3.2. Lysate Preparation

3.2.1. Lysate Preparation for Anti-Acetyl-Lysine Immunoprecipitation

1. Supplement IP Lysis Buffer with various inhibitors (see Note 15).
2. Add 2× volume of IP Lysis Buffer + Inhibitors to each cell pellet.
3. Vortex until homogenous.

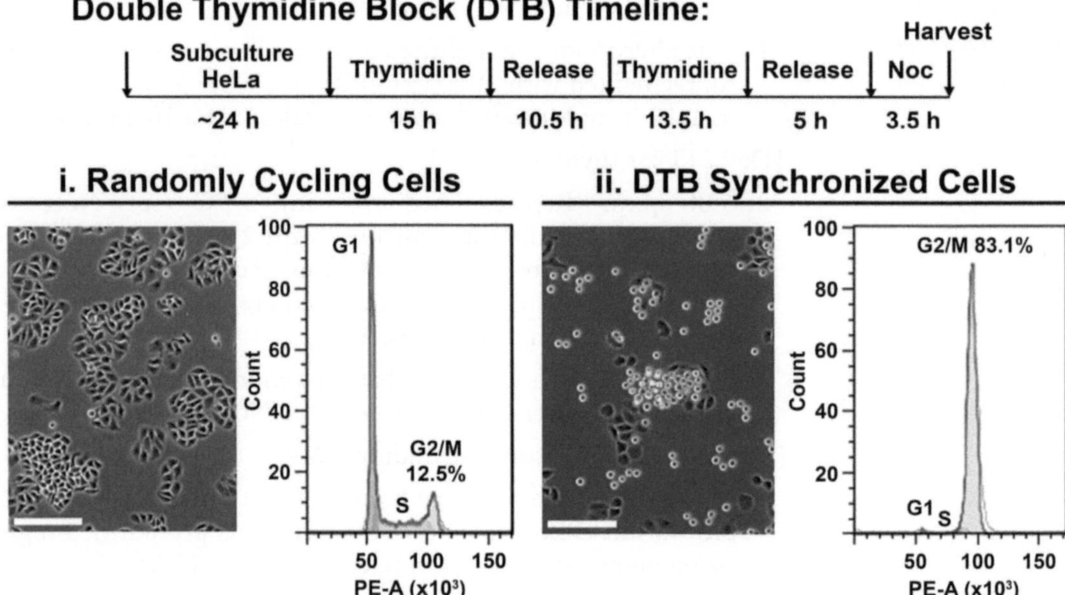

Fig. 1. Schematic of cell synchronization protocol and representative brightfield images of mitotic cells. Phase-contrast images show the mitotic population as rounded cells. Images were acquired using a Nikon TE2000 microscope system (Nikon Instruments Inc., Melville, NY, USA). Bars, 200 μm. Flow cytometry analysis shows the cell cycle distribution of (1) randomly cycling cells versus (2) mitotic cells synchronized by a double thymidine block and release (DTB) protocol. About 12.5% of the randomly cycling cells versus more than 80% of the DTB synchronized cells are in the G2/M phase of the cell cycle.

4. Use a 27 × 1/2″ needle attached to a 1 mL syringe to shear (see Note 16) the cell solution 12 times.
5. Incubate on ice for 30 min.
6. Vortex and spin down cell solution at >14,000 × *g* for 15 min at 4°C.
7. Transfer supernatant (see Note 17) into a new 1.5-mL microcentrifuge tube.
8. Determine protein concentration by Bradford Assay.
9. Proceed to anti-acetyl-Lysine immunoprecipitation.

3.2.2. Reduced Lysate Preparation for Immunoprecipitation for a Specific Protein

1. Prepare reducing buffer (see Note 18).
2. Add 200 μL reducing lysis buffer to the cell pellet.
3. Mix well by vortexing. Incubate on ice for 5 min. Meanwhile, supplement RIPA Buffer as below.
4. Supplement RIPA Buffer with various inhibitors and DNase I (see Note 19).
5. Dilute (see Note 20) the reducing lysis buffer suspension with 1.8 mL RIPA Buffer + Inhibitors + DNase I.

6. Needle shear lysed suspension with a 1 mL syringe (27 × 1/2″ needle) 12 times.
7. Incubate on ice for 5 min.
8. Proceed to reduced lysate protein-specific immunoprecipitation (see Note 21).

3.3. Immunoprecipitation

3.3.1. Anti-Acetyl-Lysine Immunoprecipitation for LC-ESI-MS/MS

1. Wash 1 mL IP beads with 1 mL IP Lysis Buffer and spin at ~300 × *g* for 30 s at 4°C. Repeat three times. After the final wash, reconstitute with 1:1 volume of beads:buffer (e.g., 500 μL beads:500 μL buffer).
2. Preclear 9 mg lysate (see Note 22) with 200 μL washed IP beads (see Note 23) in a microcentrifuge tube for 1 h on a rotator at 4°C.
3. Spin down beads at >13,000 × *g* for 5 min at 4°C.
4. Transfer 800 μL supernatant (see Note 24) to a new 1.5-mL microcentrifuge tube without disturbing the beads.
5. To each tube, add 80 μL of washed IP beads and 4.5 μL of anti-acetyl-Lysine rabbit antiserum.
6. Incubate overnight on a rotator at 4°C.
7. Spin down IP reactions at ~300 × *g* for 5 min at 4°C and use a pipet to remove the supernatant without disturbing the beads (see Note 25).
8. Wash beads with 400 μL IP Lysis Buffer supplemented with NaB at ~300 × *g* for 5 min at 4°C. Aspirate out (see Note 26) supernatant without disturbing beads. Repeat four times. On the last wash, remove as much of the IP Lysis Buffer as possible without aspirating the beads.
9. Add 70 μL of Loading Buffer (see Note 27) and boil at 95°C for 5 min.
10. Spin down beads at >14,000 × *g* for 2 min at room temperature, and load 30 μL for separation by SDS-PAGE.

3.3.2. Anti-Acetyl-Lysine Immunoprecipitation for Western Blot Analysis

1. Wash 1 mL IP beads with 1 mL IP Lysis Buffer and spin at ~300 × *g* for 30 s at 4°C. Repeat three times. After the final wash, reconstitute with 1:1 volume of beads:buffer (e.g., 500 μL beads:500 μL buffer).
2. Preclear 4.4 mg lysate (see Note 28) with 80 μL washed IP beads (see Note 23) in a microcentrifuge tube for 1 h on a rotator at 4°C.
3. Spin down beads at >14,000 × *g* for 5 min at 4°C.
4. Transfer 800 μL supernatant (see Note 24) to a new 1.5-mL microcentrifuge tube without disturbing the beads.
5. To each tube with supernatant, add 80 μL of washed IP beads and 4 μg (i.e., 4 μL of the 1 μg/μL stock) of the Millipore anti-acetyl-Lysine, Clone 4G12 mouse monoclonal antibody.

6. Incubate overnight on a rotator at 4°C.
7. Spin down IP reactions at ~300 × *g* for 5 min at 4°C and use a pipet to remove the supernatant without disturbing the beads (see Note 25).
8. Wash beads with 400 μL IP Lysis Buffer supplemented with NaB at ~300 × *g* for 5 min at 4°C. Aspirate out (see Note 26) supernatant without disturbing beads. Repeat four times. On the last wash, try to remove as much of the IP Lysis Buffer as possible without aspirating the beads.
9. Add 70 μL of Loading Buffer (see Note 27) and boil at 95°C for 5 min.
10. Spin down beads at >14,000 × *g* for 2 min at room temperature, and load 30 μL for separation by SDS-PAGE.

3.3.3. Protein Immunoprecipitation Using Reduced Cell Lysate for Western Blot Analysis

1. Wash 1 mL IP beads with 1 mL RIPA Buffer and spin at ~300 × *g* for 30 s at 4°C. Repeat three times. After the final wash, reconstitute with 1:1 volume of beads:buffer (e.g., 500 μL beads:500 μL buffer).
2. Preclear 400 μL reduced lysate with 40 μL washed IP beads (see Note 23) for 1 h on a rotator at 4°C.
3. Spin down beads at >14,000 × *g* for 5 min at 4°C.
4. Transfer 400 μL supernatant (see Note 24) to a new 1.5-mL microcentrifuge tube without disturbing the beads.
5. To each tube with supernatant, add 40 μL of washed IP beads and 1 μL of NudC G1 antibody.
6. Incubate overnight on a rotator at 4°C.
7. Spin down IP reactions at ~300 × *g* for 5 min at 4°C and pipet (see Note 25) out supernatant without disturbing beads.
8. Wash beads with 400 μL RIPA Buffer supplemented with NaB at ~300 × *g* for 5 min at 4°C. Aspirate out (see Note 26) supernatant without disturbing beads. Repeat four times. On the final wash, try to remove as much of the RIPA Buffer as possible without aspirating the beads.
9. Add 35 μL of Loading Buffer and boil at 95°C for 5 min.
10. Spin down beads at >14,000 × *g* for 2 min at room temperature and load 30 μL for separation by SDS-PAGE.

3.4. SDS-Polyacrylamide Gel Electrophoresis

1. These instructions are based on the use of an Invitrogen XCell *SureLock*™ Mini-Cell leak-free electrophoresis system. Prior to use, make sure to wash the system while wearing clean gloves with water to rinse off any possible contaminants.
2. Prepare the 1× Running Buffer (see Note 29) by diluting 50 mL of the 20× NuPAGE® MOPS Buffer with 950 mL of water in a 1 L screw cap glass bottle. Tighten cap and mix by inverting.

3. Cut open the precast gradient gel cassette pouch with scissors and drain away the gel packaging buffer to remove the precast gradient cassette from the pouch (see Note 30).
4. Peel off the tape covering the slot on the back of the gel cassette. In one fluid motion, gently pull the comb out of the gel cassette (see Note 31). This exposes the gel loading wells.
5. Set up the XCell *SureLock*™ Mini-Cell by lowering the Buffer Core into the Lower Buffer Chamber so that the negative electrode fits into the opening in the gold plate on the lower buffer chamber. Insert gels on either side of the Buffer Core (see Note 32). Lower the Gel Tension Wedge into the Lower Buffer Chamber and push the Gel Tension Wedge handle in towards the Buffer Core to secure the gels. This forms a leak-free Upper Buffer Chamber consisting of the Buffer Core flanked by the gels.
6. Fill the Upper Buffer Chamber to the rim with 1× Running Buffer and add about 200 mL of 1× Running Buffer to the Lower Buffer Chamber.
7. Use a 1-mL pipette to gently wash the gel cassette wells with 1× Running Buffer (see Note 33). Repeat.
8. Load 7 μL of the pre-stained molecular weight marker to a well and load 30 μL of each IP sample in a well.
9. Carefully top off the Upper Buffer Chamber without disturbing the samples in the gel wells.
10. Align the lid to the Buffer Core and push down until secure. The lid can only be firmly seated if the (–) electrode is aligned over the banana plug on the right. If the lid is not properly seated, no power will go through the mini-cell.
11. With the power OFF, connect the electrode cords to the power supply with the red to the (+) jack and the black to the (–) jack.
12. Turn ON the power. Run the gel at 150 V for ~1.5 h. Stop the power source when the dye fronts (blue and orange) run to the slot on the back of the gel where the tape was.
13. At the end of the run, turn off the power and disconnect the cables from the power supply. Remove the lid and unlock the Gel Tension Wedge. Proceed to stain or transfer.

3.5. Coomassie Blue Staining and In-Gel Tryptic Digestion for LC-ESI-MS/MS

3.5.1. Coomassie Blue Staining

1. Carefully remove SDS-PAGE gel from cassette and transfer into a clean gel box without touching the gel or wear clean gloves and only handle the edge of the gel (see Note 34).
2. Wash SDS-PAGE gel three times with 200 mL of water for 5 min at room temperature with gentle shaking.
3. Fix SDS-PAGE gel with Fixing Solution for 15 min at room temperature with gentle shaking.
4. Wash SDS-PAGE gel three times with 200 mL of water for 5 min at room temperature with gentle shaking.

5. Mix the GelCode Blue Stain Reagent solution immediately before use by gently swirling the bottle several times. Do not shake bottle to mix the solution.
6. Add 20 mL of GelCode Blue Stain Reagent and incubate for 1 h at room temperature with gentle shaking.
7. Destain/enhance band intensity by washing gel with 200 mL of water for 1–2 h. Change water every 30 min.

3.5.2. In-Gel Protein Tryptic Digestion for LC-ESI-MS/MS

1. Excise protein bands from SDS-PAGE gel for trypsin digestion as previously described (9–11).
2. Briefly, the gel slices were washed two times in a mixture of analytical-grade water and 1:1 v/v 0.1 M NH_4HCO_3/ACN (Sigma-Aldrich) in twice the gel volume for 15 min with agitation.
3. After removing the wash solution completely, the gel pieces were covered with ACN and left to shrink and stick together.
4. After removing ACN completely, the gel pieces were rehydrated in 0.1 M NH_4HCO_3 for 10 min and incubated in an equal volume of ACN for another 10 min for a final concentration of 1:1 v/v of 0.1 M NH_4HCO_3/ACN. This procedure is performed at room temperature.
5. Once drained of all liquids, the gel pieces were dried down in a vacuum centrifuge. After drying, the gel pieces were swelled in a solution of 10 mM DTT and 55 mM iodoacetamide in 0.1 M NH_4HCO_3.
6. The alkylating solution was then removed and the gel pieces were washed and dried as above.
7. Gel pieces were digested in trypsin for 24 h at 37°C.
8. The peptides were extracted from the gel pieces by the addition of 10 μL 25 mM NH_4HCO_3, 5 μL 5% (v/v) formic acid, and 5 μL ACN, dried down, and dissolved in a mixture containing formic acid:water:ACN:trifluoroacetic acid (0.1:95:5:0.01 v/v) for LC-MS/MS analysis.
9. Tryptic digests were separated with a reverse-phase column (C-18 PepMap 100, LC Packings/Dionex, Sunnyvale, CA, USA).
10. The column eluate was directly introduced onto a QSTAR XL mass spectrometer (Applied Biosystems, Foster City, CA, USA, and Sciex, Concord, ONT, Canada) via ESI.
11. Candidate ion selection and data collection were performed as previously described (10, 11). Half second MS scans (300–1,500 Thompson) were used to identify candidates for fragmentation during MS/MS scans. Up to five 1.5-s MS/MS scans (65–1,500 Thompson) were collected after each scan.

An ion was assigned a charge in the range of +2 to +4. The dynamic exclusion was 40.

12. Protein identifications were completed with ProteinPilot (versions 1.0 and 2.0, Applied Biosystems and Sciex), using a setting with 1.5 Da mass tolerance for both MS and MS/MS and the human "RefSeq" databases from NCBI (http:www.ncbi.nlm.nih.gov/RefSeq). ProteinPilot is the successor to ProID and ProGroup, and uses the same peptide and protein scoring method. Briefly, given a protein score, *S*, the likelihood that the protein assignment is ***incorrect*** is 10^{-S}. Scores above 2.0 require that at least two sequence-independent (unique) peptides will be identified (10, 11). An example of the list of acetylated proteins in mitosis is shown in Table 1.

3.6. Western Blotting for Acetylation of a Specific Protein

1. The following procedure is used for immunoprecipitation of a specific protein (prepared from reduced lysates) followed by blotting for acetylation of the protein using an anti-acetyl-Lysine antibody.
2. The samples that have been resolved by SDS-PAGE are electrophoretically transferred to nitrocellulose membranes. The following protocol is based on the use of a BIO-RAD Mini Trans-Blot® Electrophoretic Transfer Cell.

Table 1
Proteins identified by acetyl-lysine immunoprecipitation and LC-ESI-MS/MS in mitotic HeLa cells

Protein names	gi	Biological process	Molecular function	[a]Known Ac status
Histones				
Histone H2A.1	31980	Chromosome organization	DNA binding	(5, 6)
Histone H3	31982	Chromosome organization	DNA binding	(5, 6)
Histone H4	31995	Chromosome organization	DNA binding	(5, 6)
Translation				
40S ribosomal protein S25	51338648	Translation	Catalyzes protein synthesis	(5)
40S ribosomal protein S26	266970	Translation	Catalyzes protein synthesis	(5)
Eukaryotic initiation factor 4 gamma (eIF4G)	219613	Translation	Initiates translation	(5)
Polyadenylate-binding protein	35570	Translation	Initiates translation	(5)
Polyadenylate-binding protein II	74706522	Translation	Initiates translation	(5)

(continued)

Table 1 (continued)

Protein names	gi	Biological process	Molecular function	[a]Known Ac status
RNA binding				
DEAD (Asp-Glu-Ala-Asp) box polypeptide 3, X-linked	57209229	mRNA processing	Unwinds RNA	(5)
Heterogeneous nuclear ribonucleoprotein AO	773644	mRNA processing/transport	RNA binding	
Heterogeneous nuclear ribonucleoprotein A1	296650	mRNA processing/transport	RNA binding	(5, 6)
Heterogeneous nuclear ribonucleoprotein A/B	13528732	mRNA processing/transport	RNA binding	
Heterogeneous nuclear ribonucleoproteins A2/B1	133257	mRNA processing/transport	RNA binding	(5)
Heterogeneous nuclear ribonucleoproteins C1/C2	108935845	mRNA processing/transport	RNA binding	(5)
Heterogeneous nuclear ribonucleoproteins C3	3334899	mRNA processing/transport	RNA binding	
Heterogeneous nuclear ribonucleoprotein D0	13124489	mRNA processing/transport	RNA binding	(5)
Heterogeneous nuclear ribonucleoprotein D-like	39644771	mRNA processing/transport	RNA binding	(5)
Heterogeneous nuclear ribonucleoprotein R	2697103	mRNA processing/transport	RNA binding	(5)
Heterogeneous nuclear ribonucleoprotein U	32358	mRNA processing/transport	RNA binding	(5)
Heterogeneous nuclear ribonucleoprotein U-like	21536326	mRNA processing/transport	RNA binding	(5)
mRNA-binding protein CRDBP	7141072	RNA metabolism	RNA binding	
RNA helicase A	307383	RNA processing	Unwinds dsDNA and dsRNA	
Translocation in Liposarcoma (TLS)	386157	mRNA processing/transport	RNA binding	
Transcription				
ATP-dependent helicase SMARCA4	116242792	Transcription	Transcriptional coactivator	
Cell cycle				
Actin-binding protein anillin	8489881	Cytokinesis	Acto-myosin ring assembly	(5)
APC1 (Tsg 24 protein)	11967711	Cell cycle	Anaphase promotion	
Cip 1-interacting zinc finger protein (Ciz1)	6136800	Cell cycle	Inhibits cdks	
Cyclin B3	14275558	Meiotic prophase I	Interacts with cdk2	
NudC	619907	Cell cycle	Mitotic progression	(5)
Tripin/Shugoshin 2	23986276	Cell cycle	Protects centromeres	
Zizimin-3 (dedicator of cytokinesis, protein 10)	32469767	Cytokinesis	Guanine exchange factor	

(continued)

Table 1 (continued)

Protein names	gi	Biological process	Molecular function	[a]Known Ac status
DNA damage				
Ku70	125729	DNA damage response	Non-recombinational repair	(5)
Ku80	35038	DNA damage response	Non-recombinational repair	(5)
NF45 protein	532313	DNA damage response	Non-recombinational repair	
Mitochondrial single-stranded DNA-binding protein	188856	DNA damage response	Protects ssDNA	
Chaperones				
Stress-70 protein, mitochondrial, precursor	21264428	Stress/metabolic response	Protein folding	(5, 6)
71 kDa heat-shock cognate protein	32467	Stress/metabolic response	Protein folding	(5)
Heat-shock 70 kDa protein 5, precursor	14916999	Stress/metabolic response	Protein folding	(5)
Metabolism				
Placental alkaline phosphatase 1, precursor	130737	Basic phosphatase	Hydrolase enzyme	
Carbamoyl-phosphate synthase I, mitochondria	4033707	Urea cycle	Degrades ammonia	(6)
Carbamoyl-phosphate synthetase II, cytosol	1228049	Pyrimidine biosynthesis	Degrades glutamine	(5)
Immune response				
Bone marrow stromal cell antigen 2	506861	Humoral immune response	B cell growth and development	
Complement component 3, precursor	119370332	Inflammation	Complement activation	
Fibrinogen alpha chain, precursor	1706799	Blood clotting	Platelet aggregation cofactor	
Migration inhibitory factor-related protein 14 (MRP14)	34771	Inflammation	Calcium binding	
Structural				
Desmoplakin I	1147813	Intercellular junction	Binds desmosomes	
Ifapsoriasin (filaggrin 2)	59939295	Structural	Binds keratin	
Keratin type II	34069	Structural	Intermediate filament	
Myosin, heavy chain 9, non-muscle	3169000	Structural	Contractile protein	(5)

(continued)

Table 1 (continued)

Protein names	gi	Biological process	Molecular function	[a]Known Ac status
Miscellaneous				
Annexin A2	113950	Cell growth; signal transduction	Phospholipid binding	(5)
F-box only protein 11 isoform 1	30089926	Protein ubiquitination	Interacts with Skp1	
Granulin	31193	Cell growth	Secreted, glycosylated peptide	
FLJ00195	18676594	N/A	N/A	
Hypothetical protein	21739818	N/A	N/A	

[a]Proteins previously shown to be acetylated in response to a 24-h treatment with an HDAC inhibitor (5, 6)
Three samples prepared as in (Fig. 1) (9–15 mg each) from two independent experiments were immunoprecipitated with anti-acetyl lysine polyclonal antibody (Millipore), resolved on SDS-PAGE, and stained with GelCode Blue. 24 gel slices were processed for LC-ESI-MS/MS analysis. Protein identifications were completed with ProteinPilot (versions 1.0 and 2.0, Applied Biosystems and Sciex). Proteins identified by acetyl-lysine immunoprecipitation and LC-ESI-MS/MS in mitotic HeLa cells

3. Prepare 1× Transfer Buffer (see Note 35) by supplementing 100 mL of the 10× Transfer Buffer with 200 mL of methanol and diluting with 700 mL of water in a 1 L screw cap glass bottle. Tighten cap and mix by inverting.
4. Set up the BIO-RAD Mini Trans-Blot® Electrophoretic Transfer Cell by placing the Electrode Module (see Note 36) inside the Buffer Tank. Fill the Buffer Tank with 200 mL with 1× Transfer Buffer. Place a magnetic stir bar (see Note 37) directly between the Electrode Module.
5. A tray of 1× Transfer Buffer is prepared that is large enough to lay out a Transfer Cassette (such as the use of a glass pyrex dish). Submerge a piece of fiber pad and two 3 in. × 4 in. sheets of Whatman® chromatography paper placed on the clear and the grey side of the Transfer Cassette. A sheet of the nitrocellulose membrane, 3 in. × 3.25 in., is submerged on top of the Whatman® chromatography paper on the white side of the Transfer Cassette.
6. Remove a gel cassette from the mini-cell. Handle gel cassettes by their edges only. Carefully insert a metal spatula into the narrow gap between the two plates of the cassette. Push up and down gently on the spatula to separate the plastic plates. You will hear a cracking sound which means you have broken the bonds which hold the plates together. Rotate and repeat until you have broken the bonds on all sides and the two plates are completely separated.

7. Slowly open the cassette so that the gel may adhere to one plate. Remove and discard the plate without the gel, allowing the gel to remain on the other plate. Use a clean razor blade to cut off the gel well separations and also cut the gel so that the slot on the back of the gel where the tape was is not transferred.
8. Carefully use the spatula to separate the bottom corners of the gel from the plastic plate. Gently lay the gel on top of the Whatman® chromatography paper on the grey side of the Transfer Cassette so that it is submerged.
9. Place the presoaked nitrocellulose membrane from the clear side of the cassette on top of the gel so that it is submerged. Place the two presoaked Whatman® chromatography paper on the white side of the cassette on top of the nitrocellulose membrane. Gently roll over the Whatman® chromatography paper with a water-filled 15-mL conical tube to remove bubbles (see Note 38) between the nitrocellulose membrane and the gel.
10. Place the presoaked fiber pad from the clear side of the cassette on top of the transfer sandwich so that it is submerged. Gently roll over the fiber pad with the 15-mL conical tube to ensure that bubbles are not trapped between the nitrocellulose and the gel. Close the clear side of the Transfer Cassette on top of the sandwich and lock the cassette closed with the white latch.
11. Place the Transfer Cassette in the Electrode Module so that the grey side of the cassette is facing the black side of the Electrode Module.
12. Repeat steps 3–9 for the second gel. If there is no other gel, proceed to step 13.
13. Insert the frozen Bio-Ice cooling unit (see Note 39) in the Buffer Tank and completely fill the tank with 1× Transfer Buffer.
14. Carefully transfer the Tank Buffer and all its contents to a magnetic stir plate in a beer cooler or cold room (see Note 40). Turn on the stir plate to the maximum speed.
15. Align the lid to the Electrode Module so that the red cable matches up to the electrode protruding from the red side of the Electrode Module and the black cable matches up with the electrode protruding from the black side of the Electrode Module.
16. With the power OFF, connect the cable cords to the power supply with the red to (+) jack and the black to (−) jack.
17. Turn ON the power. Transfer the gel at 100 V for ~1.5 h.
18. At the end of the run, turn off the power and disconnect the cables from the power supply. Discard the 1× Transfer Buffer. Remove the Transfer Cassettes from the Electrode Module. Open the cassette to retrieve the nitrocellulose membrane (see Note 41). Discard the gel and Whatman® Chromatography paper.

19. Incubate the nitrocellulose membrane in 6 mL of Blocking Buffer for 1 h on a rocking platform at room temperature.
20. Discard the blocking buffer. Incubate the membrane with 1:2,000 dilution of the anti-acetyl-Lysine (rabbit polyclonal antibody) or acetylated lysine mouse mAb in blocking buffer overnight at 4°C on a rocking platform.
21. The primary antibody is then removed (see Note 42) and the membrane is washed three times for 10 min each with TBS-T.
22. The secondary antibody is freshly prepared for each experiment as a 1:20,000 dilution in blocking buffer and added to the membrane for 1.5 h at room temperature on a rocking platform.
23. The secondary antibody is discarded and the membrane is washed three times for 10 min each with TBS-T.
24. After the final wash, 2 mL aliquots of each portion of the ECL reagent are transferred onto the nitrocellulose. The nitrocellulose is first rotated by hand and then incubated in the ECL mixture for 1 min at room temperature on a rocking platform.
25. The nitrocellulose is removed from the ECL reagents, blotting the edge on a paper towel, and then placed between the leaves of a sheet protector that is adhered inside an X-ray film cassette (see Note 43).
26. Proceed to the dark room to expose a film on the blot for a suitable exposure time, typically 10 and 30 s (longer exposures may be used as needed). An example of the acetylation of NudC in mitotic cells is shown in Fig. 2.

3.7. Sodium Azide Treatment and Blotting for Specific Proteins

1. Once a satisfactory exposure for the result of anti-acetyl-Lysine reactivity has been obtained, the membrane is then treated with sodium azide to deactivate the HRP conjugate and then reprobed with a protein-specific antibody. Here we use NudC as an example. This provides a loading control for equal recovery of the IP samples through the procedure.
2. The nitrocellulose membrane is washed three times for 10 min each with TBS-T, and then incubated in TBS-T supplemented with 0.05% NaN_3 for 30 min at room temperature on a rocking platform.
3. Wash the nitrocellulose membrane three times for 10 min each with TBS-T. Block with 6-mL of ReliaBLOT® blocking buffer for 1 h at room temperature on a rocking platform.
4. The membrane is then ready to be reprobed with anti-NudC R2 antibody diluted 1:3,000 in ReliaBLOT® blocking buffer overnight at 4°C on a rocking platform.
5. The primary antibody is then removed (see Note 44) and the membrane is washed three times for 10 min each with TBS-T.

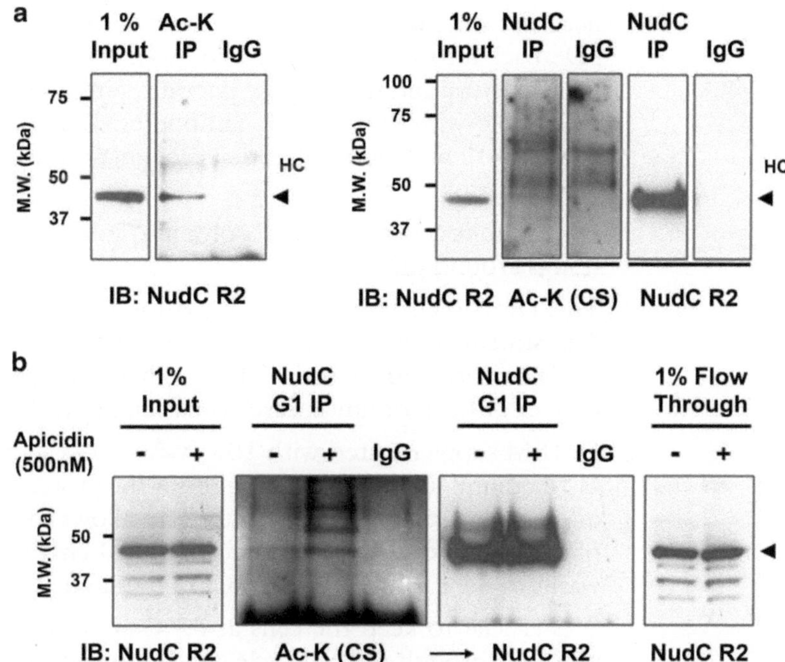

Fig. 2. Acetylation of NudC in mitosis. (**a**) Mitotic cell lysates (2 mg), prepared as in Fig. 1, were immunoprecipitated with a second anti-acetyl-Lysine antibody (Millipore, mAb) (Ac-K IP) and immunoblotted for NudC (NudC R2) (*left, arrowhead*). For confirmation, reciprocal protein-specific immunoprecipitation of NudC (*right*) was performed using reduced lysates. Western blot analysis shows acetylation through reactivity with a third anti-acetyl-Lysine antibody (Cell Signaling, mAb) (*arrowhead*), and re-immunoblotted for NudC (NudC R2) to show efficiency of immunoprecipitation and protein loading. (**b**) Mitotic HeLa cells were synchronized as in **a** and treated with or without apicidin (500 nM), a small molecule inhibitor for HDAC2/3, for 3.5 h prior to harvest to block protein deacetylation during the peak of mitosis. Reduced mitotic lysates were immunoprecipitated for NudC (NudC G1) and immunoblotted for acetylation (*arrowhead*) and re-immunoblotted for NudC as in Fig. 2b (*right*).

6. Incubate the nitrocellulose membrane in ReliaBLOT® HRP conjugate diluted at 1:3,000 in ReliaBLOT® blocking buffer for 2 h at room temperature on a rocking platform.
7. Repeat with washes and ECL detection as above. An example of the NudC acetylation result is shown in Fig. 2.

4. Notes

1. Unless stated otherwise, all solutions should be prepared in water that has a resistivity of 18.2 MΩ cm and total organic content of less than five parts per billion. This standard is referred to as “water” in this text.
2. Reducing buffer (1% (w/v) SDS, 5 mM EDTA, 10 mM DTT) is made by diluting 10% (w/v) SDS, 500 mM EDTA pH 8.0, and 100 mM DTT. Check 10% (w/v) SDS solution for precipitation before use. If SDS has precipitated out of solution,

heat and stir the solution on a heated stir plate to dissolve the SDS back into solution. Reduced lysate is used to disrupt protein structure and fully expose peptide epitopes before protein-specific IP with anti-peptide antibodies. Do not use reduced lysate for IP with antibodies that recognize epitopes of protein conformation.

3. Deoxycholic acid does not dissolve into solution until the solution is around pH 6–7.
4. This enrichment protocol is timed for HeLa cells' cell cycle. Adjustments to incubation and release times would need to be made to enrich for mitotic cells in other cell types (e.g., hepatocytes or hepatocellular carcinoma cell lines).
5. DMEM supplemented with 10% (v/v) FBS will be referred to as "medium" or "media" throughout the protocol. Also, 5 mL of media is used per 10-cm plate. This allows the conservation of media usage and allows media in the 10-cm plates to quickly warm to 37°C once in the 5% CO_2, humidified incubator.
6. It is crucial to keep the cells at 37°C at all times. HeLa cells proceed through the cell cycle at a slower rate as the temperature deviates from 37°C (12).
7. DNA synthesis is inhibited by treatment with a high concentration of thymidine (13, 14). Thus, thymidine treatment is used to block cells at the G1/S transition phase (15, 16). Do not leave drug-containing medium at 37°C for more than 1 h as the drug may lose activity. Pre-warm medium to 37°C first. Thaw 200 mM thymidine stock at 37°C for 15–30 min and vortex until precipitates go back into solution. Add thymidine to medium and mix well right before use.
8. Pre-warming the PBS and media overnight allows for immediate washing at 9 a.m. the next morning. Alternatively, come in at 8 a.m. to pre-warm the PBS and media.
9. The metal surface of the sterile hoods is a thermal conductor (i.e., readily absorbs heat). Therefore, when moving cell plates to the sterile hoods, place cell plates on a paper box or a Styrofoam/plastic conical rack (air, Styrofoam, and plastic are good insulators). Work with four to six plates at a time to quickly change media and return cell plates to the incubator. Also, have the Bunsen burner on in the sterile hood to maintain a warm temperature (> room temperature) in the hood.
10. Nocodazole is a microtubule-destabilizing drug (17). Nocodazole is used to prevent the cells from progressing through mitosis and enriches cells in a prometaphase-like state (early mitosis). Pre-warm medium to 37°C first. Thaw 500 μg/mL Nocodazole stock in the hood for 15–30 min. Add Nocodazole to medium and mix well right before use.

11. Apicidin is a selective class I HDAC inhibitor that specifically inhibits HDAC3 at 100 nM and HDAC2/3 at 500 nM (18). Pre-warm medium to 37°C first. Warm 500 μM apicidin stock in the 37°C water bath for 15–30 min. Add Apicidin to medium supplemented with 50 ng/mL Nocodazole and mix well right before use.
12. If the cells were efficiently synchronized, there should be ~80% rounded mitotic cells floating in the media. You can also wait to harvest at 6:00 p.m. to enrich for even more mitotic cells. Interphase cells are also affected by nocodazole treatment, so be *very gentle* when pipetting and collecting the mitotic cells.
13. PBS washing after harvesting cells also removes nocodazole in the medium, so nocodazole-arrested cells may start to progress through the cell cycle during the time between the wash and when the cells are finally frozen. Cold temperature destabilizes microtubules (19), so washing with cold PBS after harvesting will continue to arrest the cells in a prometaphase-like state.
14. Collect five plates of cells and media in a 50-mL conical tube. Then transfer to a 1.5-mL or a 2.0-mL microcentrifuge tube during cold PBS wash. Store the cell pellet in a 1.5-mL microcentrifuge tube for the anti-acetyl-Lysine IP and in a 2-mL microcentrifuge tube for the reduced lysate protein-specific IP.
15. Prepare enough Lysis Buffer supplemented with inhibitors for 2× volume of the cell pellet (i.e., for 100 μL pellet, add 200 μL Lysis Buffer + Inhibitors).
16. Needle shearing refers to passing the lysed suspension through the needle attached to a 1-mL syringe. Needle shearing introduces DNA fragmentation and allows chromatin-interacting proteins to come into solution. Needle shearing is performed until the viscosity is reduced to manageable levels. If the DNA is not fully sheared, it can interfere with the separation of the pellet and the supernatant after centrifugation.
17. Be careful not to disturb the pellet. Use a 200-μL gel loading pipet tip to carefully extract the supernatant and transfer it to a new 1.5-mL microcentrifuge tube.
18. Prepare enough Reducing Buffer for each cell pellet (i.e., for two pellets, prepare 450 μL Reducing Buffer).
19. Prepare enough RIPA Buffer supplemented with inhibitors and DNase I for each cell pellet (i.e., for two pellets, prepare 4 mL RIPA Buffer + Inhibitors + DNaseI).
20. The excess 1% (v/v) Triton X-100 in the RIPA Buffer quenches the SDS in the reducing buffer. This protects the antibody from being reduced during immunoprecipitation.
21. Due to the NP40 in the RIPA Buffer, one cannot accurately determine the protein concentration (20).

22. Make 10 mg lysate.
23. Transfer IP bead slurry with a 200-μL tip that has been cut off. Cut the 200-μL tips 5 mm from the tip with a clean razor blade to easily transfer beads.
24. Be careful not to disturb the beads. Use a 200-μL gel loading pipet tip to carefully extract supernatant and transfer into the new 1.5-mL microcentrifuge tube. Discard the beads.
25. Use a 200-μL gel loading pipet tip to carefully pipet out the supernatant without disturbing the beads. Save the supernatant (flow through) to run 1% flow through parallel with 1% input and IP to determine IP efficiency. Store at −20°C.
26. Connect a 200-μL gel loading pipet tip to the end of an aspirator and carefully aspirate out the supernatant without disturbing the beads.
27. This allows the loading of two gels with 30 μL each of the immunoprecipitated sample, enabling the Western blotting for two different proteins for immunoprecipitation by anti-acetyl-Lysine antibody.
28. Make 4.4 mg of proteins in a final volume of 800 μL by adding IP Lysis Buffer supplemented with NaB.
29. Unused 1× Running Buffer can be stored at room temperature (label bottle as "New MOPS"). Used 1× Running Buffer can be stored at room temperature and reused once to fill the Lower Buffer Chamber (label bottle as "Used MOPS"). Always fill the Upper Buffer Chamber with New MOPS.
30. Always handle the precast gel cassette by its edges only.
31. Hold on to the sides of the gel cassette with both hands and gently push comb up with both thumbs.
32. Insert the gels with the front (the shorter, well side) facing the Buffer Core. If only running one gel, a Buffer Dam in place of the gel can be inserted.
33. When washing, the outlines of the wells become more clearly visible. Washing removes urea in the wells to ensure that samples can be loaded onto a flat clean surface.
34. Keratin from skin cells is the most common contaminant in MS analysis. Always use clean reagents/containers and wear clean gloves when handling gels for MS analysis.
35. The methanol in the 1× Transfer Buffer may evaporate with time, so do not store the 1× Transfer Buffer for long periods and do not reuse the 1× Transfer Buffer.
36. The electrodes on the Electrode Module need to be positioned in the middle of the Buffer Tank for the lid to properly close.
37. The magnetic stir bar is to help maintain even buffer temperature and ion distribution in the tank. Set the speed as fast as possible to keep ion distribution even.

38. Bubbles interfere with electrophoretic transfer of protein samples to the nitrocellulose membrane, thus are highly undesirable. Note that miniscule bubbles can expand under the heat of transfer and block and/or distort protein transfer. Manipulations done on a completely submerged gel and filter papers facilitate the complete removal of bubbles.
39. Fill the Bio-Ice cooling unit with water and pre-freeze at –20°C for at least 1 h before transfer.
40. The beer cooler or cold room is to keep the transfer unit at a low temperature so that the SDS-PAGE gel stays solid during the high-voltage heat-generating transfer process.
41. If the pre-stained molecular weight marker is crisp and clear on the nitrocellulose filter without any remnants on the SDS-PAGE gel, this indicates that the transfer has been successful.
42. The anti-acetyl-Lysine (rabbit antibody) can be saved in a labeled 15-mL conical tube at –20°C and reused up to four times. However, the anti-acetyl-Lysine mouse mAb is not a good antibody to save for reuse.
43. Cut a sheet protector so that it fits inside an X-ray film cassette and is adhered so that it does not move. After developing, remove the nitrocellulose filter and clean the sheet protector with a wet paper towel and dry for subsequent reuse.
44. The NudC R2 antibody dilution in ReliaBLOT® blocking buffer can be saved in a labeled 15-mL conical tube at –20°C and reused four to six times for 2 months. Precipitant may develop upon longer storage at which point the diluted antibody can no longer be used.

Acknowledgements

The authors would like to thank Dr. Sue-Hwa Lin for data discussion and technical advice and Dr. Sophia Y. Tsai for support. This work was supported by NIH training grant T32 DK007696 (Carol Chuang), and grants from the NIH RO1 DK53176, U19 AI071130, and the Dan L. Duncan Cancer Center (Li-yuan Yu-Lee).

References

1. Ishii S, Kurasawa Y, Wong J et al (2008) Histone deacetylase 3 localizes to the mitotic spindle and is required for kinetochore-microtubule attachment. Proc Natl Acad Sci 105:4179–4184
2. Chuang C, Lin S-H, Huang F et al (2010) Acetylation of RNA processing proteins and cell cycle proteins in mitosis. J Proteome Res 9:4554–4564
3. Dephoure N, Zhou C, villen J et al (2008) A quantitative atlas of mitotic phosphorylation. Proc Natl Acad Sci 105:10762–10767
4. Malik R, Lenobel R, Santamaria A et al (2009) Quantitative analysis of the human spindle phosphoproteome at distinct mitotic stages. J Proteome Res 10:4553–4563
5. Choudhary C, Kumar C, Gnad F et al (2009) Lysine acetylation targets protein complexes

and co-regulates major cellular functions. Science 325:834–840

6. Kim SC, Sprung R, Chen Y et al (2006) Substrate and functional diversity of lysine acetylation revealed by a proteomics survey. Mol Cell 23:607–618
7. Kim GW and Yang XJ (2010) Comprehensive lysine acetylomes emerging from bacteria to humans. Trends Biochem Sci 36:211–220
8. Nishino M, Kurasawa Y, Evans R et al (2006) NudC is required for Plk1 targeting to the kinetochore and chromosome congression. Curr Biol 16:1414–1421
9. Josic D, Brown MK, Huang F et al (2005) Use of selective extraction and fast chromatographic separation combined with electrophoretic methods for mapping of membrane proteins. Electrophoresis 26:2809–2822
10. Huang F, Clifton J, Yang X et al (2009) SELDI-TOF as a method for biomarker discovery in the urine of aristolochic-acid-treated mice. Electrophoresis 30:1168–1174
11. Clifton JG, Huang F, Kovac S et al (2009) Proteomic characterization of plasma-derived clotting factor VIII-von Willebrand factor concentrates. Electrophoresis 30: 3636–3646
12. Rao PN and Engelberg J (1965) HeLa cells: Effects of temperature on the life cycle. Science 148:1092–1094
13. Morris NR and Fischer GA (1960) Studies concerning inhibition of the synthesis of deoxycytidine by phosphorylated derivatives of thymidine. Biochim Biophys Acta 42:183–184
14. Xeros N (1962) Deoxyriboside control and synchronization of mitosis. Nature 194:682–683
15. Studzinski GP and Lambert WC (1969) Thymidine as a synchronizing agent. I. Nucleic acid and protein formation in synchronous HeLa cultures treated with excess thymidine. J Cell Physiol 73:109–117
16. Bostock CJ, Prescott DM, Kirkpatrick JB (1971) An evaluation of the double thymidine block for synchronizing mammalian cells at the G1-S border. Exp Cell Res 68:163–168
17. Atassi G and Tagnon HJ (1975) R17934-NSC 238159: a new antitumor drug--I. Effect on experimental tumors and factors influencing effectiveness. Eur J Cancer 11:599–607
18. Khan N, Jeffers M, Kumar S et al (2008) Determination of the class and isoform selectivity of small molecule histone deacetylase inhibitors. Biochem J 409:581–589
19. Brinkley BR and Cartwright JJr (1975) Cold-labile and cold-stable microtubules in the mitotic spindle of mammalian cells. Ann New York Acad Sci 253:428–439
20. Lin SH and Guidotti G (2009) Purification of membrane proteins. Methods Enzymol 463: 619–629

Chapter 14

Comparative Proteome Analysis of a Human Liver Cell Line Stably Transfected with Hepatitis D Virus Full-Length cDNA

Celso Cunha and Ana V. Coelho

Abstract

Hepatitis D virus (HDV) is the causative agent of one of the most severe forms of virus hepatitis. HDV is a satellite virus of Hepatitis B virus (HBV) and coinfects or superinfects liver cells already infected with HBV. Investigation of HDV biology and pathogenesis has been so far impaired by the lack of an appropriate cell culture system capable of replicating and propagating the virus. This is usually partially overcome using transiently transfection systems and stably transfected cell lines. Here, we used a well-characterized human liver hepatoma cell line stably transfected with HDV cDNA (Huh7-D12) to study the changes in the host proteome induced by the expression of the virus. A 2-DE and MS approach was performed allowing the identification of 23 differentially expressed proteins when compared with the parental non-transfected Huh7 cell line. The proteomic results were validated by western blot and real-time qPCR.

Key words: Hepatitis D virus, Liver, Proteome, 2-DE, Mass spectrometry

1. Introduction

Hepatitis D virus (HDV) infection of human hepatocytes previously infected with Hepatitis B virus (HBV) substantially increases liver damage and the risk of cirrhosis and fulminant hepatitis (1, 2). The two viruses are clinically associated since they share the same envelope which consists of HBV surface antigens (HBsAgs). The HDV genome consists of a single-stranded closed circular RNA molecule of about 1.7 kb and negative polarity (3). This RNA bears a single ORF, from which two proteins are derived as a consequence of an RNA editing mechanism catalyzed by a cellular adenosine deaminase (ADAR1), the so-called small and large delta antigens (p24 S-HDAg and p27 L-HDAg, respectively).

Djuro Josic and Douglas C. Hixson (eds.), *Liver Proteomics: Methods and Protocols*, Methods in Molecular Biology, vol. 909, DOI 10.1007/978-1-61779-959-4_14,

Accordingly, the virus core includes the RNA genome and several molecules of both S-HDAg and L-HDAg that assemble into ribonucleoprotein (RNP) particles. The HDV envelope, in contrast, consists of HBsAgs encoded exclusively by the HBV genome (4). As a small and simple virus, HDV relies at a large extent on the host cell machinery for replication. However, and despite of its simplicity, the precise molecular mechanisms of HDV replication and pathogenesis remain unraveled. This was partially due to the absence of appropriate in vitro cell culture systems that could support virus replication. In an attempt to overcome this constraint, human hepatoma cell lines (Huh7) stably transfected with several copies of HDV cDNA were constructed (5). The HDV cDNA was obtained by reverse transcription of genomic RNA extracted from serum of an experimentally infected woodchuck (6). Transfected cells were selected with geneticin and subsequently cloned and expanded. One of the obtained clones (Huh7-D12) has been previously characterized and was shown to contain about 40 HDV cDNA copies inserted in a single chromosome, probably chromosome 1 (7). This clone constitutively expresses virus RNA and antigens in a molar ratio similar to that observed in HDV particles collected from experimentally infected woodchucks (8). We used the Huh7-D12 cell line to study the changes in the proteome induced by the expression of HDV RNPs. First, we separated by 2-DE, whole cell protein extracts from Huh7 and Huh7-D12 cells. The two proteomes were compared and the spots corresponding to differentially expressed proteins were excised and analyzed by mass spectrometry. After database search, 15 down-regulated and 8 up-regulated proteins were identified in Huh7-D12 cells when compared with the parental Huh7 cell line (9). The identified proteins are involved in different metabolic pathways, some of which were also reported to be altered in chronic and acute HDV patients. The results of proteome analysis were further validated by western blot analysis and real-time qPCR.

2. Materials

2.1. Cell Culture and Sample Preparation

1. RPMI 1640 Medium (Sigma) supplemented with 10% fetal bovine serum (FBS, Invitrogen, Carlsbad, CA, USA).
2. Cell scrapers (Sarstedt, Numbrecht, Germany).
3. Phosphate-buffered saline (PBS).
4. 10% (v/v) Trichloroacetic acid (see Note 1).
5. Acetone (stored at –20°C).
6. Protein quantification kit (Bradford assay; Pierce, Rockford, IL, USA).

2.2. Two-Dimensional Electrophoresis

1. Isoelectrofocusing (IEF) buffer: 7 M urea, 2 M thiourea, 4% (w/v) CHAPS, IPG buffer mix (pH 3–11/7–11; 2:1 (v/v); GE Healthcare, Pittsburgh, PA, USA), 60 mM DTE, 0.002% (w/v) bromophenol blue.
2. Immobiline™ DryStrip pH 3–11 (GE Healthcare).
3. Equilibration buffer: 0.5 M Tris–HCl, pH 8.8, 6 M urea, 30% (v/v) glycerol, 2% (w/v) SDS, 2% (w/v) DTE.
4. Blocking buffer: 0.5 M Tris–HCl, pH 8.8, 6 M urea, 30% (v/v) glycerol, 2% (w/v) SDS, 2.5% (w/v) iodoacetamide, 0.002% (w/v) bromophenol blue.
5. 0.5% (w/v) Agarose in electrophoresis buffer: 0.025 M Tris, 0.192% (w/v) glycine, 0.1% (w/v) SDS.
6. Running gel buffer (5×): 1.875 M Tris–HCl, pH 8.8, 0.5% (w/v) SDS. Store at room temperature.
7. Stacking gel buffer (5×): 0.625 M Tris–HCl, pH 6.8, 0.5% (w/v) SDS. Store at room temperature.
8. 30% Acrylamide/bis-acrylamide solution (37.5:1). Store in a dark bottle at 4°C. When unpolymerized, acrylamide is a neurotoxin. Care should be taken to avoid exposure.
9. *N,N,N,N'*-tetramethyl-ethylenediamine (TEMED, Sigma-Aldrich, St. Louis, MO, USA). Store at room temperature (see Note 2).
10. 10% (w/v) Ammonium persulfate (APS) in water. Prepare fresh, immediately before use (see Note 3).
11. Water-saturated isopropanol. Mix equal volumes of water and isopropanol, in a glass bottle, and shake. Allow to separate and use the upper phase. Store at room temperature.
12. Running buffer (5×): 125 mM Tris, 960 mM glycine, 0.5% (w/v) SDS. Store at room temperature.
13. 12% (v/v) Trichloroacetic acid.
14. 3% (v/v) Glycerol solution.
15. Staining solution: 0.12% (w/v) brilliant blue G-250, 10% (w/v) orthophosphoric acid, 10% (w/v) ammonium sulfate, 20% (v/v) methanol.

2.3. Protein Identification

2.3.1. Excision of Gel Spots

1. Standard powder-free nitrile gloves.
2. Standard 1 mL micropipette tips.
3. 1 mL Eppendorf tubes or 96-well polypropylene plate, twin. tec PCR plates (Eppendorf, Hamburg, Germany).
4. Disposable bonnet and laboratory coat.

2.3.2. In-Gel Protein Digestion

1. Destaining solution: 50% (v/v) acetonitrile.
2. Acetonitrile.
3. Reduction agent: 10 mM dithiothreitol in 100 mM NH_4HCO_3.

4. Alkylation agent: 55 mM iodoacetamide in 100 mM NH_4HCO_3.
5. Digestion buffer: 6.7 μg/mL trypsin (Sequencing Grade Modified Trypsin, Promega, Madison, WI, USA) in 50 mM NH_4HCO_3 (see Note 4).
6. Acidification solution: 5% (v/v) formic acid.

2.3.3. Desalting and Concentration of Peptide Digests

1. Acidification solution: 5% (v/v) formic acid.
2. Formic acid.
3. MALDI matrix: 3 mg/mL alpha-cyano-4-hydroxycinnamic acid matrix prepared in a solution of 50% (v/v) acetonitrile with 5% (v/v) formic acid.

2.3.4. Acquisition of MALDI-MS Spectra

Peptide calibration mixture: Cal Mix 1 (AB Sciex, Foster City, CA, USA) with m/z range 900–3,700, should be prepared according to the manufacturer's instructions.

2.4. Western Blot

2.4.1. Sample Preparation and SDS-PAGE

1. PBS.
2. Benzonase (Sigma-Aldrich).
3. Running gel buffer (5×): 1.875 M Tris–HCl, pH 8.8, 0.5% (w/v) SDS. Store at room temperature.
4. Stacking gel buffer (5×): 0.625 M Tris–HCl, pH 6.8, 0.5% (w/v) SDS. Store at room temperature.
5. 30% Acrylamide/bis-acrylamide solution (37.5: 1). Store in a dark bottle at 4°C. When unpolymerized, acrylamide is a neurotoxin. Care should be taken to avoid exposure.
6. TEMED (Sigma). Store at room temperature (see Note 2).
7. 10% (w/v) APS in water. Prepare fresh, immediately before use (see Note 3).
8. Water-saturated isopropanol. Mix equal volumes of water and isopropanol, in a glass bottle, and shake. Allow to separate and use the upper phase. Store at room temperature.
9. Running buffer (5×): 125 mM Tris, 960 mM glycine, 0.5% (w/v) SDS. Store at room temperature.

2.4.2. Protein Transfer and Detection

1. Transfer buffer: 48 mM Tris, 39 mM glycine, 0.04% (w/v) SDS, 10% (v/v) methanol.
2. Nitrocellulose membrane (Protan BA 85, Schleicher & Schuell, Dassel, Germany) and 3MM chromatography paper (Whatman, Kent, UK).
3. PBS-T; PBS with 0.05% (w/v) Tween 20.
4. Ponceau S [0.1% (w/v) in 5% acetic acid].

Table 1
Nucleotide sequences of the synthetic primers used for cDNA amplification

mRNA	Forward primer (5′→3′)	Reverse primer (5′→3′)	Accession number
β2MG	GGCTATCCAGCGTACTCCAA	TCACACGGCAGGCATACTC	NM_004048
NASP	GGAGTCTCAGCGTA-GYGGGAAT	TCTACTGGCAATCATGGAGACTG	BT006757
TPI	GCACTCAGAGAGAAGGCATGT	CAATGCAGGCGATTACTCCGA	M10036

The indicated accession number refers to the NCBI database (http://www.ncbi.nlm.nih.gov/)

5. Blocking buffer: PBS-T with 5% nonfat dry milk (w/v).
6. Wash buffer: PBS-T with 2% nonfat dry milk (w/v).
7. Primary antibodies: Mouse monoclonal anti-lamin C (Abcam, UK), mouse monoclonal anti-LA (10), and rabbit polyclonal anti-clathrin (Santa Cruz Biotechnology, Santa Cruz, CA, USA).
8. Secondary antibodies: Anti-mouse IgG and anti-rabbit IgG, both conjugated with horseradish peroxidase (BioRad, Hercules, CA, USA).
9. Enhanced chemiluminescent reagents (ECL; GE Healthcare).
10. Bio-Max ML film (Kodak, Rochester, NY, USA).

2.5. Quantitative Real-Time PCR

1. MicroPoly(A) Purist™ mRNA extraction kit (Ambion, Austin, TX, USA).
2. DNA-*free*™ kit (Ambion).
3. Revert Aid™ Moloney murine leukemia virus (M-MuLV) reverse transcriptase (Fermentas, Burlington, Canada).
4. Oligo(dT)$_{12-18}$ (Roche Applied Science, Indianapolis, IN, USA).
5. dNTP mix (Roche Applied Science).
6. Protector RNase inhibitor (Roche Applied Science).
7. Reaction buffer: 50 mM Tris–HCl, pH 8.3, 50 mM KCl, 4 mM $MgCl_2$, 10 mM DTT.
8. GFX PCR DNA and Gel Band purification kit (GE Healthcare).
9. qPCR Core kit for SYBR® Green I (Eurogentec, Seraing, Belgium).
10. Primers for gene of interest and control housekeeping gene (see Table 1).

3. Methods

HDV infects human hepatocytes and is considered to be a satellite virus of HBV. The outer envelope of the two viruses has the same protein composition and consists of HBsAgs. These antigens are coded exclusively by the HBV genome being essential for HDV and HBV propagation of infection. As a consequence, in the absence of HBsAgs HDV cannot be released from and reinfect liver cells. Different strategies and systems are currently used to study HDV replication and pathogenesis, each of which having specific advantages and disadvantages. Among them are transient transfection systems using plasmids coding for each of the HDV components (RNA or antigens) separately, and the human embryonic kidney HEK293 cell line expressing HDV antigens under tetracycline control. In 1993, Cheng et al. (5) reported the construction and characterization of human hepatoma cell lines (Huh7) stably transfected with cDNA copies of the HDV genome. Clone D12 (Huh7-D12) is perhaps the best characterized and was shown to express constitutively the virus genome and antigens in relative ratios comparable with those observed in virions released into serum of experimentally infected woodchucks. Although far from what can be considered an ideal system to study HDV biology, the Huh7-D12 cell line can still provide valuable information concerning virus replication and pathogenesis.

The clinical course of HDV infection is well documented, but the detailed and specific mechanisms of HDV pathogenesis are still poorly understood. In particular, it is necessary to identify the alterations in liver cell metabolism and protein expression that may account for the commonly observed clinical pictures. In this context, a proteomic approach based on 2D-electrophoresis followed by MS analysis was used to investigate the changes in the proteome of human liver cells that arise as a consequence of HDV replication and expression. This approach was based on the detection and identification of differentially expressed proteins in Huh7-D12 cells when compared with the parental Huh7 cell line. Spots found differentially expressed were excised for identification using MALDI-MS data. The proteins in the excised spots were subjected to trypsin digestion, which produces peptides with suitable sizes and ionization ability for these MS assays. In order to obtain enough information for search protein sequence databases, masses and sequence information of tryptic peptides was extracted, respectively, from MS and MS/MS spectra.

3.1. Sample Preparation and 2-DE

1. The HDV cDNA stably transfected Huh7-D12 cell line (European Collection of Cell Cultures no 01042712) and the parental non-transfected Huh7 cells (Japanese Collection of Research Bioresources JCRB 0403) are cultured in RPMI

1640 medium supplemented with 10% FBS. Cells are cultured as monolayers in 25 cm^2 flasks. Before reaching confluence, cultures can be passaged using trypsin/EDTA. A 1:5 split is often adequate in order to generate new confluent cultures in approximately 72 h.

2. To prepare whole cell protein samples for subsequent 2-DE separation, the cells are harvested with a cell scraper directly to the medium and collected into centrifuge tubes. Cells are then sedimented by centrifugation at 800 × *g*, for 5 min, the supernatant is rejected, and the pellet is washed three times with PBS, in the same conditions.
3. After the last washing, cell lysis and protein precipitation can be promoted in a single step by resuspending the obtained cell pellet with 10% TCA (v/v), followed by a 30-min incubation on ice. After incubation, the precipitate is collected by centrifugation at 12,000 × *g*, for 5 min, and washed with three volumes of ice-cold acetone, for 2 h, at –20°C. After washing, precipitates are sedimented by centrifugation, 5 min at 12,000 × *g*, and the supernatant is rejected. Finally, protein precipitates are allowed to dry approximately for 15 min, at room temperature. At this point, samples can be stored at –80°C, until use.
4. Before electrophoresis separation, protein concentration needs to be determined in each sample to be analyzed. This can be performed using a commercial kit (for instance from Pierce or BioRad) based on the principles of the Bradford assay (11). It is crucial to accurately measure the whole protein concentration since we will further need to compare the relative amounts of each individual protein spot after 2-DE.
5. We usually load 200 μg of total protein extracts onto the first dimension of each IEF gel. Accordingly, prior to loading 200 μg protein are diluted in IEF buffer, and incubated for 1 h, at room temperature, with gentle shaking (see Note 5). After incubation, the insoluble material is removed by two consecutive rounds of centrifugation at 16,000 × *g*, during 10 min.
6. The first dimension isoelectric focusing gel allows separating proteins according to the respective isoelectric point (pI). These gels consist of a polyacrylamide matrix containing an immobilized nonlinear pH gradient. For our purposes we use 18 cm precast Immobiline™ DrySrip pH 3–11 polyacrilamide gels and the *Ettan IPGphor Isolectric Focusing System*, both from GE Healthcare.

   Protein samples are loaded onto the Immobiline™ DrySrips, and after a 12-h rehydration with IEF buffer, IEF is carried out under the following voltage and time conditions: (a) 200 V,

5 h; (b) 500 V, 5 h; (c) 1,000 V, 3 h; (d) 2,000 V, 3 h; (e) 3,000 V, 3 h; (f) 5,000 V, 3 h; (g) 8,000 V, 8 h; and (h) 50 V, 2 h. An overall of approximately 97,500 V/h are achieved at the end of separation.

7. Before the second-dimensional SDS-PAGE, the strips are subjected to a two-step equilibration process. First, strips are equilibrated with Equilibration buffer, for 15 min, at room temperature, with gentle shaking. The buffer is replaced and strips are incubated for another 15 min in the same conditions. In a second step, strips are equilibrated with Blocking buffer, for 15 min, at room temperature with gentle shaking. The buffer is replaced and the strips are incubated again in the same conditions. After the equilibration step, strips are rinsed with bidestilated water and disposed over the SDS-PAGE gels.
8. After rinsing with water, the strips are fixed to the SDS-PAGE gel using 0.5% (w/v) agarose, dissolved in electrophoresis buffer (see Note 6). In these gels, proteins are now separated according to the respective molecular masses (see Note 7). We usually use 9% polyacrilamide gels and proteins are separated overnight at 50 V using a Hoefer Ruby SE 600 apparatus from GE Healthcare. Electrophoresis is stopped when the bromophenol blue dye reaches the end of the gel.

3.2. Protein Staining and Image Analysis

1. After electrophoresis, gels are stained according to the Silver Blue staining technique previously described by Candiano et al. (12). First, proteins in the gel are fixed in 12% TCA for 15 min, and then rinsed with water for 5 min at least three times. The gels are then stained with the Staining solution for 48 h, at room temperature, with gentle shaking.
2. After staining, gels are rinsed with water for 5 min and washed twice with water for 1 h. Finally, gels are incubated in a 3% glycerol solution for 15 min.
3. The stained two-dimensional gels (Fig. 1a, b) are then scanned in an ImageScanner (GE Healthcare) with a resolution of 300 dpi. In order to statistically validate the results, at least five gel replicas corresponding to five independent protein samples need to be digitalized and analyzed. Detection and analysis of protein spots are performed with the Phoretix PG200 software from Nonlinear Dynamics (Newcastle upon Tyne, UK). This software can detect all the spots present in a gel. However, we advise to verify manually the correct detection since some spots may be incorrectly indicated by the software and thus need to be removed.
4. The software can select a gel image as a reference image to which all the others corresponding to the same proteome will be compared with. In the next step we thus need to compare

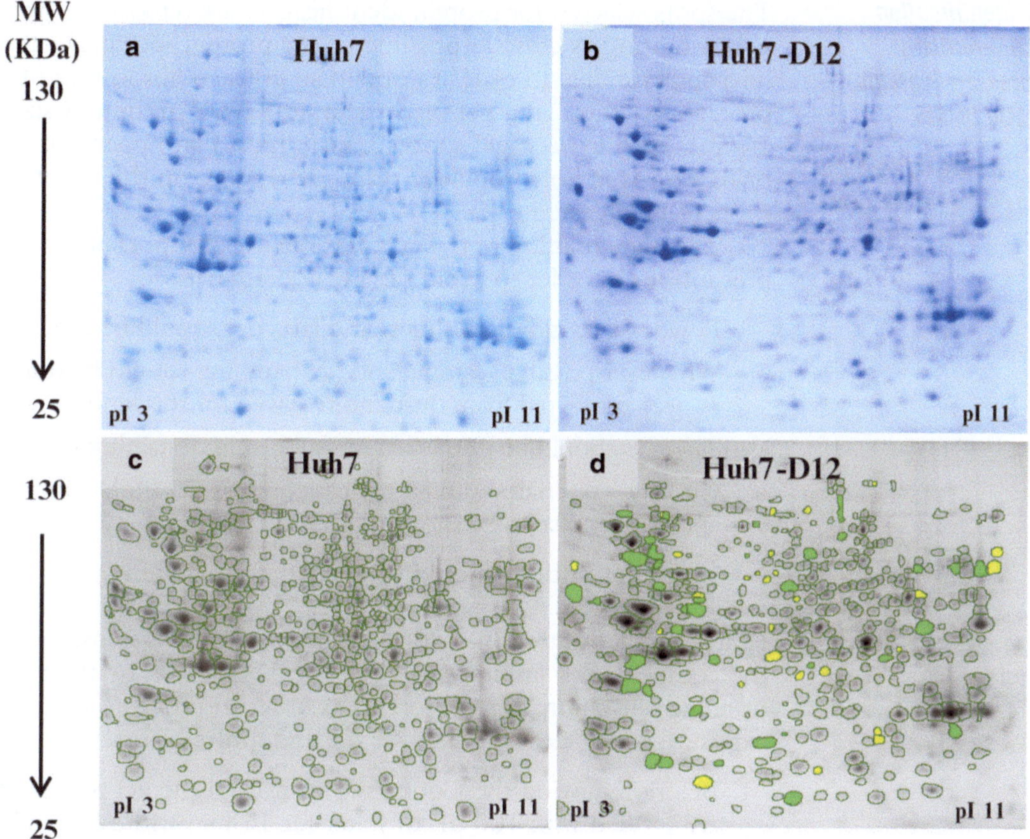

Fig. 1. 2-DE analysis of proteins extracted from the Huh7 cell line (**a**) and from the Huh7-D12 cell line (**b**). Proteins were separated on pH 3–11 NL IPG strips followed by SDS-PAGE on a 9% polyacrilamide gel. The gels were stained using Coomassie Brilliant Blue and scanned. (**c**, **d**) The digitalized images were analyzed with the Phoretix PG200 software and the differentially expressed protein spots were identified and labeled in *green* (up-regulated proteins) or *yellow* (down-regulated expression).

each gel image with the reference image previously selected by the software. The software then measures the intensity of spots in each gel and calculates the overall "spot volume." Since variations in spot intensity may be due to the experimental procedure, namely, small differences in the total amount of loaded protein in the gel, staining, etc., spot volumes need to be calibrated in order to obtain what is called the "normalized spot volume." In brief, each individual spot volume is divided by the total volume of all spots present in each gel and multiplied by 100. This "normalized spot volume" is then utilized for the statistical analysis of differentially expressed proteins. This analysis is performed only for protein spots present in at least four of the five gels in the same proteome (Fig. 1c, d). In the end, the obtained results are filtered in order to keep only those in which at least a twofold difference in expression levels between different proteome samples was obtained ($p<0.05$).

3.3. Identification of Differentially Expressed Proteins by MALDI-MS

3.3.1. Excision of Gel Spots from Polyacrylamide Gels

The spots selected for protein identification are removed from the gel by cutting as close as possible to the protein spot using 1 mL micropipette tips. In order to obtain an efficient absorption of the protease solution, the gel pieces should have a maximum size of 1 mm^3. Each excised gel spot is transferred to an Eppendorf tube or to a 96-well polypropylene plate, depending on the number of spots to be digested.

The manipulation of the gels and the excised spots needs some precautions in order to avoid sample contamination:

- Powder-free gloves, bonnet, and lab coat should be used.
- Whenever possible, sample and gel handling should be done in a space protected from dust and reserved for this kind of work.
- The gels should be immersed in water or in destaining solution.
- Cutting material should be cleaned with ethanol before use and between spot excisions.

These precautions should be applied during all the procedure.

3.3.2. Protein In-Gel Digestion

Protein in-gel digestion involves some preparative steps, namely, the complete destaining of the gel pieces, followed by the reduction and alkylation of the protein, in order to promote maximum exposure of the cleavage sites for the endoprotease reaction.

1. Before removing the Coomassie stain from the gel pieces, they are washed with 100 μL of water for 15 min with stirring. The liquid is discarded using a micropipette and 100 μL of 50% (v/v) acetonitrile is added followed by a 15-min incubation. Discard the liquid and repeat the procedure until the gel pieces are transparent. If the stain cannot be removed this way, warming of the sample up to 60°C is allowed during the destaining steps.
2. Before the reduction and alkylation steps 200 μL of acetonitrile are added to dehydrate the gel pieces, which are incubated for 15 min. The liquid is discarded and 50 μL of the reduction agent solution are added and incubated for 45 min at a minimum temperature of 56°C, in order to reduce protein disulfide bonds. The excess of solution is removed, and 50 μL of the alkylating agent solution is added followed by a 30-min incubation at room temperature in the dark, in order to alkylate the previously reduced disulfide bonds.
3. In order to maximize the absorption of the digestion buffer into the gel pieces, they are dehydrated twice with 200 μL of acetonitrile and incubated for 15 min (in between the additions of acetonitrile the liquid is discarded) and the gel pieces are dried in the sample vacuum concentration system (Thermo Savant, Whatman, MA, USA).
4. Gel pieces are rehydrated in 10–20 μL of digestion buffer, according to the volume of the gel pieces, and incubated at

4°C. After 15 min verify all samples and add more digestion buffer if the initially added volume has been completely absorbed. Incubate for 30–45 min at 4°C.

5. Discard the excess of solution and add 5–25 μL of digestion buffer without the enzyme in order to cover the gel pieces. Incubate overnight at 37°C.

   Remove the supernatant that contains the digested peptide mixture to an Eppendorf tube or to a 96-well plate. Both the supernatants and the gel pieces can be stored at −20°C for 1 year.

6. If needed, the remaining peptides still trapped in the gel pieces can be removed incubating for 15 min with stirring, first with 200 μL of acetonitrile, followed by a solution of 50% acetonitrile/5% formic acid, and to finish with 200 μL of acetonitrile. The three collected supernatants should be added in the same Eppendorf or well.

3.3.3. Desalting and Concentration of the Peptide Digests

Homemade chromatographic microcolumns (13) or commercial devices (like ZipTip pipette tips, Millipore, Billerica, MA, USA) can be used for the desalting and concentration of the peptide digests.

1. To prepare a microcolumn load a GELoader tip (Eppendorf) with a sufficient amount of a slurry of Poros R2 (Applied Biosystems, Carlsbad, CA, USA) chromatographic reverse-phase media in acetonitrile to obtain a packing of about 3–4 mm in height.
2. Wash and equilibrate the microcolumn with 20 μL of a solution of 5% formic acid. Acidify the sample adding formic acid to a final concentration of 5% and load the sample onto the microcolumn.
3. Wash the microcolumn with 20 μL of 5% formic acid. Elute the sample directly onto the MALDI plate by using 0.5–1.5 μL of alpha-cyano-4-hydroxycinnamic acid matrix.
4. Samples will dry at room temperature to promote cocrystallization of the analyte with the matrix.
5. In order to protect the components of the MALDI-TOF/TOF ion optics, the plates must be blown using compressed dry air to remove any fibers and dust before sample spotting. Until the plate is inserted into the instrument it must be protected from dust. Always use powder-free gloves and avoid touching the MALDI plate surface to avoid sample contamination.

Samples can be stored on the plate, preferably at 4°C, for a maximum period of 2 days, which allows the repetition of spectra acquisition.

3.3.4. Acquisition of MALDI-MS Spectra

After loading the MALDI plate, the acquisition of the reference MS and MS/MS spectra from calibration standard(s) is performed. Any commercially available mass spectrometry-grade peptide

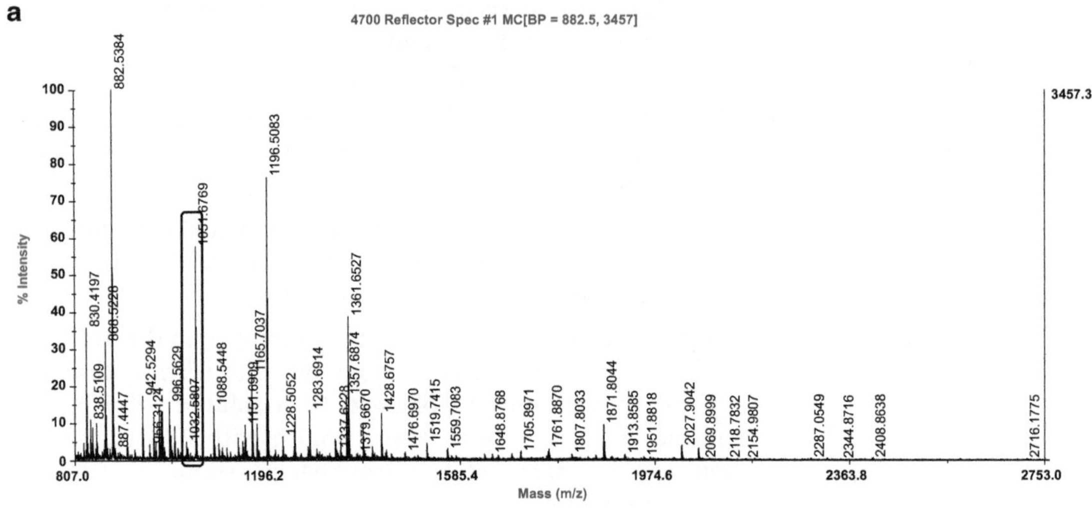

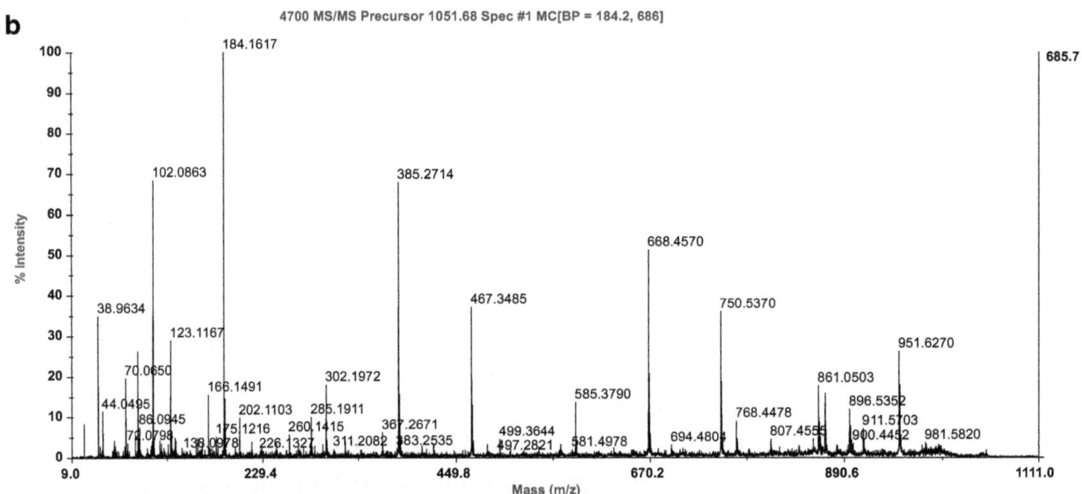

Fig. 2. Identification of the differentially expressed protein regucalcin. The molecular ion labeled in the MS spectra (**a**) was used as the precursor ion of the MS/MS spectra presented in (**b**). The combined information of both spectra was used for the identification of regucalcin (Q15493).

mixture that includes the mass values to be determined can be used for calibration. For the acquisition of MS and MS/MS spectra from unknown sample(s), laser intensity should be optimized according to the sample concentration. A MALDI-TOF/TOF 4800 plus mass spectrophotometer (Applied Biosystems) in reflectron mode is used for spectra acquisition in order to achieve maximum *m/z* resolution and accuracy (Fig. 2).

3.3.5. Protein Identification by Searching Protein Sequence Databases

The identification of a protein is determined by comparison of the peptide masses (MS) obtained experimentally with the peptide masses (MS) obtained after in silico digestion. The fragmentation masses of the daughter ions obtained (MS/MS) are also compared to the in silico fragmentation of the peptides (MS/MS) in the database. This comparison must be performed by allowing a maximum variation between the peptides mass value measured experimentally and the value determined by the in silico digestion.

Two different search software can be used for database searches. One of them is GPS Explorer (Applied Biosystems) that provides extensive Mascot (Matrix Science, Boston, MA, USA) integration and uses the Mowse® algorithm (14), enabling for both peptide mass fingerprint and MS/MS combined searches. An alternative software is Protein Pilot® (Applied Biosystems), which employs the algorithm Paragon (15) that uses a probabilistic approach to identify simultaneously hundreds of modifications, substitutions, and unexpected cleavages.

These software rank the identification results according to a determined score value. Each program calculates the minimum score values for a significant identification, according with the chosen search parameters, namely, minimum mass accuracy for the parent and fragment ions, number of missed cleavages, and fixed and variable modifications. Else than the score value other acceptance criteria are used to consider a confident identification and at least two tryptic peptides should match the respective MS/MS spectra with 95% confidence.

3.4. Validation of Proteomic Results by Western Blot

Validation of proteomic results is most commonly performed by western blot analysis of differentially expressed proteins. If the signal is not saturated, band intensities are proportional to the amount of protein. A control, not differentially expressed, usually a housekeeping protein, needs also to be detected in the same experiments in order to normalize the results. Here, we chose to validate the proteomic results for Lamin A/C and RNP LA which are differentially expressed in Huh7-D12 cells and are known to be involved in distinct metabolic pathways.

3.4.1. Preparation of Samples

1. Huh7 and Huh7-D12 cell cultures are washed with PBS, harvested with a cell scraper, and collected by centrifugation at $800\times g$ for 5 min.
2. The obtained cell pellets are immediately resuspended in electrophoresis sample buffer containing 400 U of the endonuclease benzonase (see Note 8). The cellular DNA is digested in about 20 min, at room temperature, until the protein samples lose viscosity.
3. After DNA digestion, the samples are denatured by boiling at 100°C, for 5 min. The obtained protein samples may be immediately separated by SDS-PAGE or stored at −80°C, until use.

3.4.2. SDS-PAGE

1. To separate protein samples we prepare 12% polyacrilamide minigels. The gels are 1.5 mm thick and run in a Mini Protean 3 BioRad gel system. The separation gel is prepared by mixing 4 mL of a 30% acrylamide/bis-acrylamide (29:1) solution with 2.5 mL 1.5 M Tris, 100 μL 10% SDS, 3.5 mL water, 40 μL 10% APS, and 20 μL TEMED.

2. Pour the gel leaving enough space for the stacking gel and overlay with water-saturated isopropanol.
3. After polymerization, the isopropanol is removed and the stacking gel mixture is poured on the top of the polymerized separation gel. The stacking gel is prepared by mixing 1.2 mL of a 30% acrylamide/bis-acrylamide (29:1) solution with 2.0 mL 1.5 M Tris, 100 μL 10% glycerol, 100 μL 10% SDS, 4.7 mL water, 40 μL 10% APS, and 20 μL TEMED.
4. Pour the stacking gel and insert the comb. The gels usually polymerize in about 30 min. After polymerization, the comb is carefully removed, and the wells are washed and filled with electrophoresis running buffer.
5. Protein samples are then loaded with a 1 mL syringe fitted with a 22-gauge needle. Reserve at least one well to load a molecular weight marker mixture.
6. The gel is first run at 60 V to allow concentration of the protein samples, and until they reach the separating gel. Then, electrophoresis can be carried out at 100 V for 2–3 h until the bromophenol blue dye reaches the end of the gel.

3.4.3. Western Blot

1. After separation, the gel is equilibrated during 30 min, at room temperature, with transfer buffer, and proteins in the gel are subsequently transferred to a Protran BA85 nitrocellulose membrane (Schleicher & Schuell) using a Transblot semi-dry transfer system (BioRad). Protein transfer is performed during 1 h at 100 mA and depends on the size of the gel (see Note 9).
2. To confirm if the proteins in the gel were efficiently transferred, the membrane is stained with a 0.1% Ponceau solution for 5 min at room temperature. After confirming that staining occurred efficiently, the Ponceau dye can be easily removed after extensive rinsing with PBS and/or water.
3. The membrane is next blocked with 5% low-fat milk in PBS, for 1 h with gentle shaking. Following blocking, membranes are incubated overnight, with gentle shaking, at 4°C, with the selected primary antibody diluted 1:500 in PBS containing 5% low-fat milk (see Note 10).
4. After incubation with primary antibodies, the membrane is washed three times with PBS containing 0.05% Tween 20 and 2% low-fat milk, for 15 min at room temperature, with gentle shaking. After washing, the membrane is then incubated for 1 h, at room temperature, with gentle shaking, with the corresponding secondary antibodies conjugated with horseradish peroxidase in a 1:2,500 dilution (BioRad). Secondary antibodies are also diluted in PBS containing 5% low-fat milk. Finally, the membrane is washed three times as described above.

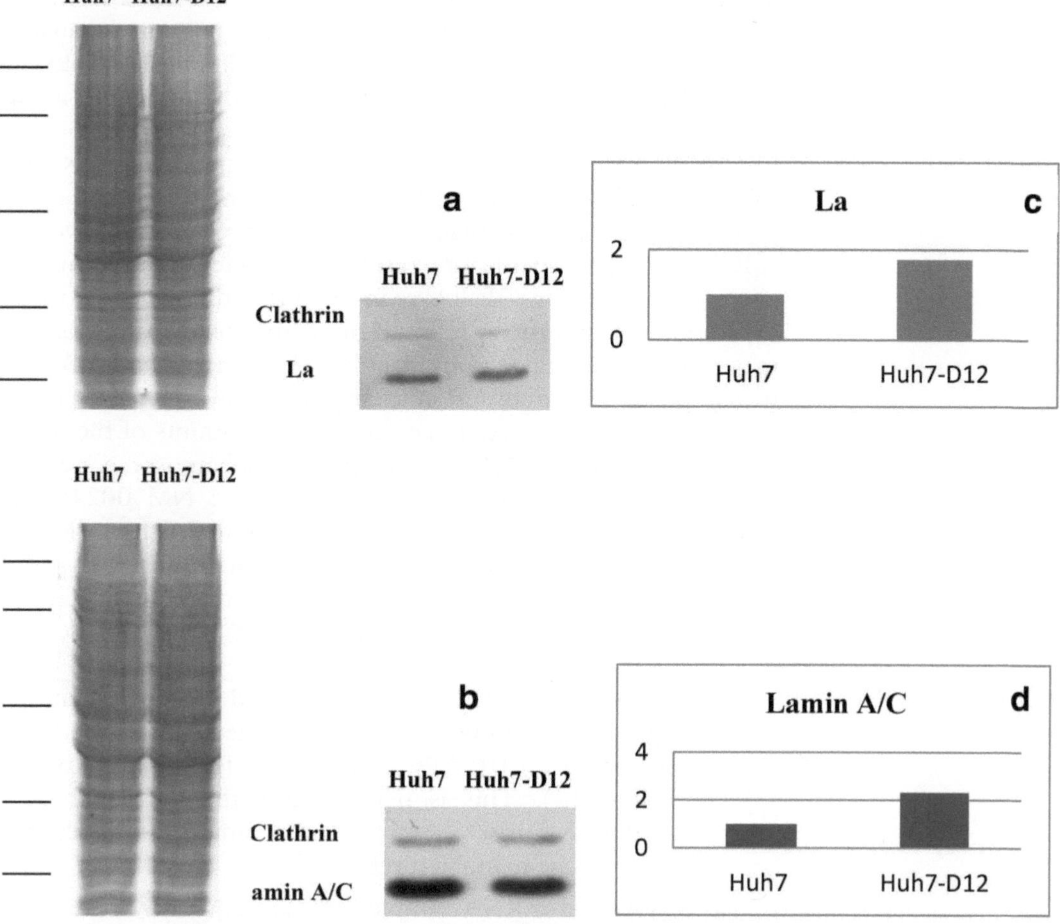

Fig. 3. Validation by western blot analysis of the overexpression of Lamin A/C and La proteins, in Huh7-D12 cells. Total Huh7 and Huh7-D12 protein extracts were separated by SDS-PAGE and transferred to a nitrocellulose membrane (*left*, the lines indicate the migration of MW markers: 120, 86, 47, 34, 26 kD, respectively). After incubation with specific anti-Lamin A/C and anti-La antibodies, membranes were developed using an ECL system (**a**, **b**), and images were analyzed with the Phoretix PG200 software. To calculate the relative expression levels the Lamin A/C and La spot volumes were normalized against the housekeeping protein clathrin ($p=0.0021$and 0.016, respectively; **c**, **d**).

5. Membrane development is performed using a commercial ECL System, according to the instructions of the manufacturer (GE Healthcare). The results obtained by western blot are digitalized with an ImageScanner (GE Healthcare) and the relative quantification of proteins is performed with the Phoretix PG200 software (Nonlinear Dynamics, see Note 11). The software removes the background and calculates the normalized spot volume of each band. Here, we used the housekeeping protein clathrin as internal control to calibrate the total amount of proteins loaded onto each well of the SDS-PAGE gel (Fig. 3).

3.5. Validation of Proteomic Results by qRT-PCR

Sometimes, western blot cannot be used to confirm differences in protein expression levels observed during proteomic analysis. This may be due to a number of reasons including the following: (1) the lack of appropriate antibodies against the protein of interest and (2) although available the antibody is not appropriate for use in western blot applications. Alternatively, the proteomic results can be validated indirectly by measuring the relative amounts of the mRNA coding for the protein of interest by real-time quantitative PCR (qRT-PCR). This does not represent a completely accurate and precise approach since we shall bear in mind that differences in mRNA expression may not totally reflect changes in protein expression. This may be due to several reasons including differences in mRNA stability and processing, transport efficiency, etc. However, we decided to use this approach to determine the relative amounts of the histone H1 binding protein (NASP) and triosephosphate isomerase (TPI) mRNA expression (NCBI accession numbers: NM_002482 and NM_000365, respectively) in Huh7 and Huh7-D12 cells when compared to a housekeeping gene that was not found, by proteomic analysis, to be differentially expressed [β-2-microglobulin (β2MG): NCBI mRNA accession number NM_004048].

3.5.1. mRNA Extraction and cDNA Synthesis

1. Huh7 and Huh7-D12 cells are collected by centrifugation at $800 \times g$ as described before. The cell pellets, corresponding to approximately 5×10^6 cells, are then used for mRNA extraction (see Note 12). This is performed with a commercial kit (MicroPoly(A)Purist™, Ambion) according to the instructions provided by the manufacturer.
2. The obtained mRNA samples are subsequently treated with DNase I to digest possible remaining contaminating DNA. This step is also performed with a commercial kit (DNA-*free*™; Ambion, Austin, TX, USA) following the manufacturer's instructions.
3. For cDNA synthesis we use the Revert Aid™ M-MuLV kit (Fermentas). To approximately 100 ng mRNA, 0.5 μg oligo$(dT)_{12-18}$ (Roche Applied Science) are added in a final volume of 12.5 μL. The mixture is heated at 70°C for 10 min and then quickly transferred to ice in order to denature RNA and remove eventual internal base-pairing secondary structures.
4. Next, reaction buffer (50 mM Tris–HCl, pH 8.3, 50 mM KCl, 4 mM $MgCl_2$, 10 mM DTT), 10 mM dNTPs, and 20 U of an RNase inhibitor are added and the mixture is incubated again at 37°C for 5 min to allow annealing of the template mRNA and the oligo$(dT)_{12-18}$.
5. Finally, 40 U of M-MuLV reverse transcriptase are added in a final volume of 20 μL. Synthesis of cDNA occurs for 1 h, at 42°C, and the reaction is stopped by heating at 70°C for 10 min.

6. After synthesis, the cDNA is purified using a commercial kit (GFX PCR DNA and Gel Band purification, GE Healthcare) according to the instructions of the manufacturer. In the last step of this purification step the cDNA is eluted from the column provided with the kit with 40 μL water. The cDNA samples can be used immediately in qPCR reactions or stored at –20°C.

3.5.2. qRT-PCR

In real-time PCR assays, the amplification products are analyzed during the exponential phase of the reaction. Accordingly, the amount of amplified DNA is proportional to the initial cDNA concentration in the samples. The methodology described here is based on the use, in reaction mixtures, of a fluorochrome (SYBR® Green I; Eurogentec) that intercalates in dsDNA. The amount of light emission upon excitation is thus proportional to the double-stranded DNA concentration present in the samples. Finally, the obtained results are analyzed using the ΔΔCt method described by Livak and Schmittgen (16).

1. The primers used in qPCR reactions to amplify NASP, TPI, and β2MG are designed using the Primer Express™ 1.5 software, and analyzed with the bioinformatics tool Oligonucleotide Properties Calculator (http://www.basic.northwestern.edu/biotools/oligocalc.html) taking into account several parameters including the melting temperature, GC content, secondary structure, and length.
2. The PCR reactions are prepared using the commercial kit qPCR Core kit for SYBR™ Green I (Eurogentec). Reaction mixtures are prepared according to the instructions of the manufacturer and contain 3.5 mM $MgCl_2$, 200 μM each dNTP, 300 nM each primer, 0.5 U HotGoldStar enzyme, and 0.6 μL SYBR™ Green I, in a final volume of 20 μL.
3. The PCR program used in amplification reactions consists of 10 min at 95°C followed by 40 cycles with 15 s at 95°C, and 1 min at 60°C. All samples are assayed in triplicate in a GeneAmp® 5700 Sequence Detector System (Applied Biosystems) and results are analyzed with the GeneAmp® 5700 SDS v1.1 software and Microsoft Excel.
4. The subsequent use of the ΔΔCt method to calculate relative expression levels implies that the efficiency of amplification of each of the selected primers is very similar and above 95%. As a consequence, primer efficiency needs to be previously determined. This is accomplished using the following formula: $E = 10^{(-1/m)}$, where m is the slope of the linear regression with the different values of Ct (see Note 13) obtained during amplification of successive serial dilutions of the cDNA with

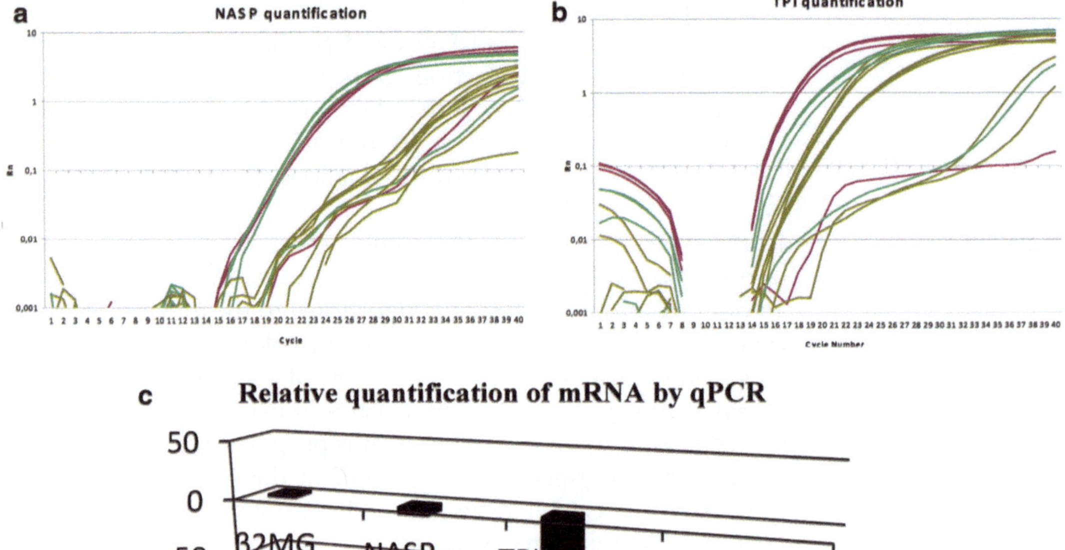

Fig. 4. Real-time PCR analysis of the histone H1 binding protein (NASP) and triosephosphate isomerase (TPI) in Huh7 and Huh7-D12 cells. (**a**, **b**) Amplification curves from real-time PCR performed in Huh7-D12 and Huh7 cells to quantify the mRNA levels of NASP and TPI. (**c**) The vertical axis represents the fold change in relative gene expression levels: 2.1-fold for NASP and 108-fold for TPI. The β-2-microglobulin gene (β2MG) was used as an internal housekeeping control. Fold changes were determined as $2^{-(\Delta Ct\ sample - \Delta Ct\ control)}$ ($p < 0.05$).

the selected primers. We routinely use 1:1, 1:10, 1:100, 1:1,000, and 1:10,000 dilutions. According to the ΔΔCt relative quantification method the Ct values obtained for different samples cannot be directly compared. Thus, the Ct values obtained for the mRNA of interest (hnRNP H) in Huh7 and Huh7-D12 cells are first compared with the Ct values obtained for the reference β2MG mRNA, in the same cDNA samples. This comparison ΔCt ($Ct_{hnRNPHRNA} - Ct\ \beta_{2MGRNA}$) is mandatory in order to normalize for the cDNA quantity obtained after reverse transcription of total mRNAs in Huh7 and Huh7-D12 cells. In the next step we compare the normalized Ct values determined for the hnRNP H mRNA in both Huh7 and Huh7-D12 cells, consequently obtaining a ΔΔCt value for each sample. The relative amount of mRNA in a given sample (N) is expressed by the formula $N = 2^{-\Delta\Delta Ct}$ (Fig. 4).

4. Notes

1. Unless stated otherwise, all solutions are prepared in water that has a resistivity of 18.2 MΩ cm and total organic content of less than 5 parts per billion.
2. TEMED bottles shall be tightly capped and stored at room temperature, preferably in a desiccator.
3. APS may be stored in aliquots at –20°C. Once thawed, do not freeze again and use immediately.
4. Add 200 μL of 0.001% (v/v) HCl to each vial containing 20 μg trypsin. After solubilization the solution is divided in 10 μL aliquots and stored at –20°C for a maximum of 6 months.
5. The protein sample volume to be loaded on the first dimension depends on the length of the strips to be used. On the other hand, the length of the strips depends also on the size of the second-dimension gels. Commercially available strips are usually of 7, 11, 13, 18, and 22 cm in length. In this system we use 18 cm IPG strips and load 400 μL of sample.
6. Before sealing the first-dimension IEF gel onto the second-dimension SDS-PAGE gel, the dissolved agarose needs to be cooled at about 42°C.
7. It is not mandatory to load molecular weight markers onto the second-dimension gel. After identification of proteins present in some spots, the respective molecular weight data can be introduced to the software which subsequently calculates the theoretical molecular weight of all spots present in the gel.
8. Benzonase is active in electrophoresis sample buffer, at room temperature, making it ideal to digest the released chromosomal DNA during preparation of protein samples.
9. High-molecular-weight proteins are usually transferred slower than low-molecular-weight proteins. Performing transfer at 2 mA/cm^2 during 30–60 min is usually a good compromise.
10. Antibody solutions shall be aliquoted and stored at –20°C. However, after thawing and dilution they may not be frozen again and need to be stored at 4°C. These antibodies can be used over several months. In order to avoid contaminations, usually 0.02% sodium azide is added from a 10% stock solution (care should be taken since sodium azide is very toxic). Note that for western blot purposes, the presence of sodium azide is not recommended. It may interfere with chemiluminescent systems producing undesirable backgrounds.
11. For an accurate quantification of the relative density of the gel bands it is very important that the signals are not saturated.

12. Work with RNA implies the use of special precautions to avoid degradation of the samples. Work on ice, and always use RNAse-free plastic and glassware. Use RNase inhibitors during manipulation and storage. If possible, reserve a dedicated, clean area on the bench solely for manipulation of RNA samples.
13. Ct represents the PCR cycle where the amplification curve intercepts a previously established threshold. In this particular case the threshold was set to 0.1 but may vary according to the experimental conditions.

Acknowledgements

This work was supported by Fundação para a Ciência e Tecnologia PPCDT/SAU-IMI/55112/2004.

References

1. Govindarajan S, Chin KP, Redeker AG, Peters RL (1984) Fulminant B viral hepatitis: role of delta agent. Gastroenterology 86:1417–1420
2. Jacobson IM, Dienstag JL, Werner BG, Brettler DB, Levine PH, Mushawar IK (1985) Epidemiology and clinical impact of hepatitis D virus infection. Hepatology 5:188–191
3. Taylor JM (2006) Structure and replication of the hepatitis delta virus RNA. Curr Top Microbiol Immunol 307:1–23
4. Bonino F, Heermann KH, Rizzetto M, Gerlich WH (1986) Hepatitis delta virus: protein composition of delta antigen and its hepatitis b virus-derived envelope. J Virol 58:945–950
5. Cheng D, Yang A, Thomas H, Monjardino J (1993) Characterization of stable hepatitis delta-expressing cell lines: effect of HDAg on cell growth. Prog Clin Biol Res 382:149–153
6. Kuo MYP, Goldberg J, Coates L, Mason W, Gerin J, Taylor J (1988) Molecular cloning of hepatitis delta virus RNA from an infected woodchuck liver: sequence, structure and applications. J Virol 62:1855–1861
7. Cunha C, Monjardino J, Chang D, Krause S, Carmo-Fonseca M (1998) Localization of hepatitis delta virus RNA in the nucleus of human cells. RNA 4:680–693
8. Gudima S, Chang J, Moraleda G, Azvolinsky A, Taylor J (2002) Parameters of human hepatitis delta virus genome replication: the quantity, quality, and intracellular distribution of viral proteins and RNA. J Virol 76:3709–3719
9. Mota S, Mendes M, Freitas N, Penque D, Coelho AV, Cunha C (2009) Proteome analysis of a human liver carcinoma cell line stably expressing hepatitis delta vírus ribonucleoproteins. J Proteomics 72:616–627
10. Pruijn GJ, Thijssen JP, Smith PR, Williams DG, Van Venrooij WJ (1995) Anti-La monoclonal antibodies recognizing epitopes within the RNA-binding domain of the La protein show differential capacities to immunoprecipitate RNA-associated La protein. Eur J Biochem 232:611–619
11. Bradford MM (1976) Rapid and sensitive method for the quantitation of microgram quantities of protein utilizing the principle of protein-dye binding. Anal Biochem 72:248–254
12. Candiano G, Bruschi M, Musante L, Santucci L, Ghiggeri GM, Carnemolla B, Orecchia P, Zardi L, Righetti PJ (2004) Blue silver: a very sensitive colloidal Coomassie G-250 staining for proteome analysis. Electrophoresis 25:1327–1333
13. Larsen MR, Cordwell SJ, Roepstorff P (2002) Graphite powder as an alternative or supplement to reversed-phase material for desalting and concentration of peptide mixtures prior to matrix-assisted laser desorption/ionization-mass spectrometry. Proteomics 2(9):1277–1287
14. Perkins DN, Pappin DJ, Creasy DM, Cottrell JS (1999) Probability-based protein identification by searching sequence databases

using mass spectrometry data. Electrophoresis 20(18):3551–3567

15. Shilov IV, Seymour SL, Patel AA, Loboda A, Tang WH, Keating SP, Hunter CL, Nuwaysir LM, Schaeffer DA (2007) The Paragon Algorithm, a next generation search engine that uses sequence temperature values and feature probabilities to identify peptides from tandem mass spectra. Mol Cell Proteomics 6(9): 1638–1655
16. Livak K, Schmittgen TD (2001) Analysis of relative gene expression data using real time quantitative PCR and the $2^{-\Delta\Delta Ct}$ method. Methods 25:402–408

Chapter 15

Identification of New Autoimmune Hepatitis-Specific Autoantigens by Using Protein Microarray Technology

Lin Wu and Guang Song

Abstract

Autoimmune hepatitis (AIH) is a chronic necroinflammatory disease of the liver with a poorly understood etiology. Detection of non-organ-specific and liver-related autoantibodies using immunoserological approaches has been widely used for diagnosis and prognosis. However, unambiguous and accurate detection of the disease requires the identification and characterization of disease-specific autoantigens. In the present study, we have used protein microarray technology to profile the autoantigen repertoire of patients with AIH versus those with other liver diseases, identifying and validating novel and highly specific biomarkers for AIH. Using two algorithms developed in the current study, we were able to demonstrate that a panel of six autoantigens could be used in combination to clearly diagnose AIH-positive serum samples.

Key words: AIH, Autoimmune diseases, Autoantibodies, Autoantigens, Protein microarray

1. Introduction

Autoimmune hepatitis (AIH) is a chronic necroinflammatory disease of the liver with a poorly understood etiology. At present, indirect immunofluorescent (IIF) staining of easily accessible rodent tissue and HEp-2 cells is the most frequently used method for detecting circulating antinuclear autoantibodies (ANAs), anti-smooth muscle autoantibodies (anti-SMAs), and anti-liver and -kidney microsomal autoantibodies (anti-LKM) (1); the antigens recognized by these autoantibodies serve as diagnostic antigens for profiling autoantibodies in patients' sera. ELISA methods

Djuro Josic and Douglas C. Hixson (eds.), *Liver Proteomics: Methods and Protocols*, Methods in Molecular Biology, vol. 909, DOI 10.1007/978-1-61779-959-4_15,

employing autoantibodies that target specific autoantigens are being used, or are under development, as a complementary approach to IIF diagnosis, with the goal of improving diagnostic accuracy (2).

Several attempts have been made to identify the targets using conventional methods of screening expression libraries, and have succeeded in identifying cytochrome P450db1 (3), a 50-kDa protein, cytosolic UGA-suppressor tRNA-associated protein (4), and asialoglycoprotein receptor (ASGP-R) (5). However, screening expression libraries by conventional methods is a time-consuming, tedious process that tends to miss proteins that are in low abundance but robust in serological assays. Here, we use protein microarray technology, a newly developed analytical method that has been shown to have great potential for identifying autoimmune disease biomarkers (6–8).

As a means of identifying novel AIH-specific autoantigens, we have screened serum samples from healthy serum controls and individuals with AIH against a protein microarray containing 5,011 human proteins. We then developed an AIH-specific autoantigen array and used it to verify the findings with patients with AIH versus those with other liver diseases. By this approach, we succeeded in identifying six AIH-specific autoantigens including three novel ones. Finally, we developed two algorithms for AIH prediction and tested them in double-blinded samples.

2. Materials

2.1. Protein Expression and Purification

1. Yeast recombinant human protein strains: Collections are stored in our laboratory at −80°C. These strains were constructed by inserting human protein ORFs downstream of a glutathione-*S*-transferase (GST) coding sequence in a yeast expression vector pEGH via in vivo recombination, and the resulting proteins have a GST tag at their N-terminal. The expression of the GST-fusion protein is under the regulation of GAL1 inducible promoter, requiring the induction by galactose. The pEGH vector also contains a URA3 gene, so a synthetic complete medium lacking uracil is used as a positive selection to maintain the expression constructs in cells.
2. SC-ura/raffinose medium: Dissolve 6.7 g YNB, 20 g raffinose, and 2 g dropout mix (URA^-) in ddH_2O, and adjust the volume to 1 L. The medium is autoclaved and can be stored at 4°C for at least half a year.

3. 40% galactose: Dissolve 40 g D-galactose in hot ddH_2O and adjust the volume to 100 mL. After autoclave, the solution can be stored at 4°C for at least 1 month.
4. Lysis Buffer: 50 mM Tris–HCl (pH 7.5), 100 mM NaCl, 1 mM EGTA, 0.1% Triton X-100. The lysis buffer can be stored at 4°C for at least half a year. Right before lysing yeast cells, add in 60 mL lysis buffer 350 μL 0.1 M PMSF, 70 μL β-mercaptoethanol, and 1 pulverized Complete Protease Inhibitor Cocktail Tablet (Roche, Basel, Switzerland) (see Note 1).
5. Glutathione sepharose 4B beads (GE HealthCare, Buckinghamshire, UK): Each 1.5 mL aliquot of beads was washed three times with 7.5 mL lysis buffer. After each wash, beads were recovered by centrifugation at 500×*g* for 5 min and the supernatant was discarded. After the last wash, each aliquot of beads was resuspended in 10 mL lysis buffer.
6. Wash Buffer I: 50 mM Tris–HCl (pH 7.5), 500 mM NaCl, 1 mM EGTA, 0.1% Triton X-100. It can be stored at 4°C for at least half a year.
7. Wash Buffer II: 50 mM HEPES (pH 7.5), 100 mM NaCl, 10% glycerol. It can be stored at 4°C for at least half a year.
8. Elute Buffer: 50 mM HEPES (pH 7.5), 100 mM NaCl, 30% glycerol, 30 mM glutathione (reduced form), 0.5% Triton X-100, 1 mM DTT. The elute buffer should be freshly made (see Note 2).
9. AcroPrep™ 96-well Filter Plates (PALL, NY, USA).

2.2. Fabrication of the Human Protein Microarrays

1. Poly-L-lysine-coated slides (Beijing Protein Innovation, LTD, Beijing, China) (see Note 3).

2.3. Serum Assays on Human Protein Chips

1. PBS: Dissolve 8 g NaCl, 0.2 g KCl, 1.44 g Na_2HPO_4, 0.24 g KH_2PO_4 in 800 mL distilled water, adjust pH to 7.4, and then add more ddH_2O to final volume of 1 L. PBS can be stored at 4°C for half a year.
2. PBST: Add 0.5 mL Tween20 to 1 L PBS. PBST can be stored at 4°C for at least 1 month.
3. Blocking Buffer: Add 1% [w/v] BSA in PBST. The blocking buffer should be freshly made.
4. Glass coverslip (LifterSlip, Erie Scientific Company, Portsmouth, NH, USA).
5. Anti-human IgG antibody labeled with Cy-5 (The Jackson Laboratory, Bar Harbor, ME, USA). This should be aliquoted and stored in a aluminum foil-covered container at −20°C. It can be stored for at least 1 year.
6. 12-hole rubber gasket (BioCapital, Beijing, China).

3. Methods

We exploited a two-phase strategy to identify AIH-specific autoantigens. In the first phase, we probed a high-content human protein microarray with serum samples from 22 AIH patients and 30 healthy controls to quickly narrow down the number of potential AIH autoantigens to a manageable number. In the second phase, we did extensive serum profiling including several types of disease controls (primary biliary cirrhosis (PBC), Hepatitis C (HC), and other diseases) on AIH-candidate protein microarray.

Data analysis is very important in microarray technology. In phase I, we used a less stringent scoring system to identify as many potential candidates as possible. We included several known autoantigens in the microarray to aid in the establishment of our scoring system. In phase II, we used a value of three times the SD above $\text{mean}_{\text{healthy}}$ as a cutoff value to compare the positive rates of each candidate among different disease groups. At the same time, we used two kinds of statistical algorithms, logistic regression and discriminant analysis, to identify candidates that significantly differentiate AIH from non-AIH groups. Logistic regression and discriminant analysis were applied to a panel of six candidates (including known autoantigens) that are qualified in all three analyses to come up with two formulas for disease prediction.

When these formulas were applied to additional AIH samples in a double-blind format, they achieved levels of specificity and sensitivity similar to those obtained with the training set.

3.1. Preparation of Samples

1. Draw blood from patients from various diseases and healthy groups, centrifuge and collect sera, aliquot, and store at −80°C.
2. Right before use for microarray hybridization, dilute the sera in PBST as required and filter them with a 0.45 μm filter.

3.2. Protein Expression and Purification

Protein expression and purification were conducted as described previously (9). The following procedure is done in 96-deep well plates with round bottom, and the reagent amount listed is for each well.

1. Activate expression yeast strains by picking single colonies in 300 μL SC-ura/raffinose medium and incubate at 30°C for 24 h with shaking at 220 rpm.
2. Transfer 20 μL activated yeast to 800 μL SC-ura/raffinose medium and incubate at 30°C with shaking at 220 rpm for about 4 h till its OD_{600} reading reaches 0.6–0.8. Then add 40% galactose to the final concentration of 2%, and incubate at 30°C overnight with shaking.

3. Collect yeast by centrifugation at 4°C with 2,500 × *g* for 3 min. Wash the pellet with 200 μL cold water, resuspend, and spin at 4°C with 2,500 × *g* for 3 min.
4. Resuspend the yeast pellet in 300 μL lysis buffer containing PMSF, β-mercaptoethanol, and protease inhibitors, and add 100 μL ceramic beads with a diameter of 0.5 mm (Zhuzhou Jintro Advanced Ceramics Co., Ltd, China). Vortex for 40 s followed by 2 min on ice. Repeat four times. Then spin at 4°C for 3 min at 2,500 × *g*. Collect supernatant.
5. Repeat step 4 and combine two supernatants. Spin again at 4°C for 5 min at 3,200 × *g*. Transfer supernatant to a new plate.
6. Add 100 μL well-suspended GST beads to each well, seal the plate, and rotate at 4°C for 2 h.
7. Spin at 800 × *g* for 2 min at 4°C, discard supernatant, and wash beads three times with 200 μL Wash Buffer I and three times with 200 μL Wash Buffer II. For each wash, vortex and spin at 800 × *g* for 2 min at 4°C and discard supernatant.
8. Add 40 μL elute buffer, seal the plate, briefly vortex, and rotate overnight at 4°C.
9. Spin at 200 × *g* from 1 min, transfer whole content (supernatant and beads) to an AcroPrep™ 96-well Filter Plate containing 10 μl ddH_2O. Put a new regular 96-well round-bottom plate underneath, and spin at 1,800 × *g* for 1 min to collect eluate (purified protein). The purified protein can be stored at −80°C for long-term storage.
10. The quality of the protein purification should be checked by SDS-PAGE.

3.3. Fabrication of the Human Protein Microarrays

1. Array 10 μL each of purified and control proteins in 384-well titer dishes using PP-125N-MS Personal Pipettor (APRICOT DESIGN, CA, USA).
2. Spot proteins in duplicate onto poly-L-lysine-coated slides using a SpotArray-72 printer (Perkin Elmer, MA, USA).
3. Keep the printed human protein microarrays horizontally at room temperature for 1 h before storage at 4°C (see Note 4).

3.4. Serum Assays on Human Protein Microarrays

1. Immerse the protein microarrays in blocking buffer, and incubate for 1 h at room temperature on a horizontal shaker.
2. Take out the protein microarrays from the blocking buffer and absorb the liquid with tissue paper. Slowly add 200 μL of serum diluted 1:1,000 in PBST (see Note 5) onto the microarray and carefully cover the microarray with a glass coverslip (avoid bubbles) to maintain the appropriate humidity. Incubate in a humidified box for 1 h at room temperature.

3. Carefully take off the coverslip, rinse the microarray briefly with PBST, and wash it three times for 10 min each at 40°C with pre-warmed PBST.
4. Take out the protein microarrays and absorb the liquid with tissue paper. Carefully add 200 μL of 1:1,000 dilution of Cy-5-labeled anti-human IgG antibody in PBST, cover with a coverslip, and incubate in a humidified box in the dark (to avoid fluorescent quenching) for 1 h at room temperature.
5. After taking off the coverslip, rinse the microarray briefly and wash three times for 10 min each with PBST.
6. Rinse the microarray twice with double-deionized water, spin briefly, and blow the microarray to dryness with compressed air.
7. Scan the microarray with a BioCapital microarray scanner (BioCapital, Beijing, China).
8. Acquire and analyze the binding signals using GenePix Pro 5.0 software (Molecular Devices, Sunnyvale, CA, USA) according to the software instruction. An example of the scan image is shown in Fig. 1.

3.5. Analysis of the Microarray Data

1. Following analysis with GenePix Pro 5.0, determine the signal intensity of each protein feature by subtracting the median background value from the median foreground value. If the resulting signals are <0, assign a minimum signal of 1. Since every protein was spotted in duplicate, the average of the duplicate can be used as the final signal intensity for a given protein.
2. Perform inter-microarray normalization using the total signal intensity. Divide the total signal intensity for a given microarray (microarray signal*n*) by the mean value for all microarrays (average (microarray signal*n*)) to obtain an intensity factor for the microarray (intensity factor (microarray*n*)). Divide each signal for a given microarray by the intensity factor (microarray*n*) to generate a normalized signal (see Note 6).
3. Calculate the mean value of a given protein for all control serum samples ($\text{average}_{\text{protein}}$ *i* control value). Determine the so-called factor *F* value by dividing the signal intensity value for each protein in the individual patient's serum by the $\text{average}_{\text{protein}}$ *i* control value. Proteins with *F* value of >2 are considered positive (10).
4. Determine the percentage of patient sera positive for each protein. Proteins with a positive rate greater than 59% among all the patient sera are considered candidates for autoantigens which in the present case are specific for AIH (see Note 7). The list of candidates and their percentage in all patients' sera is shown in Table 1.

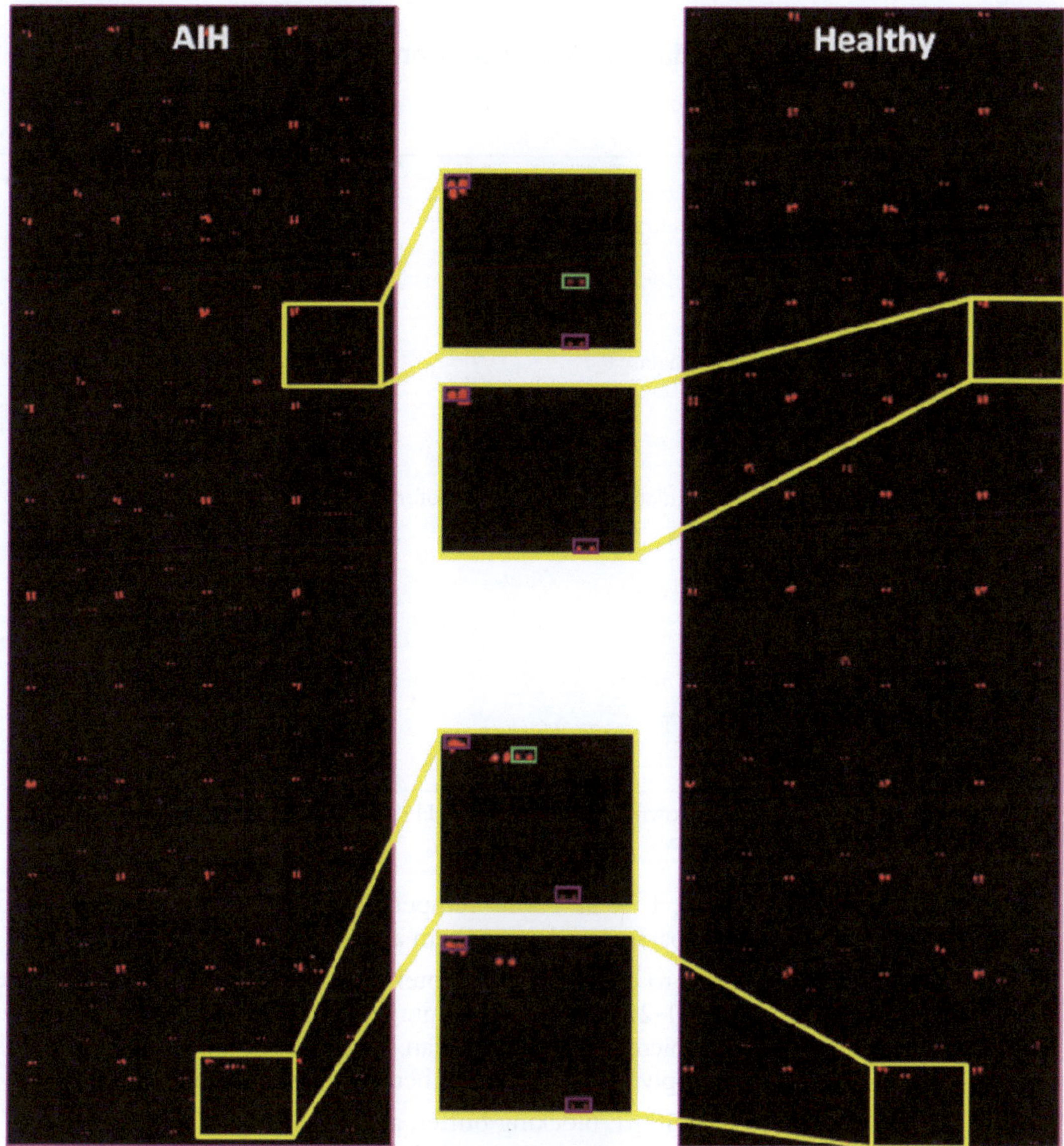

Fig. 1. Probing of the human protein microarray with AIH-positive and healthy control sera. Twenty-two AIH and 30 healthy control serum samples were diluted 1:1,000 and individually incubated with the human protein microarray, followed by the addition of the anti-human IgG antibody (Cy5-conjugated). Microarrays were dried and scanned to acquire the images. Representative images are shown, and selected areas in *yellow boxes* are enlarged and illustrated. *Green boxes* indicate the positive candidate autoantigens, and *purple boxes* indicate human IgG (the positive control).

3.6. Fabrication of an AIH-Candidate Protein Microarray and Its Probing with Human Sera

1. Print all of the AIH-specific autoantigen candidates identified on the human protein microarray together with controls (human IgG, GST, and printing buffer) in duplicate within a 2.52 mm × 1.44 mm area. Fabricate 12 identical probe areas on a single poly-L-lysine-treated microscope slide.

Table 1
AIH-specific autoantigen candidates identified by protein microarray

Protein ID	Percentage in AIH (%)
RPS20	68.2
Alba-like (hypothetical, Alba-like protein)	68.2
p62 (nucleoporin 62 kDa)	63.6
CYP2D6	63.6
PRO0245 (dUTPase, dUTP diphosphatase)	63.6
ASGPR2	59.1
GC-rich promoter binding protein 1	59.1
Similar to protein phosphatase 2A regulatory subunit delta isoform	59.1
MGC16385 protein	59.1
Glutaredoxin-1 (thioltransferase-1, TTase-1)	59.1
LOC57002 (hypothetical protein)	59.1
Hypothetical protein (ENST00000333859)	59.1
Ubiquitin-like protein SMT3B (HSMT3)	59.1
PRO3121	59.1

Notes: Italicized protein IDs are the known autoantigens of AIH

2. After 1 h at room temperature, vacuum seal the printed microarrays and store them at 4°C.
3. Prior to use, warm the protein microarray at room temperature for 10–20 min before opening the sealed bag. In order to probe the microarrays with human sera, form 12 individual chambers by applying a 12-hole rubber gasket to each microarray.
4. Add 50 μL blocking buffer, and incubate for 1 h at room temperature on a horizontal shaker. Remove the blocking buffer by carefully pipetting.
5. Add 50 μL of filtered 1:1,000 diluted human serum in each microarray chamber, and incubate in a humidified box for 1 h at room temperature (see Note 8).
6. Rinse the microarrays with PBST before carefully removing the rubber gaskets. Wash three times with pre-warmed PBST at 40°C with 10 min each.
7. Take out the protein microarrays and soak up the liquid with tissue paper. Carefully add 200 μL of 1:1,000 dilution of Cy-5-labeled anti-human IgG antibody in PBS, cover with a coverslip, and incubate in a humidified box in the dark for 1 h at room temperature.

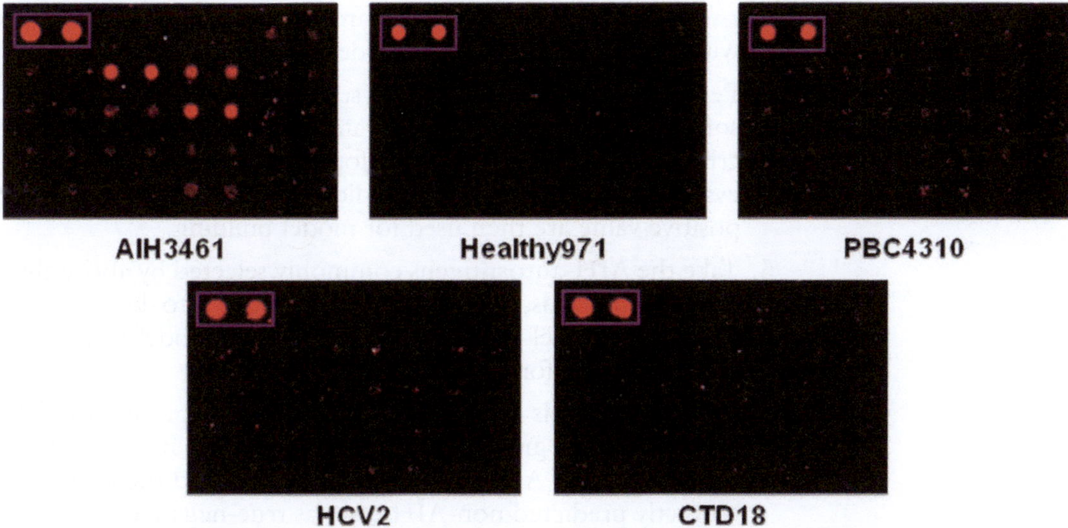

Fig. 2. Large-scale serum profiling using AIH-candidate microarrays. AIH autoantigen candidates were purified and printed onto a poly-L-lysine-coated microscope slide to form the AIH-candidate protein microarray. Shown are representative microarray images probed with sera collected from AIH patients, healthy subjects, PBC, HCV, or CTD patients. *Purple boxes* show the human IgG spots as a landmark.

8. Wash the microarray three times with PBST for 10 min.
9. Rinse the microarray twice with double-deionized water, spin briefly, and blow the microarray to dryness with compressed air. Scan the microarray with a BioCapital microarray scanner (BioCapital, Beijing, China).
10. Acquire binding signals and analyze using the GenePix Pro 5.0 software (Molecular Devices, Sunnyvale, CA, USA). A typical image for AIH-candidate protein microarray is shown in Fig. 2.

3.7. AIH-Candidate Microarray Data Analyses on Training Set to Establish the Prediction Model

1. Following image analysis with GenePix Pro 5.0, determine the signal intensity of each protein feature by subtracting the median background value from the median foreground value. If the resulting value is <0, assign a minimum signal of 1. Use the average of the two duplicates for each protein as the final signal intensity.
2. Calculate the mean signal intensities and standard deviations (SDs) for 50 healthy controls. Use the value for the $\text{mean}_{\text{healthy}} + 3 \times \text{SD}$ for each potential autoantigen as an immnoreactive cutoff value. Count the number of positives for each potential autoantigen in each group and calculate their positive rate in each group. Select those autoantigens with relatively high positive rates for further analysis.
3. Perform logistic regression analysis (see Note 9) using SPSS 12.0 software. The logistic regression analysis first performs

t-tests for each of the 14 autoantigens and then selects those with p values <0.01 for the model building.

4. Perform discriminant analysis (see Note 10) using SPSS 12.0 software. The model first generates standardized canonical discriminant function coefficients for each autoantigen in order to evaluate its importance in prediction. The autoantigens with positive value are then used for model building.
5. Take the AIH-autoantigens commonly selected by above three analysis methods, and apply their signals into both logistic regression model and discriminant analysis model to generate two prediction formulas.
6. Apply the signals into these two formulas to predict the AIH from non-AIH group. Count and record the number of correctly predicted AIH cases as true-positive, and the number of correctly predicted non-AIH cases as true-negative.
7. Calculate their sensitivity and specificity in prediction as follows:

   Sensitivity = true positive/AIH cases;

   Specificity = true negative/non-AIH cases.

3.8. Apply Prediction Model to Double-Blinded Test Set Data to Validate the Prediction Model

1. Probe double-blinded test serum samples on the AIH candidate protein microarray, and acquire signals.
2. Apply microarray signals into the two formulas to calculate their sensitivity and specificity in AIH prediction of test set data.

The prediction result for both training and double-blinded test samples is listed in Table 2.

Table 2
Prediction performance of two classification methods

Sample set (AIH/non-AIH cases)	Prediction method	True-positive	True-negative	Sensitivity[a] (%)	Specificity[b] (%)	Accuracy (%)
Training set (44/234)	Logistic regression	30	225	68.20	96.20	91.70
	Discriminant analysis	37	218	84.10	93.20	91.70
Test set (15/26)	Logistic regression	9	26	60	100	85.40
	Discriminant analysis	10	26	66.70	100	87.80

[a]Sensitivity is defined as the true positive rate
[b]Specificity is defined as true negative rate

4. Notes

1. To make 0.5 M EGTA (pH8.0), add 190.2 g EGTA in 800 mL ddH_2O and vigorously stir the mixture with a magnetic stirrer until fully dissolved. Then adjust pH value to 8.0 with NaOH (about 20 g) and add more ddH_2O to make the final volume to 1 L. After autoclave or filtration, it can be stored at 4°C for at least 1 month. To make 0.1 M PMSF, dissolve 174 mg PMSF in 10 mL isopropanol, then aliquot into small tubes covered with aluminum foil, and store at –20°C. Before use, warm the tube and make sure that it is totally liquidized.
2. The bottle of the reduced form of glutathione should be tightly sealed to prevent oxidization.
3. Microscopic slides with various surface chemistry can be used for printing proteins, such as poly-L-lysine, epoxy, etc. Precoated microscopic slides are commercially available worldwide.
4. The printed microarray can be stored at 4°C for a month or at –80°C for at least half a year.
5. The fold of serum dilution used for microarray assay needs to be tested before the real experiment. In our hands, 1:500 and 1:1,000 dilutions gave similar results in terms of signal/noise ratio. To eliminate noise and get a clearer background, it is recommended to filter all the solutions used in the microarray hybridization process.
6. There are several ways of inter-microarray normalization. We used a global normalization method to even out the total signals for each microarray. One may also use a quantile normalization method (11) to even out the median value for each microarray, or use a proCAT approach (12).
7. The criteria used for narrowing down the potential candidates in Phase I depend on the subject studied. It is better to use some known markers as a help for establishing a suitable scoring system. The purpose here is to get as many good candidates as possible, and at the same time, in a manageable number. In the case of AIH, we set the *F* value greater than 2 as positive combined with a greater than 59% positive rate among patients. With these criteria, three known markers qualified and several new candidates with similar or even better scores were identified.
8. The fold of dilution used for serum in chambered microarray needs to be tested before the real experiment. The dilution used in the regular microarray may not be suitable here because the fluid dynamics is different during shaking incubation.
9. Logistic regression is a generalized linear regression model for binary responses. The candidate features were selected by the

model using a dual-direction stepwise search with Akaike's information criteria, and the probabilities were calculated as follows: $p = \text{EXP}(\Sigma(bixi) + c)/(1 + \text{EXP}(\Sigma(bixi) + c))$, where p is the probability of each case, $i = 1$ to n; b is the regression coefficient of a given autoantigen; x is the signal intensity; and c is a constant generated by the model. For prediction, only those cases with a >0.5 probability were classified as positive.

10. Discriminant analysis is a linear regression model that predicts group membership by using a set of predictors. The formula is $y = \Sigma(bixi) + c$, where $i = 1$ to n, x is the signal intensity, b represents the unstandardized canonical discriminant function coefficient, and c is a constant provided by the model. Those cases in which the y value was >0 were considered positive.

Acknowledgements

This work is an elaboration on details and strategy of the work published by Song et al. (13). And this work was supported in part by funding from Ministry of Science and Technology of the People's Republic of China (2006AA02A311 to L.W.).

References

1. Zachou K, Rigopoulou E, Dalekos GN (2004) Autoantibodies and autoantigens in autoimmune hepatitis: important tools in clinical practice and to study pathogenesis of the disease. J Autoimmune Dis 1:2
2. Bogdanos DP, Invernizzi P, Mackay IR, Vergani D (2008) Autoimmune liver serology: current diagnostic and clinical challenges. World J Gastroenterol 14:3374–3387
3. Manns MP, Johnson EF, Griffin KJ, Tan EM, Sullivan KF (1989) Major antigen of liver kidney microsomal autoantibodies in idiopathic autoimmune hepatitis is cytochrome P450db1. J Clin Invest 83:1066–1072
4. Wies I, Brunner S, Henninger J, Herkel J, Kanzler S, Meyer Zum Büschenfelde KH, Lohse AW (2000) Identification of target antigen for SLA/LP autoantibodies in autoimmune hepatitis. Lancet 355:1510–1515
5. McFarlane BM, McSorley CG, McFarlan IG, Williams RL (1985) A radioimmunoassay for detection of circulating antibodies reacting with the hepatic asialoglycoprotein receptor protein. J Immunol Methods 77:219–228
6. Kalbas M, Lueking A, Kowald A, Muellner S (2006) New analytical tools for studying autoimmune diseases. Curr Pharm Des 12: 3735–3742
7. Chen CS, Sullivan S, Anderson T, Tan AC, Alex PJ, Brant SR, Cuffari C, Bayless TM, Talor MV, Burek CL, Wang H, Li R, Datta LW, Wu Y, Winslow RL, Zhu H, Li X (2009) Identification of novel serological biomarkers for inflammatory bowel disease using *Escherichia coli* proteome chip. Mol Cell Proteomics 8:1765–1776
8. Patwa TH, Li C, Poisson LM, Kim HY, Pal M, Ghosh D, Simeone DM, Lubman DM (2009) The identification of phophoglycerate kinase-1 and histone H4 autoantibodies in pancreatic cancer patient serum using a natural protein microarray. Electrophoresis 30: 2215–2226
9. Zhu H, Bilgin M, Bangham R, Hall D, Casamayor A, Bertone P, Lan N, Jansen R, Bidlingmaier S, Houfek T, Mitchell T, Miller P, Dean RA, Gerstein M, Snyder M (2001) Global analysis of protein activities using proteome chips. Science 293:2101–2105
10. Lueking A, Luekinga A, Huberb O, Wirthsd C, Schultea K, Stielerf KM, Blume-Peytavif U, Kowaldg A, Hensel-Wiegeld K, Tauberd R,

Lehrachg H, Meyera HE, Cahilla DJ (2005) Profiling of alopecia areata autoantigens based on protein microarray technology. Mol Cell Proteomics 4:1382–1390

11. Bolstad BM, Irizarry RA, Astrand M, Speed TP (2003) A comparison of normalization methods for high density oligonucleotide array data based on variance and bias. Bioinformatics 19:185–193

12. Zhu X, Gerstein M, Snyder M (2006) ProCAT: a data analysis approach for protein microarrays. Genome Biol 7:R110

13. Song Q, Liu G, Hu S, Zhang Y, Tao Y, Han Y, Zeng H, Huang W, Li F, Chen P, Zhu J, Hu C, Zhang S, Li Y, Zhu H, Wu L (2010) Novel autoimmune hepatitis-specific autoantigens identified using protein microarray technology. J Proteome Res 9:30–39

Chapter 16

Proteomics Analysis of Human Nonalcoholic Fatty Liver

Eva Rodriguez-Suarez, Jose M. Mato, and Felix Elortza

Abstract

Nonalcoholic fatty liver disease (NAFLD) is being increasingly recognized as a major cause of liver-related morbidity and mortality. Given the increasing prevalence of obesity in western countries, NAFLD has become an important public health problem. The principal aim of this study was to find differences in protein expression between patients with NAFLD and healthy controls. Changes in protein expression of liver samples from controls, nonalcoholic steatosis, and nonalcoholic steatohepatitis (NASH) subjects were analyzed by two-dimensional differential in-gel electrophoresis (DIGE). With this proteomic technique, hundreds of proteins can be analyzed simultaneously and their relative abundance can be calculated. Proteins showing significant changes (ratio ≥ 1.5, $p < 0.05$) were identified by MALDI TOF/TOF mass spectrometry. Western blot of tissue homogenates was then used as a complementary method to validate protein expression changes observed by DIGE. With the aim to have a noninvasive approach to detect changes produced in NAFLD-affected liver, validated proteins were further tested in serum samples of different cohorts of patients. Following this approach, we identified two candidate markers CPS1 and GRP78 that were differentially expressed between control, steatosis, and NASH. This proteomics approach demonstrates that DIGE combined with MALDI TOF/TOF and Western blot analysis of tissue and serum samples is a useful approach to identify candidate markers associated with NAFLD.

Key words: NAFLD, Liver proteomics, DIGE, MALDI-TOF/TOF, Western blot

1. Introduction

NAFLD is a term that encompasses a spectrum of abnormalities that range from simple triglyceride accumulation in the hepatocytes (hepatic steatosis) to hepatic steatosis with inflammation (nonalcoholic steatohepatitis or NASH), which may lead to hepatic fibrosis and cirrhosis, resulting in increased morbidity and mortality (1–3). Features of metabolic syndrome such as obesity, type 2 diabetes, and hyperlipidemia are all associated with nonalcoholic

Djuro Josic and Douglas C. Hixson (eds.), *Liver Proteomics: Methods and Protocols*, Methods in Molecular Biology, vol. 909, DOI 10.1007/978-1-61779-959-4_16, © Springer Science+Business Media New York 2012

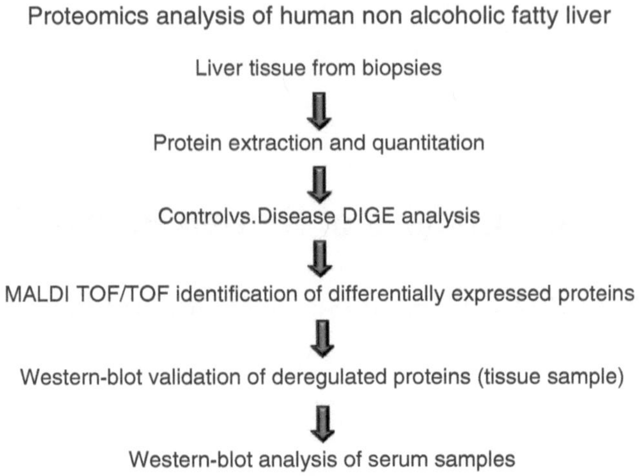

Fig. 1. Strategy followed for proteomics analysis of human non alcoholic fatty liver.

fatty liver disease (NAFLD) (4, 5). With the increasing prevalence of these metabolic conditions in the general population, NAFLD is now recognized as the most common chronic liver disease in Western World (6, 7). Numerous studies have been carried out to help to elucidate the complex molecular mechanism of NAFLD and its progression to more severe stages of the disease such as hepatocellular carcinoma (HCC) (8, 9). In fact, the poor survival of patients with HCC is related in part to the lack of reliable tools for early diagnosis. Therefore, a combination of these molecular insights and complex algorithms has been used to design a predictive test of fibrosis (10, 11) and even more, tests which may distinguish cirrhotic patients with HCC from those without (12).

Proteomics is a powerful tool for studying variations in protein expression in different populations of diseased and healthy patients. The identification of differentially expressed proteins may have a huge impact by increasing the availability of molecular markers for early diagnosis and therapy. Two-dimensional electrophoresis (2-DE) is a widespread method in proteomics useful to compare changes in protein expression (13). Hence, 2-DE analysis of cell lines, organelles, tissue, serum, or animal models (14–18) has been used to look for differences in protein expression in NAFLD and other liver-related diseases. A proteomic investigation of drug-induced steatosis in rat liver has led to the identification of protein markers (19). HCC-related biomarkers have also been identified in previous studies (20, 21).

In this chapter, we describe a strategy for using DIGE in combination with MALDI TOF/TOF to identify differentially expressed proteins in liver affected by NAFLD and validate these proteins as serum markers which are able to distinguish between steatosis and NASH. Following this approach (Fig. 1), we identified two serum proteins (CSP-1 and GRP78) that differentiate between healthy controls and NAFLD and which can also discriminate NASH from simple steatosis. To our knowledge this is the first time that DIGE has been used for a comprehensive comparative study of steatosis and early NASH in human liver (22).

2. Materials

2.1. Human Samples

For this study, we obtained surgically resected specimens from each of the following three groups of subjects (see Note 1).

We obtained six liver samples from each of the following three groups of subjects:

1. Six cholecystectomy controls with normal liver function and histology.
2. Six morbidly obese patients diagnosed with nonalcoholic hepatic steatosis. A diagnostic of hepatic steatosis in the absence of other (viral, alcohol, metabolic) causes of liver disease was established histologically.
3. Six morbidly obese patients diagnosed with early-stage NASH. A diagnostic of steatohepatitis grade 1 was established histologically (macrovesicular steatosis, lobular, and portal inflammation) in the absence of other (viral, alcohol, metabolic) causes of steatohepatitis.
   (a) Liver samples were obtained during open cholecystectomy (1) or open bariatric surgery (groups 2 and 3) using the same procedure and treated in an identical fashion.
   (b) No pre-operation regimen was used before surgery.
   (c) Biopsies were obtained at the beginning of the surgical procedure.
   (d) A single pathologist who was unaware of the clinical data of patients made the histological diagnosis.
   (e) Liver samples were divided into two groups. One group was processed for routine histology and the other group was immediately frozen in liquid nitrogen and stored at -80°C.
   (f) Serum was prepared by incubating patient venous blood in serum separator tubes for 30 min before centrifugation (2,500 × *g*, 15 min).
   (g) Aliquots of supernatants were placed into microtubes and stored at −80°C until required. Fifteen serum samples of each cohort of patients were used for this study.
   (h) Patients with other forms of liver disease, including hemochromatosis, Wilson's disease, celiac disease, and drug-induced liver disease, were excluded from the study.

(i) None of the patients included in this study was treated with insulin.

(j) The Human Research Review Committees of the participating hospitals approved the study.

(k) Informed consent was obtained from each patient included in the experimental section and the study protocol conforms to the ethical guidelines of the 1975 Declaration of Helsinki.

2.2. Protein Extraction

1. Lysis buffer: 7 M urea, 2 M thiourea, 4% v/v 3-[(3-cholamidopropyl)dimethylammonio]-1-propanesulfonate (CHAPS), 1% v/v (2S,3S)-1,4-Bis-sulfanylbutane-2,3-diol (DTT), and 0.5% v/v Bio-Lyte 3–7 ampholytes (Amersham Biosciences) (see Note 2).
2. Protein concentration of the supernatant was measured with the Bradford assay kit (Bio-Rad, Hercules, CA) using bovine serum albumin (BSA) (Sigma) as standard protein (see Note 3). Reagents: BSA 10×: 1 mg/ml in Milli-Q water. Keep at −20°C in aliquots. BSA 1× (0.1 mg/ml): dilute BSA 10× in the buffer of your sample.
3. Samples were desalted and concentrated with the 2D-Clean up kit (Amersham Biosciences, GE Healthcare) following the manufacturer's recommendations.
4. Final buffer for first-dimension electrophoresis, isoelectric focusing (IEF): 30 mM 2-Amino-2-hydroxymethyl-propane-1,3-diol (Tris), 7 M urea, 2 M thiourea, and 4% v/v CHAPS.

2.3. First-Dimension Electrophoresis: Isoelectric Focusing

1. 3–7 IPG strips, 24 cm (Bio-Rad) (see Note 4).
2. Rehydration buffer: 7 M urea, 2 M thiourea, 4% (v/v) CHAPS, 1% (v/v) DTT, and 1% (v/v) Biolyte 3–7 ampholytes (see Notes 5 and 6).
3. Equilibration buffer stock: 50 mM Tris–HCl pH 7.5, 6 M urea, 30% (v/v) glycerol, 2% (v/v) sodium dodecyl sulfate (SDS) (see Note 7).
4. Reduction buffer: 50 mM Tris–HCl pH 7.5, 6 M urea, 30% (v/v) glycerol, 2% (v/v) SDS, and 2% (v/v) DTT.
5. Alkylation buffer: 50 mM Tris–HCl pH 7.5, 6 M urea, 30% (v/v) glycerol, 2% (v/v) SDS, and 2.5% (v/v) iodoacetamide (IAA). Keep in the dark.

2.4. Second-Dimension Electrophoresis: 2D-SDS-Polyacrylamide Gel Electrophoresis (2D-PAGE) (12.5%)

1. 30% w/v Polyacrylamide (PAA, Bio-Rad): 0.8% w/v Piperazine di- Acrylamide (PDA, Bio-Rad). Stocks: 40% w/v PAA 937 ml, PDA 10 g and make up to 1,250 ml with Milli-Q water; 1.5 M Tris, pH 8.8: 181 g/l Tris, pH with HCl to 8.8; 0.5 M Tris, pH 6.8: 60.5 g/l Tris, pH with HCl to 6.8 (see Note 8).
2. 20 l of 1× Running Buffer: Tris 250 mM: 60.4 g, glycine 192 mM: 288.4 g, SDS 1% w/v: 20 g, water to 20 l.

3. Resolving gel solution: 150 mM Tris pH 8.8, 10% acrylamide–PDA, 0.1% (w/v) SDS, 0.067% (v/v) tetramethyl ethylenediamine (TEMED), and 0.25% (v/v) ammonium persulfate (APS) (all from Amersham Pharmacia).
4. Stacking gel solution: 125 mM Tris pH 6.8, 6% (v/v) acrylamide-PDA, 0.1% (w/v) SDS, 0.134% (v/v) TEMED, and 0.25% (v/v) APS.

2.5. Western Blot

1. Sample buffer: 250 mM Tris–HCl pH 6.8, 500 mM β-mercaptoethanol, 50% glycerol, 10% SDS, bromophenol blue.
2. Blocking solution: 5% nonfat dry milk in Tris-buffered saline (TBS) (pH 8.0) containing 0.1% Tween-20 (TBST-0.1%).
3. Washing solution: TBST-0.1%.

2.6. Protein Identification by Mass Spectrometry

1. Ammonium bicarbonate 50 mM solution is prepared in water. The pH of this solution is 7.8 and does not require further adjustment. This solution can be stored frozen for long periods.
2. IAA is dissolved in 50 mM Ammonium bicarbonate to a final concentration of 55 mM.
3. DTT is dissolved in 50 mM Ammonium bicarbonate to a final concentration of 10 mM.
4. α-Cyano-4-hydroxycinnamic acid (CHCA) MALDI matrix (Bruker daltonics, Bremen, Germany). This MALDI matrix is freshly prepared in 20 μg/μl in acetonitrile, 0.1% TFA, 70:30, vol/vol.
5. Proteomics grade trypsin (Roche Diagnostics, recombinant, Penzberg, Germany).
6. "Trifluoroacetic acid (TFA): TFA 0.5 ml ampoules (Pierce) were used. 1% Stock solution can be stored at room temperature for weeks.

3. Methods

Any differential proteomics analysis relies on the quality of the samples to be compared. Therefore, the clinical characterization of the samples is of paramount importance in order to classify them correctly. The sample classification parameters and collection procedures have to be rigorously followed if samples are obtained in different hospitals.

The comparison of two-dimensional (2D) gel images from different samples is an established method used to study differences in protein expression. Conventional methods rely on comparing images from at least two different gels. Due to the high variation between gels, detection and quantification of protein differences can be problematic. Two-dimensional difference gel electrophoresis (DIGE) is an advanced technique for comparative proteomics, which improves the reproducibility and reliability of differential expression analysis between samples. In the application of DIGE, different samples are labeled with spectrally resolvable fluorescent dyes prior to electrophoresis, and are then separated on the same gel. The fluorophores (Cy2, Cy3, and Cy5) have different wavelengths for excitation and emission but are structurally similar; have very similar molecular masses; are positively charged to match the charge that is replaced on the lysine residue; and undergo nucleophilic substitution with the ε-amino group of lysine residues to form an amide bond. Matching the charge and mass ensures that all the samples essentially co-migrate to the same point during electrophoresis, and the different excitation/emission wavelengths make them discernable under appropriate scanning conditions. In the labeling reaction, the dye:protein ratio is low. Several studies using DIGE for comparative proteomics have demonstrated the high sensitivity and reproducibility of the system (23). Labeling reactions are carried out under minimal labeling, such that only 20% of molecules of a particular protein are covalently modified with one Cy dye molecule. Hence, detection of proteins for excision and mass spectrometry requires post-staining of gels with a general protein stain, because the unlabeled majority of the protein will not exactly co-migrate with the labeled protein, particularly in the low molecular weight range. By default, Cy3 and Cy5 are used for labeling the samples to be compared, whereas Cy2 is exclusively used for the labeling of the so-called internal standard. Ideally this standard sample is a pool comprising equal amounts of each of the experimental samples being compared. The pooled standard therefore represents the average of all the samples being analyzed and ensures that all the proteins present on the sample are represented. The same pooled standard is loaded in each of the gels in an experiment, and it is used to normalize protein abundance measurements within each gel.

Variation in spot intensities due to the experimental factors such as protein loss during sample entry into the strip will be normalized within a DIGE gel with the use of the internal standard. Therefore, the relative amount of proteins that enters differentially will remain constant. With conventional "one sample per gel" 2D techniques, samples to be compared and separated independently in different gels, and consequently, spot migration and intensity, will differ for each gel and sample in an experimental variation. DIGE therefore increases the confidence with which quantitative protein differences can be detected and quantified (24). A further advantage of this system is that the number of gels run in an experiment is reduced.

Proteins are identified by mass spectrometry. Typically MALDI-TOF type of analysis is performed for 2D gel resolved protein identification (see Note 9).

Validation by Western blot is described for tissue and serum samples. The Western blot performance very much relies in the quality of the commercially available antibodies. Therefore, bibliography search is recommended to check if the protein of interest has been successfully detected by a given antibody in other similar studies.

3.1. Sample Preparation for 2D-PAGE Analysis

1. Wash the tissue with PBS 1×.
2. Homogenize 5 mg of liver samples in 1 ml of lysis buffer using a potter homogenizer.
3. Centrifuge cell lysates at 12,000 × *g*, 30 min, 4°C.
4. Check pH. Add HCl–NaOH 50 mM to get a pH 8–9.
5. Determine total protein in the supernatant (crude extract) with the Bradford assay kit (Bio-Rad). Read the assay in a SpectraMax microplate reader using BSA (Sigma) as standard protein (25).
6. Mix reagents, wait for 2 min, and then read within 60 min at OD 590 nm.
7. Prepare triplicates of your sample, or use different points. That is, use several dilutions that you estimate will be inside the linearity of the curve, or prepare triplicates of your sample that you estimate to be in the middle of the curve. The linearity of the assay can be improved by plotting the ratio of absorbance at 595 and 465 nm.
8. Plot OD 590 nm versus μg BSA standard. Try to use a sample concentration with OD 590 nm near the middle of the curve.

3.2. Desalt and Concentrate Samples with the 2-D Cleanup Kit

1. Add 300 μl precipitant to 1–100 μl sample (1 μg/μl). Vortex, and incubate on ice for 15 min.
2. Centrifuge for 1 min at 12,000 × *g*.
3. Remove supernatant. Centrifuge briefly to bring the remaining supernatant to the bottom of the tube. Remove the remaining supernatant with micropipette.
4. Without disturbing the pellet layer, add 40 μl of co-precipitant on top of the pellet. Let the tube sit on ice for 5 min.
5. Centrifuge for 5 min at 12,000 × *g*.
6. Add 25 μl of distilled/deionized water to the pellet. Vortex to disperse.

7. Add 1 ml chilled wash buffer and 5 μl of wash additive. Vortex for 20–30 s every 10 min. Repeat two more times.
8. Centrifuge for 5 min at 12,000 × *g*.
9. Remove and discard supernatant. Allow pellet to dry.
10. Resuspend pellet in rehydration buffer or sample solution of choice.

3.3. Sample Labeling

1. Resuspend the dyes in dimethylformamide (DMF) (see Notes 10 and 11).
2. Thaw the dyes and leave them at room temperature for 5–10 min.
3. Add 5 μl DMF/5 nmol dye (1 nmol/μl) (Dye stock solution).
4. Vortex for 30 s.
5. Centrifuge at 12,000 × *g*, 30 s.
6. Prepare working dye solution (WDS) adding DMF to a concentration equal to 400 pmol/μl.
7. Add dyes to the protein (400 pmol dye/50 μg protein) (see Note 12).
8. Aliquot 50 μg of protein in a tube.
9. Add 1 μl of WDS to the sample.
10. Vortex and spin down as in step 5.
11. Incubate for 30 min in the dark.
12. Add 1 μl of 10 mM Lys to quench the reaction.
13. Vortex and spin down.
14. Leave for 10 min on ice.
15. Mix the samples [Cy-2-labeled samples (pooled samples) were mixed with Cy-3- or Cy-5-labeled samples that were alternatively steatosis and control for the first analysis and NASH and control for the second analysis].
16. Add the same volume of sample buffer 2×.
17. Vortex for 30 s.
18. Leave for 10 min on ice.

3.4. First-Dimension Electrophoresis (IEF)

1. All strips should be from the same batch. Label strip near the acidic end (+). Remove strip protective cover from acidic end with forceps.
2. Pipette rehydration buffer (460 μl for 24 cm strips) on top of the channel. Place the IPG strip gel side down into a tray channel. Wet the strip by sliding it through the rehydration solution

as you place it into the channel of the tray. Minimize the amount of solution on top of the strip and insure that each strip is completely wetted to prevent uneven rehydration. Remove air bubbles if necessary (see Note 13).

3. Apply mineral oil or similar overlay material to each channel containing an IPG strip. Make sure that the entire strip is covered. Place the cover lid on the tray.
4. Actively rehydrate pH 3–7 IPG strips (Bio-Rad) at 50 V, 20°C for at least 12 h using a Protean IEF Cell (Bio-Rad). Make sure that the entire strip is covered by rehydration buffer. Place the cover lid on the tray. Place the tray on the Peltier cooling platform of the PROTEAN IEF cell. Be sure to align the electrodes of the focusing tray with the Peltier electrode connections on the platform and start the run (see Note 14).
5. Prewet electrode wicks in deionized water blotting off excess water. Insert electrode wicks directly on top of both the cathode and anode electrode wires in the focusing tray just prior to focusing. Carefully insert damp electrode wicks by gently lifting the IPG strip and placing it between the electrode and the strip.
6. After rehydration, add 50 μl of sample by cup-loading.
7. Place the lid on the focusing tray so that the lid pressure tabs press on the IPG strips directly over the electrodes to insure good contact between the strips and the electrodes.
8. Place the tray on the Peltier platform. Be sure that the electrodes of the focusing tray contact the color-coded electrodes of the PROTEAN IEF cell.
9. Carry out IEF at 300 V (1 h), 300–600 V (1 h), 600 V (3 h), 600–2,000 V (7 h), 2,000 V (3 h), and 2,000–3,500 V (7 h) at 20°C, with a maximum current setting 50 mA/strip in a Protean IEF Cell (see Note 15).
10. Start second dimension or store strips at –80°C.
11. Dispose of excess waste oil in "Recycled/Waste Oil" bottles.
12. Scrub re-swelling trays with 10%w/v SDS, rinse well in tap and Milli-Q water, and then air-dry.

3.5. Second-Dimension Electrophoresis (2D-PAGE)

1. Add the Tris and Milli-Q water to PAA stock and filter all under vacuum. Degas for 15 min.

   Add SDS, and degas again for 15 min. Add 10%w/v APS and TEMED (see Note 16).
2. Assemble the Iso-Dalt Tank casting chamber.

   Fill tank with Milli-Q water to check for leaks. Clean plates. Place plastic sheet between plate pairs and assemble with spacer plates. Ensure that tank is leveled. Pour 140 ml 50%v/v glycerol into side reservoir. Add acrylamide through the bottom hose until the solution reaches 3 cm from the top of the chamber. Remove the glycerol and overlay gels gently with water or water-saturated butanol. Allow ~1 h to set.

3. Assembling the Bio-Rad Protean II Casting Stand.

   Clean and assemble plates and spacers. Apply Vaseline to surfaces of the spacers to avoid current leakage. Use 0.5% w/v agarose to plug plates along bottom edges. Fill with acrylamide to 1.5 cm from the top. Remove bubbles by tapping plates, and overlay gently with water-saturated butanol. Allow gel to set ON at 25°C. Wash out water-saturated butanol and cover gels with Milli-Q water. Gels can be stored at 4°C for up to 2 weeks wrapped in plastic, with a water overlay. Stacking gels should be applied immediately prior to use.

4. Iso-Dalt running buffer (stable at RT).

   Iso-Dalt Tank: for Dalt plates 18–19 l of 1× running buffer; for BioRad plates 15–16 l of 1× buffer. Alternatively, add solids and water to tank and allow to dissolve. The pH should be 8.3. Top up or siphon off buffer to ensure that buffer just covers plates.

5. Running conditions: Gently run samples into gel at 20 mA/gel. Resolve at ~60 mA/gel at 4°C for 4–6 h or ~20 mA/gel, overnight at 20°C (see Note 17).

6. Wipe all clamps and stands down and put away.

3.6. Image Analysis

Efficient analysis of changes in protein expression by using 2-DE relies on the use of automated image processing techniques. A standard analysis of 2-DE gels includes at least the following three basic steps: protein spot detection, spot quantitation, and gel-to-gel matching of spot patterns.

Automatic spot detection, manual editing, creation of reference gels, average background subtraction, matching, warping, normalization, and creation of difference maps were performed to identify changes based on the profile of each spot volume between NASH, steatosis, and control. A threshold of 1.5 was chosen for displaying the difference map.

1. DIGE gels were visualized using the Typhoon TRIO (Amersham Pharmacia) with excitation at 553 nm (Cy3), 648 nm (Cy5), and 491 nm (Cy2) and emission at 572 nm (Cy3), 669 nm (Cy5), and 506 nm (Cy2) (see Notes 18 and 19).

2. With the Typhoon imager it is possible to adjust sensitivity by tuning the voltage setting of the photomultiplier tube that captures the fluorescent image. The starting point for the photomultiplier tube was around 600 V. All gels were scanned at 100 mm resolution.

3. Images were cropped to remove areas extraneous to the gel image using Image Quant (Amersham Pharmacia) prior to analysis. Gel images were stored as .gel file.
4. Images are imported to DeCyder workflow using the "Image Loader" application. All images in an experiment are grouped in one Project.
5. Images were processed in the Batch Processor by Differential In-gel Analysis (DIA) and Biological Variation Analysis (BVA) modules.

   DIA: DeCyder Differential Analysis Software DIA processes images from a single gel, performing spot detection and quantitation.

   DeCyder Differential Analysis Software BVA processes multiple Ettan DIGE system gel images, performing gel-to-gel matching of spots, allowing quantitative comparisons of protein expression across multiple gels.

   Some parameters must be specified before batch processing.
6. Image item selection:

   An estimated number of the spots to be detected in the gels must be given (10,000).

   Some properties of the spots to be detected must be considered in order to filter out non-spot events (spot vol. > 30,000).
7. BVA item settings:

   Images belonging to the same sample type (control, treated, etc.) must be grouped in defined experimental groups.

   Match images from separate gels with the BVA module of the DeCyder software using the Cy-2-labeled image on each gel for normalization.
8. Compare experimental groups according to normalized spot volumes. Calculate the statistical significance of the differences using Student's *t*-test and accept when the value was $p < 0.05$. For the study at least a 1.5-fold change was confirmed in five independent experiments.

   Using the DeCyder software (v6.5, Amersham Biosciences), Cy-2 images (pooled samples) were compared with Cy-3 or Cy-5 images that were alternatively steatosis and control for the first analysis and NASH and control for the second analysis.

3.7. SYPRO Ruby Staining

1. Wash the gel twice, 10 min with Milli-Q water.
2. Stain for 2 h with SYPRO Ruby staining.
3. Destain three times with 10% v/v methanol and 7% v/v acetic acid.

4. Wash the gel three times, 15 min, with Milli-Q water (see Note 20).

3.8. Spot Picking

1. Picking can be done in both the DIA and BVA modules. Usually a preparative gel with a bigger protein load than the analytical ones (about 300 μg, SYPRO Ruby stained gel) is run for this purpose.
2. The preparative gel must be processed in the DIA module for spot and pick reference detection and for defining the position (X and Y coordinates) of the spots to be picked. This DIA workspace must then be loaded to the BVA workspace of the experiment, and the pick image must be landmarked and matched to the analytical gels, so that the spots deregulated (spots showing a fold change higher than 1.5 comparing steatosis vs. control or NASH vs. control) in the image analysis have a correspondence in the pick gel.
3. All the information about the pick references and the spots to be picked is exported to an rtf file and loaded to the Ettan Picker. Spot picking is automatically performed by the robot based on the information given. Picked spots are placed in 96-well plates and further trypsinized for their identification. Procedure is described in Subheading 3.9.

3.9. Western Blot of Tissue and Serum

1. Boil tissue protein extracts at 95°C for 5 min in SDS-polyacrylamide gel electrophoresis (PAGE) sample buffer.
2. Separate 20 μg (for total extracts) or 80 μg of protein (for serum) by SDS-PAGE in 8%, 11%, or 15% acrylamide gels (according to the size of the protein) using Mini-PROTEAN Electrophoresis System (Bio-Rad).
3. Transfer gels onto nitrocellulose membranes by electroblotting using Mini Trans-Blot cell (Bio-Rad). Block membranes with 5% nonfat dry milk in TBS (pH 8.0) containing 0.1% Tween-20 (TBST-0.1%), for 1 h at room temperature.
4. Wash membranes three times with TBST-0.1%.
5. Incubate overnight at 4°C with commercial primary antibodies:
   - Alpha 1 acid glycoprotein from Abcam (Cambridge, UK); used dilution: 0.5 μg/ml.
   - Beta actin (Abcam); used dilution: 1:5,000.
   - Calumenin from Santa Cruz Biotechnology, Inc. (Santa Cruz, California); used dilution: 1:200.
   - Calreticulin from Abcam; used dilution: 0.1 μg/ml.
   - Carbamoyl phosphate synthase 1 (CPS1) from Proteintech Group, Inc.(Chicago, USA); used dilution: 1:100.

- 14-3-3 Epsilon protein from IBL (Gunma, Japan); used dilution: 1 μg/ml.
- BiP/GRP78 from BD Biosciences (NJ, USA); used dilution: 1:250.
- Grp94/Endoplasmin from Abcam; used dilution: 1:500.
- Heat shock protein 90 (Hsp 90) from BD Biosciences; used dilution: 1:1,000.
- Peroxiredoxin 4 from Abcam; used dilution: 1:10,000.
- Glyceraldehyde 3-phosphate dehydrogenase (GAPDH) from Abcam was used as control loading. Used dilution: 0.5 μg/ml.

6. Wash membranes three times with TBST-0.1% and incubate for 1 h at room temperature (RT) in blocking solution containing secondary anti-rabbit or anti-mouse antibodies conjugated to horseradish peroxidase.
7. Detect immunoreactive proteins by Western lightning Chemiluminescence Reagent ECL (Amersham) and expose to X-ray films (Amersham) in a Curix 60 Developer (Agfa HealthCare NV, Mortsel, Belgium).
8. Quantify bands by densitometry using the free image processing program ImageJ (http://rsbweb.nih.gov/ij/).
9. Measure serum protein concentration in serum by Bradford assay. Resolve equal amounts of protein (80 mg) in 12% SDS-PAGE. Perform Western blot analysis with the same set of antibodies used for successful validation of the potential markers in tissue.

3.10. Protein Identification by Mass Spectrometry

1. Excise bands in automatic fashion by using an Ettan spot picker (General Electrics). To avoid sample contamination, special care should be taken during in-gel tryptic digestion (see Note 21). Transfer selected spots to Eppendorf brand 1.5 ml tubes (see Note 22).
2. Reduce gel pieces with 30 μl of 10 mM DTT at 56°C for 20 min. Discard the DTT solution and alkylate in 30 μl of 50 mM iodoacetamide for 20 min at room temperature in darkness (see Note 23). Discard the supernatant and shrink the gel spots by adding 30 μl of acetonitrile for 5 min. Discard the acetonitrile.
3. Swell gel pieces on ice in 15 μl digestion buffer containing 50 mM ammonium bicarbonate and 12.5 ng/μl of trypsin. After 30 min, remove and discard the supernatant. Add 10 μl of 50 mM ammonium bicarbonate to the gel piece and the digestion was allowed to proceed at 37°C overnight.

4. Transfer the supernatant to an empty Eppendorf tube, and dry the samples in a SpeedVac (Christ, Osterode am Harz, Germany).
5. Dissolve the peptides in a solution containing 0.1%v/v TFA and desalt and concentrate on a homemade POROS-R2 nano-column (PerSeptive Biosystems, Framingham, MA, USA) according to Gobom et al. (26). This procedure can also be performed with the commercially available ZipTip columns (Millipore, Bedford, MA). Equilibrate the column with 10 μl 0.1%v/v TFA and after loading the sample, wash with 10 μl of 0.1%v/v TFA. Elute peptides directly onto the MALDI target with 0.5 μl CHCA matrix solution and air-dry for 5 min.
6. Identify proteins by mass spectrometry. Mass spectrometry analysis (MS) was performed with an Ultraflex TOF/TOF (Bruker) spectrometer equipped with an LIFT ion selector and a Reflectron ion reflector. The instrument was calibrated with a peptide calibration standard in the mass range of 500–3,500 kDa (Bruker). Typically, 1,000 spectra for peptide mass fingerprint (PMF) and 1,500 spectra for peptide fragment fingerprinting (PFF) were acquired (see Note 24).
7. Protein identities were obtained by using Mascot searching engine (Matrix Science) against SwissProt nonredundant databases selected for human taxonomy. A precursor mass tolerance of 30 ppm for MS and 0.7 Da for MS/MS was allowed. Up to two missed cleavages were allowed; oxidation of methionine was set as variable modification and carboxyamidomethylation of cysteine as fixed modification (see Note 25).

4. Notes

1. When working with human samples it is important to have a record of physical and healthy characteristics in order to characterize the samples under study correctly. Moreover, in this particular study, the pathologist subjectivity is crucial for the characterization of the samples. In order to help with the characterization of the samples, the following analytical data were obtained: the age, sex, Body Mass Index (BMI), and levels of triglycerides, glucose, cholesterol, aspartate aminotransferase (AST), and alanine aminotransferase (ALT).
2. All solutions were prepared in Milli-Q water.
3. The Coomassie blue G250 dye used in the Bradford assay appears to bind most readily to arginyl and lysyl residues of proteins (not to the free amino acids). This specificity can lead to variation in the response of the assay to different proteins, which is the main disadvantage of the method. Keep in the

dark at 4°C. Do not use directly from the bottle. Transfer the volume you are going to use to another recipient.

4. Always wear gloves when handling IPG strips to prevent contamination.
5. IPG strips must be rehydrated prior to the IEF run. The focusing tray must be used for active rehydration. Passive rehydration of the IPG strips can take place in the disposable rehydration tray or in the IPG focusing tray.
6. Composition of the IEF buffer can vary; however it should be prepared in a buffer containing urea, nonionic, and/or zwitterionic detergents, carrier ampholytes, and a reducing agent.
7. SDS is an irritant, so use a facemask during handling.
8. Acrylamide is a neurotoxin. Waste acrylamide can be left to polymerize, before discarding.
9. Regarding protein identification by mass spectrometry, it is recommended to desalt and concentrate peptides after in-gel digestion and prior to MALDI-TOF/TOF analysis by, for example, reverse phase-based microcolumns. This procedure will ensure higher quality spectra (e.g., better signal-to-noise ratio), thus enabling better protein identification.
10. Primary amines interfere with the labeling reaction. Therefore, reducing agents or ampholytes should be added after the labeling. It is important to use a 30 mM Tris buffer at pH 8–9.
11. Dyes resuspended in Dye Stock Solution can be kept at −20°C for two months and in Working Solution at −20°C for 2 weeks. Once the sample is labeled it can be used or kept at −80°C (maximum 3 months). The sample should be run immediately after mixing the dyes and adding 2× sample buffer.
12. For sample labeling, run steatosis and NASH experiments separately but in identical fashion. Label liver samples (50 μg) from patients with steatosis and their control counterparts with Cy-3 or Cy-5 in six different gels. Label a pool of both samples (25 μg each) with Cy-2 for normalization using 400 pmol of fluorochrome per 50 μg of protein. In parallel, NASH samples were compared against control samples.
13. Avoid trapping bubbles between the strip and the re-swelling tray when rehydrating the strips. It is important to rehydrate the strips homogeneously in order to get reproducible and reliable results.
14. The rehydration program in the IEF Cell may include a pause step after rehydration in order to add the sample and to place electrode wicks before the IEF begins (or to add mineral oil).
15. Running parameters of the IEF must be experimentally determined for any given sample/pH gradient/protein load/salt conditions, etc.

16. After adding the polymerizing agents (APS and TEMED), pour the gels (1.5 mm thickness) and add a layer of water that will allow to remove the bubbles, create a flat surface, and prevent the gel from drying out. Polymerize for a minimum of 1 h at room temperature. Remove the water from the top of the gel. Degas the stacking gel before loading it to help in polymerization. After degassing, load it over the gel leaving a gap of 2 cm between the upper stacking gel surface and the top of the gel glass for loading the IPG strip. Store the polymerized gels overnight at 4°C.
17. Prepare fresh running buffer each time. Reusing running buffer will affect gel resolution. Distilled water can be used to prepare the running buffer.
18. Clean and dry thoroughly the glass plates and the scanner surface before acquiring images.
19. Acquire images of the gels using lasers with excitation at 553 nm (Cy3), 648 nm (Cy5), and 491 nm (Cy2) and emission at 572 nm (Cy3), 669 nm (Cy5), and 506 nm (Cy2). Information on filter sets can be obtained from the online Handbook at www.probes.com.
20. Gels can be stored in water at 4°C.
21. Most common contaminants in proteomics are the keratins, which are introduced during different steps in management of the samples such as handling for scanning, spot picking, and in-gel digestion. Therefore, it is highly recommended to have the scanner and the spot picker in a clean room. On top of that, white clothes and gloves must be worn. Pipette tips do not need to be sterilized but care should be taken to avoid contamination.
22. Perform in-gel digestion reactions and peptide extraction procedures in Eppendorf brand tubes. We found plastic-derived contaminants by mass spectrometry when using other tube brands. These contaminants hamper the correct analysis of the peptides and even the identification of the protein.
23. IAA solution should be freshly prepared and stored in the dark prior to use and also during incubation for alkylation.
24. The parameters used for the mass spectrometry analysis may vary from instrument to instrument. Typically each model has its own methods optimized to achieve the best performance.
25. Because obtained precursor ion and fragment ions mass accuracy may be different depending on the calibration method and the type of instrument used, mass accuracy tolerance may be variable for database search. If trypsin autolysis peptides can be observed, they can be used for internal calibration. Once internal calibration is performed higher mass accuracy can be achieved and therefore, better quality identification is generally obtained.

Acknowledgements

We would like to acknowledge Estefania Fernandez, Carol Gomara for the DIGE and tissue Western blot analyses, Nere Alkorta for the MALDI TOF/TOF analyses, Begoña Rodriguez for the serum Western blot analyses, and our collaborators Maria Luz Martinez-Chantar, Antonio M. Duce, Juan Caballeria, and Shely C. Lu.

References

1. Marchesini G, Bugianesi E, Forlani G et al (2003) Nonalcoholic fatty liver, steatohepatitis, and the metabolic syndrome. Hepatology 37:917–923
2. Reid AE (2001) Nonalcoholic steatohepatitis. Gastroenterology 121:710–723
3. Starley BQ, Calcagno CJ, Harrison SA (2010) Nonalcoholic fatty liver disease and hepatocellular carcinoma: a weighty connection. Hepatology 51:1820–1832
4. Clark JM, Diehl AM (2002) Hepatic steatosis and type 2 diabetes mellitus. Curr Diab Rep 2:210–215
5. Yki-Jarvinen H (2010) Liver fat in the pathogenesis of insulin resistance and type 2 diabetes. Dig Dis 28:203–209
6. Adams LA, Lindor KD (2007) Nonalcoholic fatty liver disease. Ann Epidemiol 17: 863–869
7. Angulo P (2007) Obesity and nonalcoholic fatty liver disease. Nutr Rev 65:S57–S63
8. Charlton M, Viker K, Krishnan A et al (2009) Differential expression of lumican and fatty acid binding protein-1: new insights into the histologic spectrum of nonalcoholic fatty liver disease. Hepatology 49:1375–1384
9. Rubio A, Guruceaga E, Vazquez-Chantada M et al (2007) Identification of a gene-pathway associated with non-alcoholic steatohepatitis. J Hepatol 46:708–718
10. Imbert-Bismut F, Ratziu V, Pieroni L et al (2001) Biochemical markers of liver fibrosis in patients with hepatitis C virus infection: a prospective study. Lancet 357:1069–1075
11. Poordad FF (2004) FIBROSpect II: a potential noninvasive test to assess hepatic fibrosis. Expert Rev Mol Diagn 4:593–597
12. Vanderschaeghe D, Laroy W, Sablon E et al (2009) GlycoFibroTest is a highly performant liver fibrosis biomarker derived from DNA sequencer-based serum protein glycomics. Mol Cell Proteomics 8:986–994
13. Gorg A, Weiss W, Dunn MJ (2004) Current two-dimensional electrophoresis technology for proteomics. Proteomics 4:3665–3685
14. Ding SJ, Li Y, Shao XX et al (2004) Proteome analysis of hepatocellular carcinoma cell strains, MHCC97-H and MHCC97-L, with different metastasis potentials. Proteomics 4:982–994
15. Douette P, Navet R, Gerkens P et al (2005) Steatosis-induced proteomic changes in liver mitochondria evidenced by two-dimensional differential in-gel electrophoresis. J Proteome Res 4:2024–2031
16. Gray J, Chattopadhyay D, Beale GS et al (2009) A proteomic strategy to identify novel serum biomarkers for liver cirrhosis and hepatocellular cancer in individuals with fatty liver disease. BMC Cancer 9:271
17. Lee IN, Chen CH, Sheu JC et al (2005) Identification of human hepatocellular carcinoma-related biomarkers by two-dimensional difference gel electrophoresis and mass spectrometry. J Proteome Res 4:2062–2069
18. Santamaria E, Avila MA, Latasa MU et al (2003) Functional proteomics of nonalcoholic steatohepatitis: mitochondrial proteins as targets of S-adenosylmethionine. Proc Natl Acad Sci USA 100:3065–3070
19. Meneses-Lorente G, Guest PC, Lawrence J et al (2004) A proteomic investigation of drug-induced steatosis in rat liver. Chem Res Toxicol 17:605–612
20. Orimo T, Ojima H, Hiraoka N et al (2008) Proteomic profiling reveals the prognostic value of adenomatous polyposis coli-end-binding protein 1 in hepatocellular carcinoma. Hepatology 48:1851–1863
21. Santamaria E, Munoz J, Fernandez-Irigoyen J et al (2006) Molecular profiling of hepatocellular carcinoma in mice with a chronic deficiency of hepatic s-adenosylmethionine: relevance in human liver diseases. J Proteome Res 5:944–953
22. Rodriguez-Suarez E, Duce AM, Caballeria J et al (2010) Non-alcoholic fatty liver disease proteomics. Proteomics Clin Appl 4:362–371
23. Gharbi S, Gaffney P, Yang A et al (2002) Evaluation of two-dimensional differential gel electrophoresis for proteomic expression analysis

of a model breast cancer cell system. Mol Cell Proteomics 1:91–98

24. Alban A, David SO, Bjorkesten L et al (2003) A novel experimental design for comparative two-dimensional gel analysis: two-dimensional difference gel electrophoresis incorporating a pooled internal standard. Proteomics 3:36–44
25. Bradford MM (1976) A rapid and sensitive method for the quantitation of microgram quantities of protein utilizing the principle of protein-dye binding. Anal Biochem 72:248–254
26. Gobom J, Nordhoff E, Mirgorodskaya E et al (1999) Sample purification and preparation technique based on nano-scale reversed-phase columns for the sensitive analysis of complex peptide mixtures by matrix-assisted laser desorption/ionization mass spectrometry. J Mass Spectrom 34:105–116

Chapter 17

Proteomic Methods for Biomarker Discovery in a Rat Model of Alcohol Steatosis

Billy W. Newton, William K. Russell, David H. Russell, Shashi K. Ramaiah, and Arul Jayaraman

Abstract

Understanding changes in the expression of specific proteins and/or alterations in their posttranslational modifications is crucial to elucidating the molecular mechanisms underlying disease states such as alcoholic liver disease. Protein separation and analysis techniques such as two-dimensional electrophoresis and mass spectrometry can be used for identifying biomarker proteins that are altered during progression of alcoholic liver disease. In this chapter, we outline methods for resolving liver tissue proteins from a rodent model of alcoholic liver disease using two-dimensional electrophoresis and identifying differentially expressed proteins using mass spectrometry. In addition, since oxidative stress strongly correlates with alcoholic liver disease, we also describe methods for identifying oxidatively modified proteins from liver tissue. We specifically focus on identifying proteins that are carbonylated as protein carbonylation is a permanent modification and considered deleterious to cells. The combination of two-dimensional electrophoresis for protein resolution, mass spectrometry for protein identification, and affinity-based methods for enriching and identifying carbonylated proteins is a powerful methodology for protein biomarker identification.

Key words: Proteomics, Two-dimensional electrophoresis, Alcoholic liver disease, Steatosis, Mass spectrometry, Carbonylated proteins, Biomarkers

1. Introduction

The term proteome refers to the entire complement of proteins produced by an organism, analogous to the organisms genome, and proteomics is the systematic and large-scale study of proteins (1). Several qualitative and quantitative proteomic studies have had a direct impact on our understanding of disease states. For example,

Djuro Josic and Douglas C. Hixson (eds.), *Liver Proteomics: Methods and Protocols*, Methods in Molecular Biology, vol. 909, DOI 10.1007/978-1-61779-959-4_17, © Springer Science+Business Media New York 2012

proteomics has been used to identify biomarkers in cerebral palsy (2), immunodeficiency (3) as well as Alzheimer disease (4, 5). Rieske et al. found that treatment with simvastatin reduced levels of the Alzheimer's marker, phosphorylated tau protein in hypercholesterolemic patients (5). Proteomics has also been used to develop a fundamental understanding of disease mechanisms. Examples of these are the studies by Witzman et al. who found differentially expressed proteins in the brains of alcohol- and non-alcohol-preferring rats using two-dimensional electrophoresis and mass spectrometry (6), and the work by Moon et al. who identified nitrosylated proteins in alcohol fed rats (7).

Many proteomic studies have focused on understanding liver diseases or pathologies arising from a broad range of insults or injury (e.g., toxic exposure, burn injury, hepatic ischemia) (8–14). The focus on the liver is pertinent as it is the metabolic hub of the body and carries out several important functions such as detoxification and energy metabolism. The detoxification function of the liver is especially important as it is often exposed to several toxic metabolites and/or metabolic intermediates during the degradation and clearance of endogenous and exogenous substances, which in turn, affects its function (15, 16). Xenobiotic compounds such as ethanol or pharmaceuticals (17, 18) are known to induce enzymes belonging to the cytochrome P450 (CYP) family, which are monooxygenases involved in metabolizing these compounds. CYP enzymes metabolize drugs primarily by hydroxylating compounds (19). Hydroxylation is known as phase one reaction, and often produces a toxic intermediate, such as *N*-acetyl-*p*-benzoquinone imine (NAPQI) from paracetamol (16, 20). Such intermediates are further conjugated with glutathione or sulfate (phase 2 reaction), which neutralizes and allows for excretion. It has been estimated that a small percentage (~0.1–1.0%) of oxygen entering such catalytic cycles exits the cycle prematurely as superoxide ion ($O_2^{\cdot-}$) (19). Superoxide generation is a main source of cellular reaction oxygen species (ROS) (21, 22), and its downstream products, especially the hydroxyl radical are extremely reactive and can cause oxidative damage to all classes of biomolecules including proteins (22, 23). Thus, there is significant interest in investigating the effects of different substances on liver protein expression and function.

In the case of ethanol metabolism, long-term consumption of alcohol leads to upregulation of CYP2E1 (24, 25), and this has been correlated to a state of oxidative stress, where the generation and effects of ROS begin to overwhelm cellular antioxidant defenses (26, 27) leading to a state of oxidative stress. A common consequence of oxidative stress is the oxidative modification of different cellular proteins (7, 23, 28), and carbonylation of proteins is a deleterious and permanent protein modification (4, 28, 29) that is often correlated to oxidative stress.

While changes in liver tissue protein expression can be used to identify metabolic pathways and mechanisms involved in disease, liver proteomics requires the use of invasive sampling techniques such as biopsies. Therefore, a complementary approach has been used where changes in the serum proteome, where invasive sampling is not a concern, are monitored as liver disease progresses (13, 30). Since several serum proteins are synthesized in the liver (e.g., albumin), serum proteomics enables discovery and identification of biomarkers instead of using clinical tissue biopsies.

One of the standard techniques used in proteomics studies is two-dimensional electrophoresis (2DE). The popularity of 2DE methods is primarily due to its ability to resolve proteins based on two orthogonal properties—charge and size and the lack of dependence on costly separation equipment. However, 2DE methods have their limitations (reproducibility, inability to separate high molecular weight and basic proteins), which has led to it being used more in conjunction with mass spectrometry (MS) (30, 31). In a typical 2DE/MS strategy, proteins are separated by 2DE, followed by excision and identification of protein spots by MS (31, 32).

This paper covers methods developed for using 2DE and affinity chromatography followed by mass spectrometry for the discovery of biomarkers in alcoholic steatosis. The protocol describes sample preparation from liver tissue from a rodent model of alcoholic steatosis, 2DE separation of proteins, image analysis for determination of differentially expressed 2DE protein spots, and mass spectrometry analysis to identify the differentially expressed protein spots. In addition, we also describe the use of affinity chromatography for the enrichment of carbonylated proteins in liver tissue.

2. Materials

2.1. Liver Excision and Homogenization

1. Krebs–Henseleit Buffer: 6.9 g/l NaCl, 0.2 g/l KCl, 1.44 g/l Na_2HPO_4, 0.24 g/l KH_2PO_4, 0.41 g/l $MgSO_4{\cdot}6H_2O$, 0.37 $CaCl_2{\cdot}2H_2O$, 1 g/l glucose, pH 7.3, sterilize by autoclaving or filtration.
2. DE homogenization buffer: 2 M thiourea, 7 M urea, 4% CHAPS, 50 mM dithiothreitol (DTT), 0.5% ampholytes (Bio-Rad, Hercules, CA).
3. Bio-Lyte 3/10 ampholyte, item no. 163-1112 (Bio-Rad).
4. Protease inhibitor cocktail, item no. P8340 (Sigma, St.Louis, MO).
5. Deoxyribonuclease (DNase), item no. 0210057520 (MP Biomedicals, Solon, OH).

6. 50× DNase stock: 150 mM Tris, pH 7.5, 50 mM $MgCl_2$, DNase 5 mg/ml.
7. Ribonuclease (RNase), item no. 0210107691 (MP Biomedicals).
8. 50× RNase stock: 150 mM Tris, pH 7.5, 50 mM $MgCl_2$, RNase 2.5 mg/ml.
9. WMP-THP115 homogenizer (White Mountain Process, Boston, MA).
10. Disposable test tube (10 mm × 75 mm).
11. Portable liquid N_2 dewar (approximately 500–2,500 ml).
12. Ceramic mortar and pestle.

2.2. Bradford Protein Assay

1. Pierce 660 nm Protein Assay (Pierce, Rockwell, IN).
2. Bovine serum albumin (BSA), 2 mg/ml protein standard (Thermo-Fisher), serially diluted to 1.5, 1.0, 0.75, 0.50, 0.25, and 0.125 mg/ml.
3. 96-Well clear bottom assay plates.

2.3. Ready-Prep 2DE Cleanup Kit

1. Ready-Prep 2D Cleanup Kit, item no. 163-2140 (Bio-Rad).
2. 9 M Urea rehydration buffer bromophenol Blue Free: 9 M urea, 2%w/v CHAPS, 18 mM DTT, 0.5% ampholytes
3. 9 M Urea rehydration buffer: 9 M urea, 2% CHAPS, 18 mM DTT, 0.5% ampholytes, 0.05% bromophenol blue.

2.4. Isoelectric Focusing and IPG Equilibration

1. Protean IEF cell (Bio-Rad).
2. IEF 12 strip Rehydration Tray (Bio-Rad).
3. Readystrip IPG strips, 7, 11, 17 cm, pH 5–8, item nos. 163-2004, 163-2018, 163-2011 (Bio-Rad).
4. Mineral oil.
5. Filter paper cut into 4 mm × 8 mm strips.
6. Equilibration buffer: 50 mM Tris–HCl, pH 8.8, 6 M urea, 30%v/v glycerol, 2%w/v SDS, 0.05%w/v bromophenol blue.
7. Equilibration buffer with 0.1%w/v DTT: add 100 mg DTT to 10 ml equilibration buffer just prior to use.
8. Equilibration buffer with 0.25%w/v iodoacetamide: add 250 mg iodoacetamide to 10 ml equilibration buffer just prior to use.

2.5. SDS-PAGE Gel Casting

1. Mini-PROTEAN 3 Multi-Casting Chamber (Bio-Rad).
2. Spacer plates with 1.0 mm Integrated Spacers, item no. 165-3311 (Bio-Rad).
3. Short plates, item no. 165-3308 (Bio-Rad).

4. Separation sheets, item no. 165-4165 (Bio-Rad).
5. Acrylamide: Bis-acrylamide 40%w/v solution, add 158 ml ddH_2O directly to 100 g acrylamide/bis-acrylamide (37.5:1) (Thermo-Fisher) dry powder in original container, mix store in the dark at 4°C.
6. Sodium dodecyl sulfate (SDS) 10%w/v.
7. Ammonium persulfate 10%w/v, store in the dark at 4°C.
8. Tetramethylethylenediamine (TEMED).
9. Tris–HCl 1.5 M, pH 8.8: 182 g/l Tris, adjust to pH 8.8 with 10 M HCl, store at 4°C.
10. Butanol saturated with H_2O, butanol with 20%v/v water, butanol phase will rise to top.

2.6. SDS-PAGE with IPG Strips

1. Mini-PROTEAN 3 Cell, gel electrophoresis unit (Bio-Rad).
2. PowerPac Basic 300 V power supply (Bio-Rad).
3. 10× Tris/glycine/SDS running buffer: 3.0%w/v Tris, 14.4%w/v glycine, 1.0%w/v SDS, pH 8.3. Store at room temperature.

2.7. Gel Fixing and Staining

1. Fixing/washing solution: 20%v/v methanol, 7%v/v acetic acid.
2. GelCode Blue Stain (Pierce).
3. Sypro Ruby Stain (Invitrogen, Carlsbad, CA).

2.8. Gel Imaging, 2DE Analysis Software, Protein Spot Excision

1. VersaDoc 3000 Gel Imaging System (Bio-Rad).
2. PDQuest 2-Dimensional gel analysis software (Bio-Rad) or equivalent.
3. Wide bore 200 μl pipette tips.

2.9. Protein Gel Sample Preparation and Digestion

1. ProGest automated gel digestion system (Genomic Solutions, Ann Arbor, MI) or any equivalent gel digestion system.
2. ProGest/ProMS (Genomic Solutions); sample handling and storage: ProPrep-blue collection plates, ProGest-pink plates pierced, ProPrep sealing mat, item nos. PRO10005, PRO10006, CP000636 or any equivalent model.
3. Trypsin, sequencing grade (Promega, Madison, WI).
4. Ammonium bicarbonate (ABC), 25 mM.
5. Acetonitrile (ACN).
6. Dithiothreitol 10 mM.
7. Iodoacetamide 50 mM.
8. Ziptips, C18 (Millipore, Billerica, MA).

2.10. MALDI Matrix Spotting and Mass Spectrometry

1. ProMS automated target spotting system (Genomic Solutions) or equivalent system.
2. MALDI matrix solution: alpha-cyano-4-hydroxycinnamic acid 5 mg/l, 50%v/v ACN, ammonium monobasic phosphate 10 mM, 0.1%v/v trifluoroacetic acid (TFA).
3. 4700 Proteomics MALDI TOF/TOF Analyzer (Applied Biosystems, Foster City, CA).
4. GPS Explorer V2.1 (Applied Biosystems).
5. Mascot search engine (Matrix Science, UK).

2.11. Carbonylated Protein Biotin Tagging and Affinity Chromatography

1. Carbonylation lysis buffer: 0.1%w/v SDS, 0.5%w/v sodium deoxycholate, 1.0%w/v CHAPS, 0.1 M NaCl, 0.1 M sodium phosphate, 1 mM EDTA (pH 7.5), and protease inhibitor cocktail 0.05%v/v.
2. Biotin hydrazide (Pierce).
3. Sodium cyanoborohydride (Sigma).
4. Slidealyzer dialysis cassettes 1.5–3 ml, 7500 MWCO (Pierce).
5. Disposable polypropylene columns, 5 ml, item no. PI-29922 (Pierce).
6. Monomeric Avidin Resin, item no. PI-20228 (Pierce).

2.12. Carbonylated Protein Gel Sample Preparation

1. SDS-PAGE sample buffer: 100 mM Tris, pH 6.8, 2%w/v SDS, 5%v/v 2-mercaptoethanol, 15%v/v glycerol, 0.05%w/v bromophenol blue.
2. Ready Gel Precast Gel 4–20%, item no. 161-1159 (Bio-Rad).
3. Precision Plus Dual Color SDS-PAGE MW standards (Bio-Rad).

3. Methods

The experimental workflow used is shown in Fig. 1 (see Notes 1 and 2).

3.1. Liver Tissue Homogenization

1. Fill approximately 300 ml of liquid N_2 in a portable dewar. Safety glasses or goggles should be worn while handling liquid N_2.
2. Take the tube containing the frozen liver tissue and place it in an ice bucket containing liquid N_2.
3. Add 3 ml of 2DE homogenation buffer supplemented with 10 μl of protease inhibitor cocktail and 60 μl each of 50× DNase stock and RNase stock. Place solution in a 10 mm × 75 mm test tube on ice.

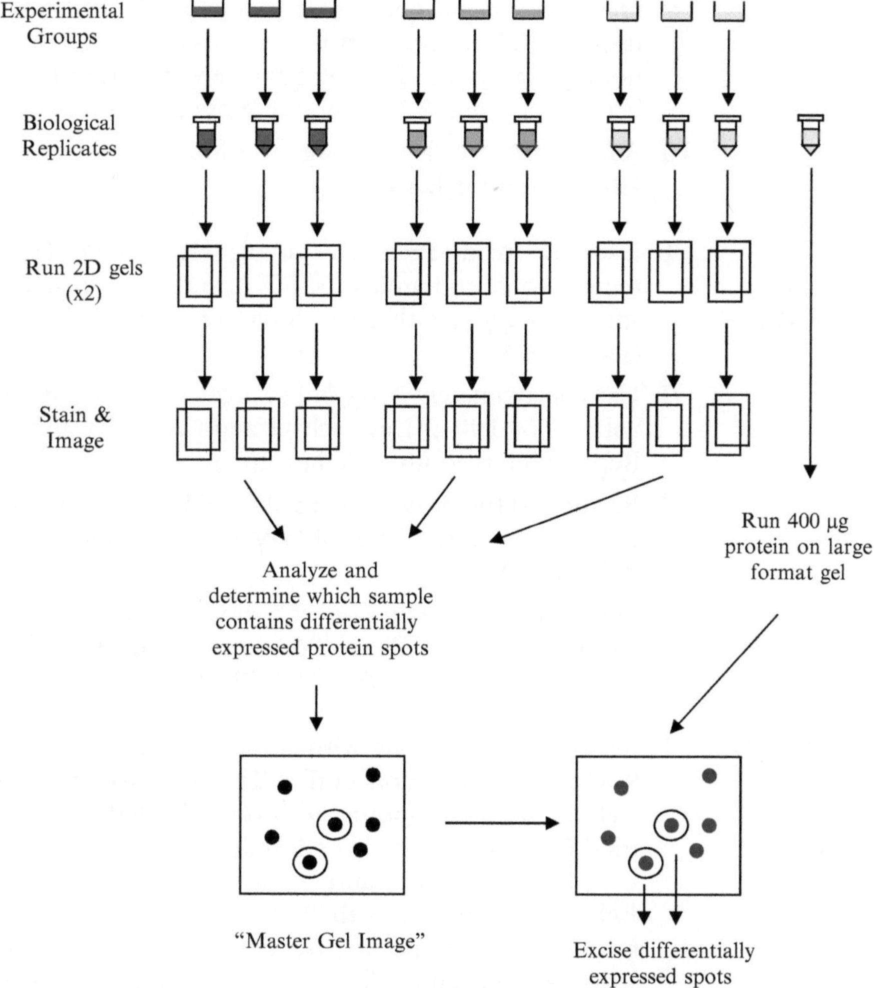

Fig. 1. Flowchart representing workflow for biomarker discovery using two-dimensional electrophoresis and mass spectrometry. Samples for use in large format gels are chosen after the determination of differentially expressed protein spots from 2D gel image analysis. Protein spots are excised from large format gel and identified for MS-based identification.

4. Cool the mortar and pestle with liquid N_2, and add several ml of liquid N_2 to the pestle.
5. Place a sample of frozen liver tissue in the mortar, and crush into millimeter sized pieces with the pestle.
6. Cool a small weigh boat with liquid N_2 and carefully place it on a balance to weigh ~0.2 g of liver tissue. Place liver tissue in the test tube containing homogenation buffer.
7. Keeping the test tube in an ice bath, homogenize the liver tissue with three 12 s pulses.
8. Transfer the homogenate to 1.5 ml tubes and centrifuge at 13,000 × *g* for 15 min at 4°C.

9. Quantify protein concentration in the clarified homogenates using the Pierce 660 nm Protein Assay kit (Pierce). Typically, homogenates from 200 mg tissue have a protein concentration between 1.5 and 2 mg/ml (see Note 3).
10. Aliquot the clarified homogenate into new tubes and store at −80°C until further use.

3.2. Removal of Contaminants from Tissue Homogenates

1. Use the Ready-Prep 2D sample cleanup kit (Bio-Rad) to remove ionic contaminants, DNA, and lipids from the liver tissue homogenates that will interfere with isoelectric focusing (see Note 4).
2. Use approximately 2 mg/ml of protein (or a maximum sample volume of 100 μl) in each cleanup reaction and follow the Ready-Prep recommended protocol.
3. Resuspend the protein pellet obtained in step 13 of the Ready-Prep protocol in 150 μl of bromophenol blue-free 9 M Urea Rehydration Buffer.
4. Quantify the protein concentration of the processed protein samples using the Pierce 660 nm Protein Assay kit (Pierce). The typical protein recovery is 70–90%.

3.3. Rehydration of IPG Strips for First Dimension Isoelectric Focusing

1. Rehydrate IPG strips with protein samples reconstituted in 9 M urea rehydration buffer. The total volume of buffer, as well as the total amount of protein loaded is determined by the IPG strip length. Therefore, protein samples must diluted to proper volume and concentration with 9 M Urea Rehydration Buffer with 0.05% bromophenol blue prior to rehydration. Failure to use proper volume or overloading strips with protein will result in poor isoelectric focusing (see Notes 5 and 6).
2. Carefully pipet the protein sample into the middle of a rehydration well of a clean IEF rehydration tray. The protein sample is pipetted such that a strip of liquid about 2 cm shorter than the actual length if the IPG strip is formed. This allows for easier and more reliable, bubble-free placement of the IPG strips.
3. Remove the plastic backing of the IPG strip and carefully set the IPG strip gel side down upon the liquid sample. This is done in a slow, gradual manner so to prevent splashing and trapping of air bubbles.
4. Carefully apply mineral oil to the well and overlay the IPG strip. This prevents moisture loss and keeps dust out.
5. Ensure that the rehydration tray is on a level surface.
6. Rehydrate samples for 12–16 h (see Note 7).

3.4. Isoelectric Focusing

1. Prepare the isoelectric focusing tray by placing a small piece of moist filter paper in the well directly over the wire electrodes at each end. Press the wetted paper and shape it with tweezers to make contact with the electrode wire.
2. Drain the IPG strip of excess oil and place the strip gel side down in the focusing tray, making sure that polarity of strip matches the polarity of the tray. The gel will make contact on top of the wetted filter paper. The filter paper pads help the IPG gel make contact with the wire electrodes and also allow any residual salts to migrate out of the gel.
3. Gently press the strip to assure contact between the gel and the filter paper. Overlay the gel with mineral oil and place the cover of the focusing tray on top of the strips.
4. Place the focusing tray in the IEF unit with proper polarity. Choose the appropriate focusing program. For our experiments, we used the following program: 15 min conditioning period at 250 V, linear ramping of voltage from 250 to 4,000 V (for 7 cm IPG strips) over 2 h; focused for 20,000 V-h; and held at 4°C indefinitely (see Note 8).
5. At the end of the IEF run, IPG strips may be stored at –80°C until use or prepared for the second-dimension electrophoresis by dabbing the IPG strips to remove excess mineral oil and placing them sequentially in 10 ml equilibration buffer with 0.1%w/v DTT for 20 min, and in 10 ml equilibration buffer with 0.25%w/v iodoacetamide for 20 min.

3.5. Second Dimension Electrophoresis

1. Thoroughly clean and dry glass plates, separation sheets, and casting chamber before assembly (see Note 9).
2. Assemble the glass plates in the gel casting unit following the manufacturer's protocol.
3. Place the chamber on a level bench and make sure all spacer plates are flat and evenly aligned.
4. It is important that Tris–HCl and acrylamide solutions, and ddH_2O all be stored at 4°C prior to starting the gel making process. This ensures that polymerization will not happen too rapidly.
5. Prepare the acrylamide gel solution for 10% gels by combining 6 ml Tris–HCl 1.5 M, pH 8.8, 15 ml of 40% acrylamide stock, 12 ml of 10% SDS, 27 ml ddH_2O, 120 μl of ammonium persulfate, and 60 μl of TEMED in 150 ml vacuum flask. Add a stir magnet and stir for 2 min. Sixty ml of acrylamide gel solution is enough to make 12 gels.
6. Degas the solution in the vacuum flask for 10 min.
7. Pour the acrylamide gel mixture into the assembled casting chamber until the liquid level is 3–4 mm below the top of the

short plate, ensuring that each spacer plate gets about the same amount. Tilt the chamber gently a few times to ensure that the solution is evenly distributed. Approximately 6–8 ml of water saturated butanol is added dropped wise across the top of each gel. Tap the chamber to even it out and allow to polymerize at room temperature for at least 3 h.

8. Pour butanol from the casting unit after polymerization and rinse out residual butanol with 2–3 washes of SDS-PAGE running buffer. The entire casting unit or individual gels may be stored in the dark in an airtight container at 4°C.
9. Assemble gel cassettes in the Mini-PROTEAN 3 Cell or equivalent gel electrophoresis unit according to the manufactures instructions.
10. Drain gel cassettes of excess liquid and place one on each side of the electrode assembly, with spacer plates facing outward. Place the cassette-assembly in the clamping frame, which is then locked in closed position forcing the gasket to seal against short plate forming upper electrophoresis reservoir. Place the complete assembly in the dry buffer tank.
11. Carefully place focused and equilibrated IPG strips in the center of the gel cassette well. The IPG strip should not be too wet or too dry. Orientation of the strip should be noted in the gel cassette. Typically the plastic backing side of the strip is placed against the spacer plate with the positive end towards the left.
12. Gently press the IPG down with thin spatulas or tweezers on each of the side of the strip so that the strip is in contact with the top of gel. A flat and even interface should be observed along union of the strip and gel. This must be free from gaps and bubbles. If gaps are observed because the top of the gel is not flat, then the focused proteins will not enter the gel properly and the resulting gel will be poor with much streaking.
13. Add SDS-PAGE running buffer until it completely covers the IPG strips and the tank lid is place on top of the electrode assembly with the proper polarity. Electrophoresis is conducted at 100 V until the bromophenol dye front is 1–2 mm from the bottom of gel (60–70 min).

3.6. Gel Fixation and Staining

1. Carefully remove gels from the gel cassette. A notch can be cut one of the corners of the gel to designate orientation of the gel.
2. Place the 7 cm gel in a tray containing 40–50 ml of gel fixing/rinsing solution. Cover and incubate on a rocker for 15 min. Replace with fresh fixing solution and repeat the incubation step. Rinse the gel in ddH_2O and incubate for twice for 15 min. For 17 cm gels, use 250 ml of water.

3. For identifying differentially expressed proteins, stain 7 cm gels with 30–40 ml Sypro Ruby stain solution on a rocker according to the manufacturer's protocol. Incubate gel a minimum of 3 h to as long as overnight (see Note 10). For excising protein spots, stain 17 cm gels with 200 ml of GelCode Blue Stain Reagent (Pierce) or equivalent (G-250 colloidal Coomassie) for 12–16 h on a rocker. Destain gels by incubating with 300 ml of ddH_2O for 30 min three times before imaging.

3.7. Image Analysis

1. Wipe the imaging platform with water to remove any traces of dirt or dust that may lead to imaging artifacts. Set the zoom level so that all of the gel is captured and a sharp digital image is obtained. The same zoom level should be used for all gels.
2. Set exposure time between 30 and 60 s for Sypro Ruby and less than 1 s for Gel Code Blue. The gel images should not be over exposed as this will make analysis more difficult. Place the gel on the imaging deck so that it is in the proper orientation (this should be same for all gels) and so that no bubbles are under the gel. After imaging, return gels to 4°C, in darkness for storage (see Note 11).
3. When all the gels have been imaged, crop them identically using the PDQuest crop tool to remove blank space around the gel.
4. Open all the raw gel images that are to be analyzed in PDQuest (Bio-Rad) or comparable 2DE analysis software. The steps given below are based on using PDQuest.
5. Start the gel spot detection wizard in the detection menu. The detection wizard begins by asking manual guidance by highlighting the faintest spot, smallest spot, and then largest single spot on the gel image. Based on this software then calculates the sensitivity, size scale, and minimum peak to be used in detecting spots. Select the gaussian option, horizontal and vertical streak removal, background removal, and smoothing. The ruby speckle filter is enabled on the default setting. When selected, the spot detection wizard will then begin to detect spots on all of the open gels using these criteria. Identified proteins in the resulting scan set will be saved as a *.gsp file.
6. After spot detection assemble all the gels into a "matchset". This set will eventually contain all information about the gels used in a single experiment. After creation of a matchset, the user will be prompted to determine which gel on which to base the master gel. The "master" gel is a single synthetic gel created by PDQuest that best represents the gel images in a matchset. The gel chosen for this is usually the highest quality gel image in the control group. Assemble gel images from the same experimental group into replicate groups at this time

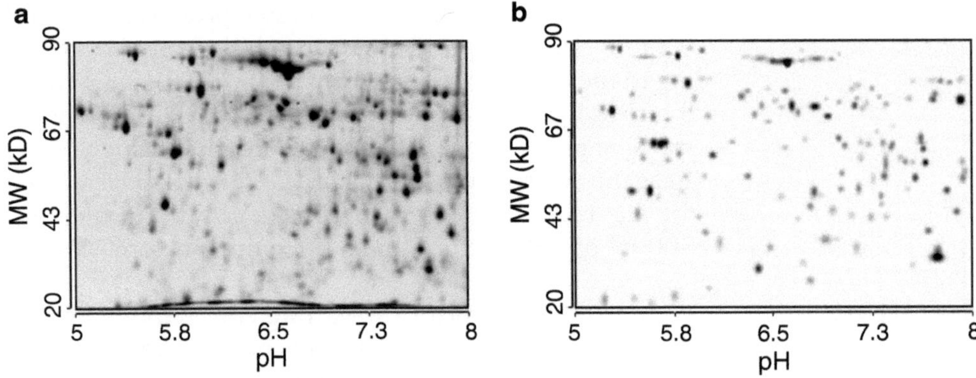

Fig. 2. Comparison of (**a**) a representative 2D gel image with (**b**) "master gel" image generated from all gels by the PDQuest image analysis software.

using a tab under the analyze menu. Automatic matching can be conducted this time with or without using landmarks. Landmarks are spots that consistently appear in all gels in the matchset, and are manually selected. Select 4–5 landmark spots distributed all over the gel to reduce the number of mismatched spots. A representative gel image and the corresponding "master" gel appear in Fig. 2.

7. Start "automatic matching" in the default mode using the extended matching tab. The success of automatic matching can be viewed under the group consensus tab under the analyze menu. This will show the number of spots detected to be common across each gel in replicate group and the entire matchset. Not all spots will be correctly matched across all of the gels, in many cases software will assign a spot where clearly there is no spot, or that a spot is clearly visible but not assigned. There are automated tools to remove or add spots to gels in a consensus group. But, the most effective way to handle mismatches is to go through each manually, viewing the same spot or mismatch simultaneously across all gels and manually assigning or removing it. Ultimately, every protein spot has to be reconciled across all gels in a matchset.
8. After completing manual matching, normalize the gels to total spot volume (i.e., volume of each spot is normalized to the total spot volume in the gel). The spot review tool in the analysis menu allows all spots and their calculated volumes to be viewed simultaneously.
9. Next, open the analysis set manager and create a quantitative analysis set to identify differentially expressed spots between replicate groups according to criteria such as minimum fold-change. It is highly recommended that the results of any

analysis set be verified by checking the spot review. Annotations for differentially expressed proteins can be viewed on any gel including the master gel (see Note 12).

3.8. Protein Spot Excision and Automated In-Gel Digestion

1. Excise differentially expressed proteins spots from a large format gel stained with Coomassie (see Note 1). Determine the correct spots for excision by examining a printout of the master gel containing annotations mentioned in the analysis section.
2. Create a gel plug extraction tool by trimming a 1,000 μl pipette tip with a scalpel to an inside diameter of approximately 2 mm.
3. Place a drop of ddH_2O on a protein spot before excision.
4. Set the pipette to volume of approximately 200 μl. Depress the pipette plunger and then puncture the gel spot with the pipette tip, completely severing it. Release plunger and the gel plug will be extracted (see Note 13).
5. Deposit the gel plug in a clean labeled microcentrifuge tube.
6. Digest gel plugs by in-gel trypsin digestion according to standard protocols (33–35). This can be done either manually or automatically using a system such as the ProGest system (Genomic Solutions). Gel plugs are washed with 300 μl 25-mM ammonium bicarbonate (ABC) and then dehydrated by washing with 100 μl of 50% 25-mM ABC (ammonium bicarbonate) and 50% acetonitrile (ACN) three times, followed by one wash with 100% ACN. The sample is reduced with 10 mM DTT at 60°C for 30 min followed by alkylation with 50 mM iodoacetamide for 45 min at room temperature. The gel spots are washed and dehydrated as before, prior to digestion with 15 μl of 0.05 mg/ml trypsin at 37°C for 4 h.

3.9. MALDI MS/MS of Protein Spot Digests

1. Desalt the samples manually or with an automated system such as the ProMS system using C18 Ziptips prior to spotting. Prepare the MALDI matrix with 5 mg/ml of α-cyano-4-hydroxy cinnamic acid prepared in 50:50 ACN/ddH20 containing 10 mM dihydrogen ammonium phosphate and 0.1% (by volume) trifluoroacetic acid. Two 10 μl sample loadings are used followed by washing steps and elution onto the MALDI target with the MALDI matrix.
2. MALDI MS-MS analysis can be performed using a 4700 Proteomics Analyzer (Applied Biosystems) or equivalent. MS data acquisition is using the reflectron detector in positive mode (700–4,500 Da, 1,900 Da focus mass) using 800 laser shots (40 shots per subspectrum) with internal calibration. The collision gas is air at the medium pressure setting, with 1 kV of collision energy applied across the collision cell.

3. Search the MS data against the latest Swiss-Prot protein sequence database using the GPS Explorer V2.1 software. A score of greater than 65 obtained with the Mascot search engine (Matrix Science, UK) is considered as significant ($p > 0.05$) (36). The parameters for this work are as follows: species, *Rattus norvegicus*; enzyme, trypsin; maximum missed cleavages, 1; variable modifications, oxidation (Met); peptide tolerance, 85 ppm; and MS/MS fragment tolerance, 0.3 Da.
4. Manually reanalyze several spots to confirm the generated MS data.

3.10. Biotin Tagging and Affinity Chromatography of Carbonylated Proteins

Carbonylated protein labeling with biotin hydrazide and isolation was performed following the protocol described by Mirzaei et al. (29, 37).

1. Prepare protein extracts from liver tissue by homogenizing whole rat liver tissue in manner described in Subheading 3.1, except that the lysis buffer contains 0.1% SDS, 0.5% sodium deoxycholate, 1.0% CHAPS, 0.1 M NaCl, 0.1 M sodium phosphate, 1 mM EDTA (pH 7.5), and protease inhibitor cocktail (Sigma). These compounds are free of primary amines which form Schiff bases with carbonyl groups and can interfere with the biotin hydrazide tagging (38), and are also known to be dialyzable. In addition, add biotin hydrazide to a concentration of 5 mM before homogenation. Incubate the homegenate for 30 min and then add sodium cyanoborohydride.to concentration of 15 μM and vortex (see Note 14).
2. Incubate the lysate for 30 min, centrifuge 13,000 × *g* for 15 min at 4°C, and collect lysate.
3. Add 1.5 ml of lysate to 3 ml dialysis cassette (Pierce) and dialyze with three 300 ml buffer changes (at least 200× sample volume) for 4 h each at 4°C to remove detergents and excess reagents. Determine the concentration of protein in the dialyzed lysate (see Note 14).
4. Fill disposable spin columns (5 ml size) with 0.6 ml settled monomeric avidin beads (1.2 ml 50% slurry). Wash the column with 3 ml of elution buffer containing 2 μM biotin and 25 mM ABC, vortex, and then centrifuge at 1,000 × *g* for 30 s, in order to bind nonreversible binding sites. Wash the column with 3 ml of 25 mM ABC twice, and regenerate the column by washing three times with 3 ml of regeneration buffer containing 0.1 M glycine, pH 2.8, followed by four washed with 3 ml of 25 mM ABC.
5. Add the dialyzed sample containing biotinylated carbonylated proteins to the column and incubate at room temperature for 20 min and then centrifuge 13,000 × *g* for 15 min at 4°C. Wash the column four times with 3 ml of 25 mM PBS to remove

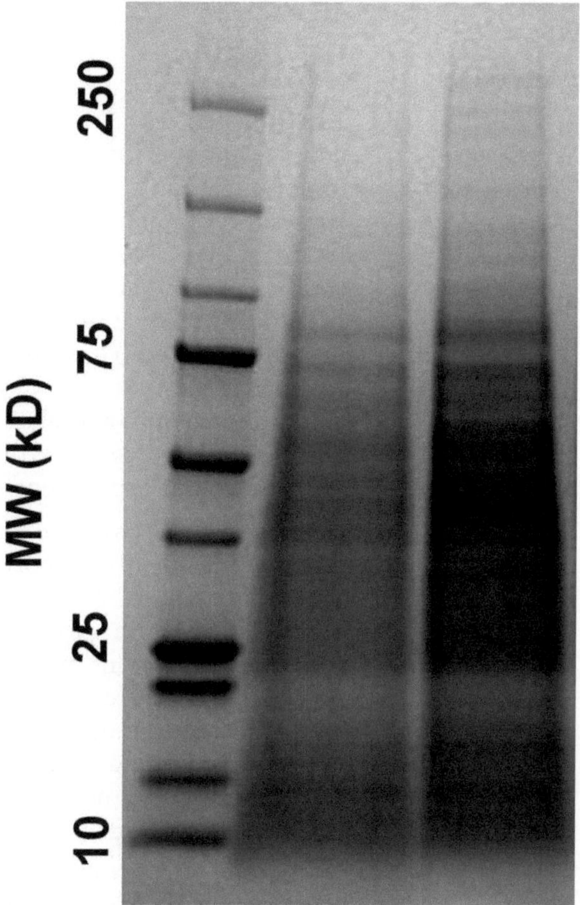

Fig. 3. Gel image of carbonylated protein isolated from two different experimental samples. Samples were resolved on a 7 cm, 8–16% gradient gel and stained with GelCode Blue.

nonspecifically bound proteins, followed by washing twice with 25 mM ABC to remove salt. Elute the biotinylated carbonylated proteins off the column by washing the beads four times with 350 μl of the elution buffer and collecting the flow-through each time in the same tube (see Note 15). The eluate is lyophilized and stored at −80°C.

3.11. Carbonylated Protein Detection and MS/MS Analysis

1. Rehydrate purified carbonylated protein samples to desired concentration in reducing SDS-PAGE buffer and resolve on a 7 cm 4–20% precast gel (Fig. 3). Gradient gels are used to give greater retention of lower molecular weight proteins. Fix and stain gels with Coomassie as described in Subheading 3.6. After imaging, gel lanes are manually excised into 35 gel slices for robotic in-gel digestion as described in Subheading 3.8. MS analysis and database searching for protein identification are performed as described earlier in Subheading 3.9.

4. Notes

1. Experimental workflow—the number of experimental groups and the number of biological samples in each group is very important when considering the design of a 2DE experiment. This affects the number of protein samples that must be collected and processed as well as the number of gels that must be run, stained and analyzed. In our experience (12), using liver tissue from three animals in each experimental group and two independent 2DE runs for each sample is the minimum number of gels needed to derive statistically significant conclusions. This translates into a total of 18 gels (3 experimental groups × 3 liver tissue samples per group × 2 independent 2DE runs per sample) for our experiment. Refer to Fig. 1.
2. To identify differentially expressed proteins, we used 7 cm IEF strips and Sypro Ruby fluorescent stain. Since this stain is very sensitive, it allows us to use lower amounts of protein. For excision of protein spots for MS identification, each protein spot must be present in sufficient quantities to be excised and detected. Therefore, a large format gel (17 cm IEF strip and compatible second dimension gel) is recommended. Moreover, the use of a large gel will ensure that only the desired spot is excised and contamination from neighboring spots is minimized. Refer to Fig. 1.
3. Pierce 660 nm Protein Assay is a modified protein binding Bradford based dye. It gives good repeatability, is shelf stable, and is less sensitive to salt, detergents, and pH than earlier dyes.
4. Use of the Ready-Prep (Bio-Rad) two-dimensional sample preparation kit improves isoelectric focusing by removing salt and reduces overall background in 2DE gels allowing for sharper delineation of protein spots. In a sample with low salt and contaminants, the dye front starts moving immediately upon starting IEF and disappears after 15–30 min into the IEF run.
5. It is recommended that multiple IEF belonging to replicates of an experimental group be run simultaneously to minimize variations in first dimension separation.
6. The manufacturer recommends that IEF strips of lengths 7, 11, 17 cm are be rehydrated with 140, 220, and 340 μl of sample with buffer. Recommended protein loads for the different strips are 50, 200, and 400 μg of protein, respectively.
7. Rehydration of IEF strips can be conducted passively (as described in Subheading 3.3), or actively where strips are placed directly in the IEF tray and the tray is placed in the IEF cell 50 V is applied to the IEF strip for a minimum of 6–8 h. We

found no advantage to using active rehydration other than possible time and labor savings. It is easier to physically manipulate the IEF strips and prevent or remove bubbles in a specially designed passive rehydration tray. Rehydrated strips can also be examined when being moved from the rehydration tray to the IEF tray to ensure even rehydration and no strip defects.

8. During the IEF, the maximum voltage in the second step is 4,000, 8,000, or 10,000 V for strips 7, 11, and 17 cm in length. The IPG strip voltages should reach 250 in the conditioning period, and then reach the maximum voltage at the end of the ramping period. If this does not occur, the resistance is likely to low due presence of excess salts in the sample. After the ramping period, 7, 11, and 17 cm lengths are focused for 20,000, 35,000, or 60,000 V-h, respectively.
9. It is recommended that a Mini-PROTEAN 3 Multi-Casting Chamber (Bio-Rad) or equivalent unit be used so that replicate gels can be run simultaneously and inter-gel variability minimized.
10. Sypro Ruby staining is preferred over silver staining because it is an equilibrium stain that avoids the notorious overstaining problems of silver stain, while delivering almost the same sensitivity but with far less background and artifacts. Overnight staining has produced the most consistent results.
11. The imaging surface should be clean to minimize contamination during spot excision for MS analysis. Always wear gloves and ensure that glass plates are clean to minimize keratin contamination.
12. In PDQuest, the spot review tool can be used to check if spots across all gels are congruent. Additionally, it provides individual spot volume and can be used to manually calculate spot quantitations.
13. During excision the gel can be placed on a light box with a smooth translucent top and illuminated from underneath. This greatly eases the manual excision process.
14. Excess biotin hydrazide is added to mixture before homogenization to immediately begin tagging aldehyde groups and reduce possible cross-linking between aldehydes and amine groups. Sodium cyanoborohydride reduces and stabilizes the hydrazone bond formed between hydrazide and aldehyde groups (38). Adding BSA increases the osmolarity of the dialysis buffer, and decreases the amount of water that will diffuse into the dialysis cassette, reducing the chance of excess swelling and cassette failure.
15. In addition to elution by binding competition with the 2 mM biotin solution, the glycine regeneration buffer can be used to elute biotinylated carbonylated proteins; however, pH may need to be adjusted as it is very acidic.

References

1. Wilkins MR, Pasquali C, Appel RD, Ou K, Golaz O, Sanchez JC, Yan JX, Gooley AA, Hughes G, Humphery-Smith I, Williams KL, Hochstrasser DF (1996) From proteins to proteomes: large scale protein identification by two-dimensional electrophoresis and amino acid analysis. Nat Biotechnol 14:61–65
2. Paintlia MK, Paintlia AS, Barbosa E, Singh I, Singh AK (2004) *N*-acetylcysteine prevents endotoxin-induced degeneration of oligodendrocyte progenitors and hypomyelination in developing rat brain. J Neurosci Res 78: 347–361
3. Reynolds JL, Mahajan SD, Sykes DE, Schwartz SA, Nair MP (2007) Proteomic analyses of methamphetamine (METH)-induced differential protein expression by immature dendritic cells (IDC). Biochim Biophys Acta 1774: 433–442
4. Soreghan BA, Yang F, Thomas SN, Hsu J, Yang AJ (2003) High-throughput proteomic-based identification of oxidatively induced protein carbonylation in mouse brain. Pharm Res 20:1713–1720
5. Riekse RG, Li G, Petrie EC, Leverenz JB, Vavrek D, Vuletic S, Albers JJ, Montine TJ, Lee VM, Lee M, Seubert P, Galasko D, Schellenberg GD, Hazzard WR, Peskind ER (2006) Effect of statins on Alzheimer's disease biomarkers in cerebrospinal fluid. J Alzheimers Dis 10:399–406
6. Witzmann FA, Strother WN (2004) Proteomics and alcoholism. Int Rev Neurobiol 61: 189–214
7. Moon KH, Hood BL, Kim BJ, Hardwick JP, Conrads TP, Veenstra TD, Song BJ (2006) Inactivation of oxidized and S-nitrosylated mitochondrial proteins in alcoholic fatty liver of rats. Hepatology 44:1218–1230
8. Henkel C, Roderfeld M, Weiskirchen R, Berres ML, Hillebrandt S, Lammert F, Meyer HE, Stuhler K, Graf J, Roeb E (2006) Changes of the hepatic proteome in murine models for toxically induced fibrogenesis and sclerosing cholangitis. Proteomics 6:6538–6548
9. Sul D (2006) Evaluation of toxicological monitoring markers using proteomic analysis. J Proteome Res 5:2525–2526
10. Duan X, Berthiaume F, Yarmush DM, Yarmush ML (2009) Dissimilar hepatic protein expression profiles during the acute and flow phases following experimental thermal injury. Proteomics 9:636–647
11. Vascotto C, Cesaratto L, D'Ambrosio C, Scaloni A, Avellini C, Paron I, Baccarani U, Adani GL, Tiribelli C, Quadrifoglio F, Tell G (2006) Proteomic analysis of liver tissues subjected to early ischemia/reperfusion injury during human orthotopic liver transplantation. Proteomics 6:3455–3465
12. Newton BW, Russell WK, Russell DH, Ramaiah SK, Jayaraman A (2009) Liver proteome analysis in a rodent model of alcoholic steatosis. J Proteome Res 8:1663–1671
13. Gray J, Chattopadhyay D, Beale GS, Patman GL, Miele L, King BP, Stewart S, Hudson M, Day CP, Manas DM, Reeves HL (2009) A proteomic strategy to identify novel serum biomarkers for liver cirrhosis and hepatocellular cancer in individuals with fatty liver disease. BMC Cancer 9:271
14. Ma Y, Yu J, Chan HL, Chen YC, Wang H, Chen Y, Chan CY, Go MY, Tsai SN, Ngai SM, To KF, Tong JH, He QY, Sung JJ, Kung HF, Cheng CH, He ML (2009) Glucose-regulated protein 78 is an intracellular antiviral factor against hepatitis B virus. Mol Cell Proteomics 8:2582–2594
15. Acosta D, Anuforo DC, Smith RV (1980) Cytotoxicity of acetaminophen and papaverine in primary cultures of rat hepatocytes. Toxicol Appl Pharmacol 53:306–314
16. Ioannides C, Parke DV (1990) The cytochrome P450 I gene family of microsomal hemoproteins and their role in the metabolic activation of chemicals. Drug Metab Rev 22:1–85
17. Greuet J, Pichard L, Bonfils C, Domergue J, Maurel P (1996) The fetal specific gene CYP3A7 is inducible by rifampicin in adult human hepatocytes in primary culture. Biochem Biophys Res Commun 225:689–694
18. Savas U, Griffin KJ, Johnson EF (1999) Molecular mechanisms of cytochrome P-450 induction by xenobiotics: an expanded role for nuclear hormone receptors. Mol Pharmacol 56:851–857
19. Fridovich I (2004) Mitochondria: are they the seat of senescence? Aging Cell 3:13–16
20. Bachmann KA (1996) The cytochrome P450 enzymes of hepatic drug metabolism: how are their activities assessed in vivo, and what is their clinical relevance? Am J Ther 3:150–171
21. Fridovich I (1995) Superoxide radical and superoxide dismutases. Annu Rev Biochem 64:97–112
22. Stadtman ER (1992) Protein oxidation and aging. Science 257:1220–1224
23. Levine RL, Moskovitz J, Stadtman ER (2000) Oxidation of methionine in proteins: roles in

antioxidant defense and cellular regulation. IUBMB Life 50:301–307

24. Perrot N, Nalpas B, Yang CS, Beaune PH (1989) Modulation of cytochrome P450 isozymes in human liver, by ethanol and drug intake. Eur J Clin Invest 19:549–555
25. Lieber CS (1993) Biochemical factors in alcoholic liver disease. Semin Liver Dis 13:136–153
26. Cederbaum AI (1998) Ethanol-related cytotoxicity catalyzed by CYP2E1-dependent generation of reactive oxygen intermediates in transduced HepG2 cells. Biofactors 8:93–96
27. Ramaiah S, Rivera C, Arteel G (2004) Early-phase alcoholic liver disease: an update on animal models, pathology, and pathogenesis. Int J Toxicol 23:217–231
28. Dalle-Donne I, Aldini G, Carini M, Colombo R, Rossi R, Milzani A (2006) Protein carbonylation, cellular dysfunction, and disease progression. J Cell Mol Med 10:389–406
29. Mirzaei H, Regnier F (2005) Affinity chromatographic selection of carbonylated proteins followed by identification of oxidation sites using tandem mass spectrometry. Anal Chem 77:2386–2392
30. Duan X, Yarmush D, Berthiaume F, Jayaraman A, Yarmush ML (2005) Immunodepletion of albumin for two-dimensional gel detection of new mouse acute-phase protein and other plasma proteins. Proteomics 5:3991–4000
31. Duan X, Yarmush DM, Berthiaume F, Jayaraman A, Yarmush ML (2004) A mouse serum two-dimensional gel map: application to profiling burn injury and infection. Electrophoresis 25:3055–3065
32. Welsh GI, Griffiths MR, Webster KJ, Page MJ, Tavare JM (2004) Proteome analysis of adipogenesis. Proteomics 4:1042–1051
33. Shevchenko A, Wilm M, Vorm O, Mann M (1996) Mass spectrometric sequencing of proteins silver-stained polyacrylamide gels. Anal Chem 68:850–858
34. Jensen ON, Wilm M, Shevchenko A, Mann M (1999) Peptide sequencing of 2-DE gel-isolated proteins by nanoelectrospray tandem mass spectrometry. Methods Mol Biol 112:571–588
35. Lopez MF, Berggren K, Chernokalskaya E, Lazarev A, Robinson M, Patton WF (2000) A comparison of silver stain and SYPRO Ruby Protein Gel Stain with respect to protein detection in two-dimensional gels and identification by peptide mass profiling. Electrophoresis 21:3673–3683
36. Perkins DN, Pappin DJ, Creasy DM, Cottrell JS (1999) Probability-based protein identification by searching sequence databases using mass spectrometry data. Electrophoresis 20:3551–3567
37. Mirzaei H, Regnier F (2007) Identification of yeast oxidized proteins: chromatographic top-down approach for identification of carbonylated, fragmented and cross-linked proteins in yeast. J Chromatogr A 1141:22–31
38. Hermanson GT (2008) Bioconjugate techniques. Maryland Heights Academic Press, pp 200–201

Chapter 18

Targeted Mass Spectrometry-Based Metabolomic Profiling Through Multiple Reaction Monitoring of Liver and Other Biological Matrices

Angelo D'Alessandro, Federica Gevi, and Lello Zolla

Abstract

In a systemic viewpoint, relevant biological information on living systems can be grasped from the study of small, albeit pivotal molecules which constitute the fundamental bricks of metabolic pathways. This holds true for liver which plays, among its unique functions, a key role in metabolism. The nonbiased analysis of all this small-molecule complement in its entirety is known as metabolomics. However, no practical approach currently exists to investigate all metabolic species simultaneously without including a technical bias towards acidic or basic compounds, especially when performing mass spectrometry-based investigations. Technical aspects of rapid resolution reversed phase HPLC online with mass spectrometry are hereby described. Such an approach allows to discriminate and quantify a wide array of metabolites with extreme specificity and sensitivity, thus enabling to perform complex investigations even on extremely low quantities of biological material. The advantages also include the possibility to perform targeted investigations on a single (or a handful of) metabolite(s) simoultaneously through single (multiple) reaction monitoring, which further improves the dynamic range of concentrations to be monitored.

Such an approach has already proven to represent a valid tool in the direct (on the liver) or indirect (on human red blood cell metabolism which is hereby presented as a representative model, but also on blood plasma or other biological fluids) assessment of metabolic poise modulation and pharmacokinetics for drug development.

Key words: Metabolomics, MRM, Mass spectrometry, Rapid resolution reverse phase HPLC, Extraction methods, Liver

1. Introduction

Liver is a vital organ playing a wide range of functions, including protein synthesis, and production of biochemicals necessary for digestion. Liver also plays a key role in metabolism and detoxification, which makes it a critical target for those studies aiming at the

Djuro Josic and Douglas C. Hixson (eds.), *Liver Proteomics: Methods and Protocols*, Methods in Molecular Biology, vol. 909, DOI 10.1007/978-1-61779-959-4_18, © Springer Science+Business Media New York 2012

determination of pathological conditions or at the assessment of metabolization/adverse effects upon drug assumption. Indeed, hepato- and nephrotoxicity are major attrition factors in preclinical drug development and thus investigative studies on the alterations of the small-molecule complement in the liver represent the basis of a recently expanding discipline which goes under the name of predictive toxicology (1). The liver is the major site of synthesis of endogenous metabolites, and the alterations in the profiles of endogenous metabolites ("the metabolome") may precede development of clinically overt drug-induced liver injury (2). This could be either assessed directly on liver, as in the case of animal trials on drugs (3), or indirectly on other biological fluids, like blood plasma or urine. The latter option is more practical and informative, especially in human clinical trials on drugs, such as in the recent cases of acetaminophen (APAP) and ximelagatran (2).

On the other hand, direct on liver analyses in mice models have already proven useful to improve comprehension of the physiological alterations taking place in obese individuals, paving the way for characterization of optimal diets for those patients seeking health improvements through scientifically validated dietary regimens (4, 5). In all the cases mentioned above, metabolomics has been proposed as a precious tool to produce fundamental data to shed light into these hot biological questions.

1.1. From Clinical Biochemistry to "Omic" Sciences and Metabolomics

However, the root of this discipline shares consistent traits with clinical biochemistry, which has historically pursued determination of standard and anomalous parameters (i.e., absolute concentration, relative abundance, etc.) of small molecular compounds in blood and its components (plasma/serum and cellular fractions). Recent advancements in the field of clinical biochemistry, especially in transfusion medicine and immunohematology closely related fields (6, 7), are mainly tied to the big technical strides in "omic" disciplines, including transcriptomics, proteomics, and metabolomics. During the last decades, "omic"-oriented strategies have constantly gained momentum, which delve into biological complexity as a whole (e.g., proteins in proteomics, mRNAs in transcriptomics) rather than dissecting biological samples through targeted analysis of single molecules (8).

Metabolomics is "the nonbiased quantification and identification of all metabolites present in a biological system," although the term metabolomics is routinely used in a broader perception as to include global identification of as many small molecule (MW lower than 1,500 Da) metabolites as possible or of a subset of them (acidic compounds; basic compounds; sugar phosphates; just to mention few). While the dawn of metabolomics dates back to 1960s, it was only in 1971 that Pauling, Robinson et al. conceived the core idea that information-rich data reflecting the functional status of a complex biological system resides in the quantitative and qualitative pattern of metabolites in body fluids (9).

Part of the ambiguity in the use of the term metabolomics is a result of the fact that truly nonbiased quantification and identification of all metabolites present in biological systems is currently not obtainable, due to technical limitations (10). The main technical obstacles hindering the way to an omni-comprehensive metabolome portrait stem both from the optimization of sample extraction efficiency of a series of metabolites as broad as possible and from the approach used to perform metabolic analyses, either nuclear magnetic resonance (NMR) or mass spectrometry (MS). In this respect, literature has recently flourished around these topics (11–16).

1.2. Technical Evolution of Metabolomics: From NMR to MS

Earliest approaches to metabolomic investigations mainly relied on NMR which was favored by machine accessibility, established data handling, and the nondestructive nature of the analysis (16). Nonetheless, MS has gradually replaced NMR due to the higher sensitivity, improved metabolite discrimination, coverage of the metabolome space, and modularity to perform compound-class-specific analyses, other than to a dramatically reduced demand for starting material necessary to perform an extensive analysis (17). MS also offers the advantage to perform targeted analyses, thus to follow one (or a handful) of metabolites through isolation and fragmentation of precursor ion and subsequent isolation of the product ions or features of interest, through selected/multiple reaction monitoring (SRM or MRM) (18). Such an approach holds the advantage to monitor metabolites through a wide spread range of linear concentrations (from millimolar to nanomolar, down to picomole quantities, depending on the characteristics of the MS instruments), while allowing to directly test the levels of metabolites of interest even in samples as low as 0.5 μl (as in the case of blastocoele fluid) (18).

On the other hand, MS could be also exploited to perform a nontargeted strategy which potentially enables de novo target discovery since the exploration of the chemical space is only limited by the sample preparation and the characteristics of the analytical technique (sensitivity and coverage). However, it is often difficult to process huge amounts of raw data to unequivocally elucidate the chemical identity of the potential targets, mainly due to platform-dependent software limitations or by the elevated noise of the recorded m/z signals, especially when performing analyses on very low amounts of samples or on extremely low-abundance metabolites. Another level of complexity is added by the necessity to include all the possible adducts of the species of interest depending on the ion mode in which the experiments are performed (e.g., H^+, K^+, and Na^+ adducts in positive ion mode), which requires time-consuming manual data handling or post hoc bioinformatic elaboration of raw data.

Targeted strategies offer the potential to lower both limits of detection (LOD) and limits of quantification (LOQ), which eases quantification of low-abundance metabolic species also in scarcely available samples. Nevertheless, absolute quantification, which is the ultimate goal in biomarker individuation and testing, can be only performed through monitoring metabolite concentrations against signals of calibration standards, either internal (added to the sample before extraction) or external standards (added to the sample after extraction). In so doing, individual variance between the preparation of the samples and, more importantly, matrix effects and other inferences in the sample are minimized (19).

1.3. The Role of Sample Extraction Efficiency

A successful analytical strategy for untargeted metabolomics workflows ideally should be rapid, robust and follow an extraction and separation protocol that gives adequate consideration to variables such as the nature of extraction solvent, quenching of metabolic turnover and inclusion of internal standards that helps gauge the success of the extraction procedure (11).

On the other hand, sample-handling steps should be kept to a minimum as they introduce uncontrolled analyte loss, and by their very nature are selective and thus discriminating in a global analysis. Such a bias can be purposedly introduced when performing targeted analyses, as to enrich specific classes of compounds (for example water or hydrophilic solvents represent the eligible choice for extraction of sugar-phosphates and other hydrophilic species) (16).

There are also a series of additional parameters which should be considered during sample preparation for MS-based metabolomic analyses, viz., sample storage temperature, protein precipitation methods, and processing-time considerations. The extraction solvent pH is but a minor parameter to be adjusted in order to bring analytes to a state where they can be extracted from one of the matrices, maximizing selectivity at a particular pH with minimal loss of recovery (20).

Recent investigations have compared a wide series of solvents for the optimization of metabolite extraction in model biological matrices, such as red blood cells (18, 21). Projection to latent structure of the GC/MS and LC/MS data suggested that the most efficient solution for the extraction of metabolites from wet erythrocytes (50 mg) could be a methanol–chloroform–water mixture (950 μl, 700:200:50, v/v/v) (21). Independent investigators came to the same conclusion, although they pointed out that the time sequence in which the solvents are used is essential to the optimal outcome of the extraction (11). A methanol–chloroform–water extraction protocol is but an evolution of the Bligh/Dyer protocol for lipid extraction (22). In 1959, when studying lipid deterioration in frozen seafood, Bligh and Dyer optimized a chloroform–methanol–water phase diagram, based on the hypothesis that "optimum lipid extraction should result when the tissue is homogenized with a mixture of chloroform and methanol which, when mixed with the water in the tissue, would have yielded a monophasic solution" (22). The resulting homogenate could then be diluted with water and/or chloroform to produce a biphasic system, the chloroform layer of which should contain the lipids and the methanol–water layer the nonlipids (22). These conclusions are still widely accepted by the scientific community and corroborated by further experimentation on optimum extraction protocol setting up (11).

1.4. HPLC Settings: Recent Advancements in Liquid Chromatography Columns and Chemistry

Three main pre-MS analytical approaches have been proposed over the years, namely, gas chromatography (GC), liquid chromatography (LC), and capillary electrophoresis (CE). GC is dampened by the poor discrimination against large intermediates such as nucleotides, flavines, and coenzyme A derivatives. On the other hand, LC holds several advantages including widespread coverage, sensitivity, ease of use, robustness to matrix, and robustness in routine operation. CE is equivalent to LC in terms of separation and sensitivity, although it lacks in robustness, which is pivotal for routine analysis of biological extracts (23).

However, LC suffers from some drawbacks as well, especially concerning the analysis of very polar compounds, such as a wide array of anionic metabolites of primary metabolism. Several solutions have been proposed to overcome this hurdle, including (a) the use of ion exchange chromatography, (b) the addition of postcolumn sodium-proton exchanger, (c) hydrophilic interaction chromatography (HILIC) with an aminopropyl stationary phase or ion pairing-reversed phase chromatography in the analysis of phosphorylated compounds, carboxylic acids, nucleotides, and coenzyme A esters, (d) normal-phase chromatography on silica hydride (24–26).

Nonetheless, all of these methods have some detrimental pitfalls as well, including reduction of peak width and sensitivity, overexposure of the MS to ion-pairing coupling agents (12).

Most of the recent HPLC-MS-based studies mainly relied on the use of RP-HPLC with C18 columns (11, 12, 18). As elegantly described by Buescher et al. (12), the interaction with the end-capped C18 phase depends on inherent and ion pairing mediated hydrophobic properties, both influencing separation of isomers and coverage of many different compounds. It is thus possible to perform compound class specific analyses as they tend to have similar retention time, in a chronological order which also depends on the mobile phases (water for phase A, methanol or, more often, acetonitrile for phase B) and the gradient (either linear or multistep, reducing the initial slope to obtain an improved discrimination of early-eluting polar compounds). A frequent RP-HPLC gradient on C18 columns tends to elute (a) sugars and aliphatic compounds with a positive net charge earlier, then (b) purines and pyrimidines, (c) acetic and aromatic amino acids, (d) sugar phosphates and monocarboxylic acids, (e) dicarboxylic acids and nucleotide monophosphates, (f) nucleotide diphosphates and redox cofactors and sugar diphosphates, and (g) nucleotide triphosphates and aromatics and coenzyme A esters (12). Hydrophobic characteristics or selective interaction with the end-capping groups allow further separating even identically charged compounds (i.e., dicarboxylic acids and sugar phosphates, respectively) (12). Charged groups could attract ion pairing molecules thus to render molecules more hydrophobic, which results in delayed retention times (8).

Recent advancements in the field of HPLC not only involve an in-depth understanding of the chemistry behind small-molecule interaction with novel sub-2 μm or fused silica stationary phases, but also the introduction of technical innovation, such as rapid resolution (RR) and ultra-HPLC (UHPLC). In RR and UHPLC short columns are used which are packed with 2.7 μm fused-core silica particles that are made by fusing a 0.5 μm layer of porous silica onto a solid silica particle (27). These unique particles enable very rapid chromatographic separation at the expense of very high backpressures, which can be easily handled with UHPLC and RR-RP-HPLC, with the advantage to obtain very sharp separation profiles in 10–25 min runs (10–12, 18).

1.5. MS Settings and Physicochemical Properties of Metabolic Species

Ion pairing agents (formic acid, trifluoroacetic acid) are introduced in HPLC phases and do influence retention times of analyzed molecules. Nonetheless, their main function is to enhance ionization of the analytes in order to improve MS detection and quantitation. Theoretically, anionic compounds at physiological pH such as several acids of the Krebs cycle and sugar phosphates can be better monitored in negative ion mode, as they naturally tend to loose protons in solution. This is the main reason why most of the common MS-based metabolomics studies have been performed in negative ion mode or, at least, switching between negative and positive modes (10–12). Indeed, ionization in positive mode is in principle possible with excellent results in sensitivity, which allows improving the dynamic range of linear concentrations and thus to reduce the requirements for the quantities of biological materials needed to perform the analysis (18). Unfortunately, routine operation in positive mode drastically increases maintenance and cleaning on the initial stages of the mass spectrometer to remove the sediments of ion pairing agent (12).

Ion traps, Fourier transform mass spectroscopy (FT-ICR-MS), Orbitrap instruments and triple quadrupole MS instruments are routinely used in MS-based metabolomics. In the case of the analysis of tissue metabolomes, extraction is a necessary prerequisite, except in cases where analysis is to be performed directly on the tissue, as is the case with matrix-assisted laser desorption/ionization (MALDI) imaging or desorption electrospray ionization (DESI) (17).

The advantages of one mass analyzer over the other have been resumed by Griffiths et al. (17), which described in detail how the analyzers can be arranged in series in space, such as on hybrid instruments (for example, tandem quadrupole, Q-TOF instruments), or in time, such as with ion traps. Ion traps are particularly suited to record MS^3 and further MS^n spectra. Triple quadrupole instruments offer other advantages such as SRM "scans," where MS1 is "parked" on an m/z value of interest and MS^2 on the m/z value(s) of a known fragment ion (SRM) or multiple fragment ions (MRM) (17).

1.6. Data Analysis: Online Databases and In Silico Elaboration

One of the main issues to deal with when performing untargeted metabolomics analyses is data filtering and interpretation, as signals over the whole retention time range are often noisy and difficult (or time-consuming) to interpret manually. A joint effort is currently underway to put into place a public database of tandem MS spectra. This is not an easy task as results obtained through different instruments (or from the same instruments from different vendors) tend to differ in fragmentation spectra (10). Despite these issues, it is now possible to rely on freely accessible databases such as HMDB (28), MassBank, Metlin (29), LipidMaps, ChemACX, and ChemSpider which have been built as to contain millions of m/z profiles and chemical structures. However, they only represent just a preliminary tool to ease the data interpretation steps, since they are far from containing all relevant structures and. For example, small molecules in biological systems are subject to phase I and II metabolism (glucuronidation, reduction, oxidation, sulfation, amino acid conjugation, etc.), and many of these modified small molecules are not covered in current databases (10). Independent laboratories have thus sought to build up their personal databases for high-throughput untargeted purposes, although at the expenses of lengthy and expensive investments on instruments, in-house software and trained personnel (10).

On the other hand, targeted metabolomics analyses through SRM or MRM also suffer from minor bioinformatic issues, which include partial drift of HPLC peaks over different HPLC-MS runs, standardization of peak picking criteria for optimal quantitation of MRM spectra and normalization of signal-to-noise (S/N) ratio against technical variables (intra- and interday reproducibility).

In order to overcome these obstacles, a series of open source valid programs for peak alignment, normalization, and peak picking have been realized and freely distributed. These software packages ease standardization of data handling in proteomics (but also metabolomics), such as ToppView OpenMS, InsilicosViewers and other tools for in silico elaboration of mass spectra, 3D visualization of MS outputs (retention times, m/z, and MS-detected counts on each axis) (30, 31).

2. Materials

2.1. HPLC and Metabolite Extraction

Acetonitrile, formic acid, and HPLC-grade water for phase A (0.1% formic acid in water) and phase B (acetonitrile, 0.1% formic acid) were purchased from Sigma Aldrich (Milano, Italy).

An Ultimate 3000 Rapid Resolution HPLC system [LC Packings DIONEX (Sunnyvale, CA, USA)] was used to perform metabolite separation. The system featured a binary pump and vacuum degasser, well-plate autosampler with a six-port micro-switching valve, a thermostated column compartment. A Dionex Acclaim RSLC 120 C18 column 2.1 mm × 150 mm, 2.2 μm was used to separate the extracted metabolites.

2.2. Mass Spectrometry

Metabolites were directly eluted into a High Capacity ion Trap HCTplus [Bruker-Daltonik (Bremen, Germany)].

2.3. Metabolite Standards

Methanol and chloroform and HPLC-grade water for metabolite extraction were purchased from Sigma Aldrich (Milano, Italy).

Standards (equal or greater than 98% chemical purity) were purchased from Sigma Aldrich (Milan, Italy) and included ATP, L-lactic acid, phosphogluconic acid, NADH, D-fructose 1,6 biphosphate, D-fructose 6-phosphate, glyceraldehyde phosphate, phosphoenolpyruvic acid, L-malic acid, L-glutamic acid, oxidized glutathione, α-ketoglutarate.

Alternative internal standards were 1-naphthylamine and 2-(methylthio)benzothiazole, which were purchased from Sigma Aldrich (St. Louis, MO, USA) and Acros organics (Morris Plains, NJ, USA), respectively.

2.4. Standard Storage and Preparation

Standards were stored either at −25°C, 4°C or room temperature, following manufacturer's instructions. Each standard compound was weighted and dissolved in nanopure water. Starting at a concentration of 1 mg/ml of the original standard solution, a dilution series of steps (in 18 MΩ, 5% formic acid) was performed for each of the standards in order to reach the limit of detection (LOD) and limit of quantification (LOQ).

3. Methods

3.1. Red Blood Cells

RBC units were drawn from healthy human volunteers according to the policy of the Italian Blood Transfusion Service for donated blood and all the volunteers provided their informed consent in accordance with the declaration of Helsinki. RBC units were collected from three donors [male = 2, female = 1, age 44 ± 6.5

(mean ± SD)] in Latium (Italy). Saline–adenine–glucose–mannitol (SAGM) erythrocyte concentrates were removed aseptically for the analysis within the first day of storage. For each sample, 0.5 ml from the pooled erythrocyte stock was transferred into a microcentrifuge tube and processed for metabolite extraction, as described below. Erythrocyte samples were then centrifuged at 1,000 × *g* for 2 min at 4°C as to remove SAGM and any further contaminant. Tubes were then placed on ice while supernatants were carefully aspirated, paying attention not to remove any erythrocyte at the interface. Samples were further processed for metabolite extraction.

3.2. Sample Extraction

Samples were extracted following the protocol proposed by Sana et al. (11), with minor modifications (18) (see Note 1).

1. The sample was resuspended by adding 0.15 ml of ice-cold ultrapure water (18 MΩ—see Note 2) to lyse cells.
2. The tubes [2 ml original Eppendorf (see Note 3)] were plunged into dry ice or a circulating bath at −25°C for 0.5 min and then into a water bath at 37°C for 0.5 min (see Note 4).
3. To each tube was added first 0.6 ml of −20°C methanol containing L-malic acid for red blood cells or 1-naphthylamine and 2-(methylthio)benzothiazole as internal standards(see Note 5).
4. 0.45 ml of −20°C chloroform was added (see Notes 6 and 7).
5. The tubes were mixed every 5 min for 30 min.
6. 0.15 ml of ice-cold pH adjusted ultrapure water (18 MΩ) was added to each tube.
7. The tubes were centrifuged at 1,000 × *g* for 1 min at 4°C (see Note 8).
8. The tubes were transferred to −20°C for 2–8 h (see Note 9). An equivalent volume of acetonitrile was added to precipitate any proteins and then the tubes were transferred to refrigerator (4°C) for 20 min.
9. Each tube was centrifuged at 10,000 × *g* for 10 min at 4°C and the supernatant was recovered into a 2 ml tube.
10. Collected supernatants were dried as to obtain visible pellets (see Notes 10 and 11).
11. Finally, the dried samples were resuspended in 1 ml of 5% formic acid in water and transferred to glass autosampler vials for LC/MS analysis (see Note 12).

3.3. Determination of Extraction Efficiency: The Red Blood Cell as a Model

Internal standards were added in order to assess extraction efficiency. L-Malic acid was used as internal standard in red blood cell extracts, through exogenous addition at step 3 of the extraction protocol at different concentrations (0; 1; 5; 10 mg/ml—Fig. 1) (18).

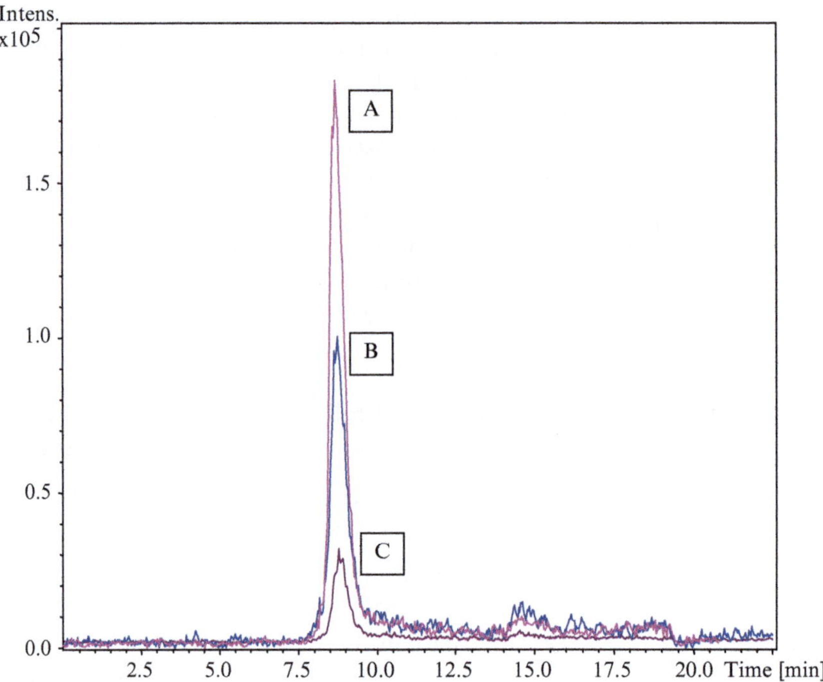

Fig. 1. MRM spectra for exogenously added L-malic acid, as internal standard to test sample extraction efficiency in three independent extractions from red blood cells at different concentrations of the standard metabolite: 1 mg (A), 5 mg (B), and 10 mg (C).

This allowed testing the linearity and reproducibility (four technical replicates for each extraction) of the extraction method. Since mature red blood cells are devoid of nuclei and mitochondria, erythrocytes are incapable of generating energy via the (oxidative) Krebs cycle. Therefore, erythrocytes mainly rely on four main metabolic pathways:

1. The Embden–Meyerhof pathway (glycolysis), in which most of the red blood cell adenosine triphosphate (ATP) is generated through the anaerobic breakdown of glucose.
2. The hexose monophosphate shunt (HMS), which produces NADPH to protect red blood cells from oxidative injury.
3. The Rapoport–Luebering shunt, responsible for the production of 2,3-diphosphoglycerate (DPG) for the control of Hb oxygen affinity.
4. Finally, the methemoglobin (met-Hb) reduction pathway, which reduces ferric heme iron to the ferrous form to prevent Hb denaturation (32, 33). Red blood cells do maintain a number of proteins which have been demonstrated to be potentially enzymatically active, such as malate dehydrogenase, although they represent but a functionless remainder after the

dedifferentiation of reticulocytes into the mature red blood cells (34). The exogenously introduced L-malic acid could be thus useful to test the efficiency of the extraction protocol and thus calculate a coefficient to derivitize the absolute concentration of the monitored metabolite in the original sample (18).

3.4. Rapid Resolution Reverse-Phase HPLC Settings

LC parameters: injection volume, 20 μl; column temperature, 25°C; and flow-rate of 0.2 ml/min. The LC solvent gradient and timetable were identical during the whole period of the analyses. A 0–95% linear gradient of solvent A [0.1% (v/v) formic acid in water] to B [0.1% (v/v) formic acid in acetonitrile] was employed over 15 min followed by a solvent B hold of 2 min, returning to 100% A in 2 min and a 6-min post-time solvent A hold (see Notes 12–15) (18).

3.5. ESI Mass Spectrometry Settings

Mass spectra for metabolite extracted samples were acquired in positive ion mode. ESI capillary voltage was set at 3,000 V (+) ion mode. The liquid nebulizer was set to 30 psig and the nitrogen drying gas was set to a flow rate of 9 l/min. Dry gas temperature was maintained at 300°C. Data was stored in centroid mode. Internal reference ions were used to continuously maintain mass accuracy. Data was acquired at a rate of 5 spectra/s with a stored mass range of m/z 50–1,500. Data was collected using Bruker Esquire Control (v. 5.3—build 11) data acquisition software. In MRM analysis, m/z of interest were isolated and monitored throughout the whole RT range. Validation of HPLC online MS-eluted metabolites was performed by comparing transitions fingerprint, upon fragmentation and matching against the standards metabolites through direct infusion with a syringe pump (infusion rate 4 μl/min).

3.6. Data Elaboration and Statistical Analysis

LC/MS data files were processed by Bruker DataAnalysis 4.0 (build 234) software. Files from each run were either analyzed as .d files or exported as mzXML files, to be further elaborated for spectra alignment, peak picking and quantitation with InSilicos Viewer 1.5.4 [Insilicos LLC (Seattle, WA, USA)]. For Total Ion Current (TIC) analyses, all compounds and compound-related components (i.e., features) in a spectrum were considered for quantitation. In positive-ion mode this included adducts (H+, Na+, and K+), isotopes and dimers. These related ions were treated as a single compound or feature for preliminary qualiquantitative analysis of metabolites of interest (untargeted analysis). Absolute quantitative analyses of standard compounds were performed on MRM data (targeted analyses). Each standard metabolite was run in triplicate, at incremental dilution until LOD and LOQ were reached. Precursor ions, fragmentation energies and transition features to be isolated and monitored were determined through direct infusion through a syringe pump (4 μl/min). The limit of detection for each compound was calculated as the minimum amount injected which gave a detector response higher than three times the S/N ratio. Basic compounds were tested in positive ion mode, while acidic compounds, including sugar phosphates were preferentially monitored in negative ion mode. However, positive ion mode was preferred to negative ion mode also for the latter group, especially when performing analyses on samples displaying a high dynamic range of metabolite concentrations, since positive ion mode guaranteed a broader range of linearity for MS signals (18). For scarce samples, negative ion mode represented the method of choice, due to improved S/N ratio.

To evaluate the potential of the method for quantitative analysis of selected metabolites, intra- and interday repeatability of retention times, and linearity of the RR-RP-HPLC-ESI-MS method were tested. Intraday repeatability was measured by injecting the same standard solution (standard metabolite at a concentration of 1 μg/ml) three times in a single day. Interday repeatability was measured by analyzing the same standard solution over six different days. Intra- and interday repeatability of retention times using our method gave relative standard deviations (RSD) of less than 2%. The linearity of the RR-RP-HPLC-ESI-MS response (LOQ) was measured for each compound by recording the responses at different concentrations (Fig. 2), over the range of at least 1 mg/ml to 1 μg/ml (down to 10 ng/ml for glutamic acid, from 6.8 mM to ≈68 nM, corresponding to a minimum of 1.35 pmol). Five-point standard curves were established by plotting

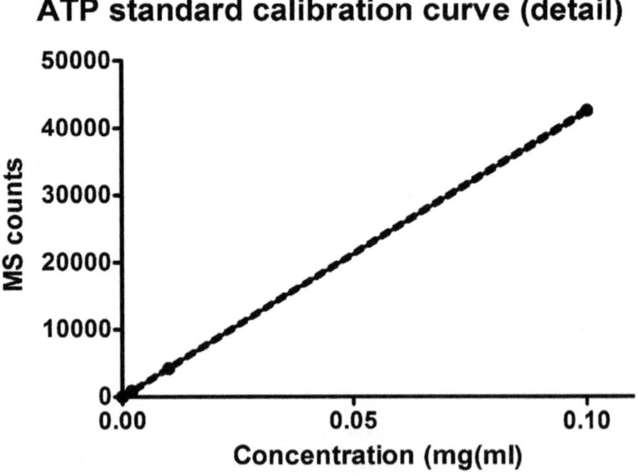

Fig. 2. A detail of the ATP standard calibration curve in the low concentration range (from 0.1 mg/ml to 0.1 μg/ml). Higher concentration points are not graphed in this curve due to space limitations. The gaped lines indicate the 99% confidence interval for linear regression calculated for the independent points in this graph. *X*-axis indicates concentrations while *Y*-axis graphs MS-counts.

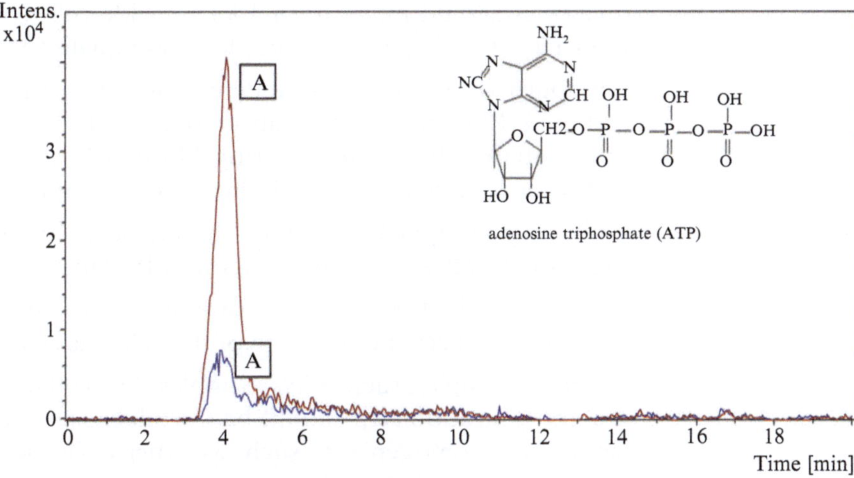

Fig. 3. Differential levels of ATP as detected from measurements on TAp73$\alpha^{+/-}$ (A) versus TAp73$\alpha^{-/-}$ (A) MEF cells through MRM (positive ion mode, 508 *m/z* → 410 *m/z*).

integrated peak areas versus concentrations. Each point on the calibration curve is the mean value of three independent measurements using the RR-RP-HPLC-ESI-MS method (Fig. 2). Linearity of the observed quantities, slope, intercept, and linear correlation values were all calculated via Microsoft Excel (Redmond, WA, USA). Data were further refined and plotted with GraphPad Prism 5.0 [GraphPad Software Inc. (San Diego, CA, USA)].

For absolute quantification in all tested biological matrices (Fig. 3), detected *m/z* signals were used to calculate quantities against calibration curves and normalized against internal standard signals, as to include any influence of the sample extraction protocol efficiency.

4. Notes

1. This method is robust as it can be applied to a wide array of biological matrices. Consistency in sample handling is very important. When collecting, it is important to minimize operational variation (e.g., collection technique, time of sampling, time to freezer, etc.). Materials collected for other experimental work and stored at −80°C can be used for metabolomic studies as long as all of the samples were treated in a consistent way during the collection process. When performing analyses on tissues (liver samples), 20 (liver) to 100 mg (other) of tissue samples are the routine requirement for metabolomics preparations; for cell samples (hepatocarcinoma cell lines, red blood cells),

quantities should range from 1×10^7 viable cells (optimal) to $0.5 \times 10^5 - 1 \times 10^6$ cells (average) for good quality MS results.

2. Ultrapurβ 18 M water is fundamental during the extraction and the HPLC-phases preparation. In the latter case, nonultrapure water might end up covering MS signals or contributing to Na^+ or K^+ adducts of metabolite species of interest.
3. Only original eppendorf tubes should be used, since other tubes might release polymers altering MS signals.
4. Check water bath temperature carefully as the thermal shock directly influences the outcome of the cell lysis step.
5. For tissue samples, such as liver samples (20–100 mg of tissue), a homogenization step should be introduced with a common laboratory homogenizer, such as Micro Dismembrator S (Sartorius, Goettingen, Germany) by using glass beads (0.5–0.75 mm) and $1500 \times g$ for 3 min.
6. Metanol and chloroform are toxic. Be cautious. Prepare methanol and chloroform stock aliquots and store them at –20°C. Low temperatures positively affect the metabolite extraction efficiency and protein precipitation steps.
7. Low temperatures are important as to prevent any metabolite degradation. In general, all sample preparation and processing steps should be as fast as possible and performed at low temperature, in order to exclude any metabolite degradation due to residual enzymatic contamination of the samples.
8. After this step, a three phase system can be observed, with liquid upper and lower phases and an intermediate cellular phase. This is particularly evident for extractions of erythrocytes, where the intermediate band is completely red. When extracting metabolites from biological fluids (such as urine or plasma), only two main bands will be visible.
9. This step can be prolonged overnight, if necessary.
10. Sample drying could be effectively performed with a common rotavapor (Rotray evaporator) at 50°C for at least 3 h.
11. After this step, a pellet should be visible at the bottom of the eppendorf tube. The absence of this pellet might be a symptom of an inefficient extraction.
12. Although HPLC separation could be also performed with classic HPLC, fast (rapid resolution) HPLC holds the advantage to perform faster separation while maintaining high resolution of metabolites over the whole retention time range. This is particularly time-saving when performing targeted analyses on multiple biological samples and elevated numbers of technical replicates, as it ends up cutting analytical times at least by a factor 2–4. For example, the chromatographic flow-rate can be increased up to 0.6 ml/min when working with RR-RP-HPLC

and specific columns (such as in the present case), which can sustain far higher backpressures (up until 800 bars), thus reducing by a factor 3 the length of the run. However, this can be done only when working online with ion trap MS, which can receive such a consistent volume per minute, while other MS need the flux to be splitted prior to direct injections into the source of the spectrometer.

13. Remember to activate and condition columns with mobile phases by performing at least two blank runs prior to the first analysis. This helps regularizing pressures over the whole retention time range.
14. Column temperature could be raised to improve chromatographic peak separation and reduce backpressure. Depending on the column used, temperature should not be raised above 50°C, as it may result in shortening of column half-lives.
15. Formic acid (FA) is the optimal coupling agent, as it works best in enhancing MS signals. However, overexposure to FA might result in increased need for maintenance of the MS instrument. The operator might think of switching to negative ion mode for the analysis of anionic compounds, especially when handling abundant starting biological material.

References

1. Suter L, Schroeder S, Meyer K et al (2011) EU Framework 6 Project: predictive toxicology (PredTox)-overview and outcome. Toxicol Appl Pharmacol 252(2):73–84. doi:10.1016/j.taap. 2010.10.008
2. O'Connell TM, Watkins PB (2010) The application of metabonomics to predict drug-induced liver injury. Clin Pharmacol Ther 88:394–399
3. Harris SR, Zhang GF, Sadhukhan S et al (2011) Metabolism of levulinate in perfused rat livers and live rats: conversion to the drug of abuse 4-hydroxy-pentanoate. J Biol Chem 286(7): 5895–904. doi:10.1074/jbc.M110.196808
4. Kim HJ, Kim JH, Noh S et al (2010) Metabolomic analysis of livers and serum from high-fat diet induced obese mice. J Proteome Res 10:722–731
5. Pilvi TK, Seppanen-Laakso T, Simolin H et al (2008) Metabolomic changes in fatty liver can be modified by dietary protein and calcium during energy restriction. World J Gastroenterol 14:4462–4472
6. D'Alessandro A, Zolla L (2010) Pharmacoproteomics: a chess game on a protein field. Drug Discov Today 15:1015–1023
7. D'Alessandro A, Zolla L (2010) Proteomics for quality-control processes in transfusion medicine. Anal Bioanal Chem 398:111–124
8. Vinayavekhin N, Homan EA, Saghatelian A (2010) Exploring disease through metabolomics. ACS Chem Biol 15:91–103
9. Pauling L, Robinson AB, Teranishi R, Cary P (1971) Quantitative analysis of urine vapor and breath by gas–liquid partition chromatography. Proc Natl Acad Sci USA 68:2374–2376
10. Evans AM, DeHaven CD, Barrett T et al (2009) Integrated, non targeted ultrahigh performance liquid chromatography/electrospray ionization tandem mass spectrometry platform for the identification and relative quantification of the small-molecule complement of biological systems. Anal Chem 81:6656–6667
11. Sana TR, Waddell K, Fischer SM (2008) A sample extraction and chromatographic strategy for increasing LC/MS detection coverage of the erythrocyte metabolome. J Chromatogr B Analyt Technol Biomed Life Sci 871:314–321
12. Buescher JM, Moco S, Sauer U, Zamboni N (2010) Ultrahigh performance liquid chromatography–tandem mass spectrometry method for fast and robust quantification of anionic

and aromatic metabolites. Anal Chem 82: 4403–4412

13. Lee Do Y, Bowen BP, Northen TR (2010) Mass spectrometry-based metabolomics, analysis of metabolite–protein interactions, and imaging. Biotechniques 49:557–565
14. Michopoulos F, Lai L, Gika H et al (2009) UPLC-MS-based analysis of human plasma for metabonomics using solvent precipitation or solid phasenextraction. J Proteome Res 8: 2114–2121
15. Bruce SJ, Tavazzi I, Parisod V et al (2009) Investigation of human blood plasma sample preparation for performing metabolomics using ultrahigh performance liquid chromatography/mass spectrometry. Anal Chem 81:3285–3296
16. Parab GS, Rao R, Lakshminarayanan S et al (2009) Data-driven optimization of metabolomics methods using rat liver samples. Anal Chem 81:1315–1323
17. Griffiths WJ, Koal T, Wang Y et al (2010) Targeted metabolomics for biomarker discovery. Angew Chem Int Ed Engl 49:5426–5445
18. D'Alessandro A, Gevi F, Zolla L (2011) A robust high resolution reversed-phase HPLC strategy to investigate various metabolic species in different biological models. Mol Biosyst 7(4):1024–32. doi:10.1039/C0MB00274G
19. Burnum KE, Cornett DS, Puolitaival SM et al (2009) Spatial and temporal alterations of phospholipids determined by mass spectrometry during mouse embryo implantation. J Lipid Res 50:2290–2298
20. Hendriks G, Uges DRA, Franke JP (2007) Reconsideration of sample pH adjustment in bioanalytical liquid–liquid extraction of ionisable compounds. J Chromatogr B 853:234–241
21. Zhang Y, Wang G, Huang Q et al (2009) Organic solvent extraction and metabonomic profiling of the metabolites in erythrocytes. J Chromatogr B Analyt Technol Biomed Life Sci 877:1751–1757
22. Bligh EG, Dyer WJ (1959) A rapid method of total lipid extraction and purification. Can J Biochem Physiol 37:911–917
23. Buscher JM, Czernik D, Ewald JC (2009) Cross-platform comparison of methods for quantitative metabolomics of primary metabolism. Anal Chem 81:2135–2143
24. Cai X, Zou L, Dong J et al (2009) Analysis of highly polar metabolites in human plasma by ultra-performance hydrophilic interaction liquid chromatography coupled with quadrupole-time of flight mass spectrometry. Anal Chim Acta 650:10–15
25. Coulier L, Bas R, Jespersen S et al (2006) Simultaneous quantitative analysis of metabolites using ion-pair liquid chromatography–electrospray ionization mass spectrometry. Anal Chem 78:6573–6658
26. Taymaz-Nikerel H, Mey M, Ras C et al (2009) Development and application of a differential method for reliable metabolome analysis in *Escherichia coli*. Anal Biochem 386:9–19
27. Hsieh Y, Duncan CJ, Brisson JM (2007) Fused-core silica column high-performance liquid chromatography/tandem mass spectrometric determination of rimonabant in mouse plasma. Anal Chem 79:5668–5673
28. Wishart DS, Tzur D, Knox C et al (2007) HMDB: the human metabolome database. Nucleic Acids Res 35:521–526
29. Smith CA, O'Maille G, Want EJ et al (2005) METLIN: a metabolite mass spectral database. Ther Drug Monit 27:747–751
30. Bertsch A, Gröpl C, Reinert K, Kohlbacher O (2011) OpenMS and TOPP: open source software for LC-MS data analysis. Methods Mol Biol 696:353–367
31. Nasso S, Silvestri F, Tisiot F et al (2010) An optimized data structure for high-throughput 3D proteomics data: mzRTree. J Proteomics 73:1176–1182
32. Schmaier AH, Petruzzelli LM (2003) Hematology for the medical student. Lippincott Williams & Wilikins, Baltimore, pp 22–23, 3
33. Leskovac V, Jerance D, Burany E (1975) Evidence for a histidine and a cysteine residue in the substrate-binding site of yeast alcohol dehydrogenase. Int J Biochem 6:563–568
34. Warburg O (1956) On the origin of cancer cells. Science 123:309–314

Chapter 19

Discovery of Lamin B1 and Vimentin as Circulating Biomarkers for Early Hepatocellular Carcinoma

Kwong-Fai Wong and John M. Luk

Abstract

The recent advancements in proteomic technologies have reconstituted our research strategies over different type of liver diseases including hepatocellular carcinoma (HCC). Combined analyses on HCC proteome and clinicopathological data of patients have allowed identification of many promising biomarkers that can be further developed into noninvasive diagnostic assays for cancer surveillance. Capitalizing our established proteomic platform primarily based on two-dimensional polyacrylamide gel electrophoresis (2DE) and MALDI-TOF/TOF mass spectrometry, our groups have identified lamin B1 (LMNB1) and vimentin (VIM) as promising biomarkers for detection of early HCC. Protein levels of both biomarkers were significantly elevated in cancerous tissues when compared to the controls in disease-free and cirrhotic liver subjects. Further investigation of the circulating LMNB1 mRNA level in patients' blood samples by standard PCR showed 76% sensitivity and 82% specificity for detection of early HCC. In parallel, an ELISA assay for measuring circulating vimentin level in patients' serum samples could detect small HCC at 40.91% sensitivity and 87.5% specificity. The candidate biomarkers were evaluated with the diagnostic performance of α-fetoprotein (AFP) for HCC. In this article, we address the current protocols for HCC biomarker discovery, ranging from clinical sample preparation, 2DE proteomic profiling and informatics analysis, and assay development and clinical validation study. Focus is emphasized on the methods for sample preservation and low-abundance protein enrichment.

Key words: Hepatocellular carcinoma, Circulating biomarker in blood, Lamin B1, Vimentin, Early diagnosis, Proteomics

1. Introduction

Liver cancer, or hepatocellular carcinoma (HCC), causes nearly 600,000 deaths annually worldwide because of late detection and ineffective treatment of this malignancy. HCC is asymptomatic by nature; therefore, many patients at the time of diagnosis are inflicted

Djuro Josic and Douglas C. Hixson (eds.), *Liver Proteomics: Methods and Protocols*, Methods in Molecular Biology, vol. 909, DOI 10.1007/978-1-61779-959-4_19, © Springer Science+Business Media New York 2012

of advanced tumors and distant metastasis and thus become unsuitable for surgical resection and liver transplantation—the two possible curative treatment approaches today. More importantly, patients who received surgical treatment often suffer from tumor recurrence and/or metastasis having short disease-free survival (1). To cope with these unmet medical needs, we have begun to identify new biomarkers for early HCC detection and better predicting tumor recurrence.

Molecular targeted therapies using small-molecule kinase inhibitors and monoclonal antibodies have imparted significant interest in drug discovery and screening for putative targets against HCC. Their antitumor benefits remain to be unequivocally established in multicenter clinical studies. However, one major pitfall of the targeted therapies is the activation of compensatory pathways and feedback loops that allow cancer cells to survive in spite of specific inhibition on targeted oncogenes (2). Conceivably, the problem is attributed to the heterogeneity of HCC, and the heterogeneous nature of HCC is twofold: (1) Etiologically, HCC occurrence is linked to a wide spectrum of risk factors such as hepatitis viral infection (e.g., HBV, HCV), exposure to aflatoxin B1, alcoholism, fatty liver disease, and hemochromatosis. (2) Genetically, HCC tumors show diverse backgrounds of chromosomal abnormality and epigenetics, resulting in differential activation and suppression on oncogenic and tumor-suppressing genes, respectively (3, 4).

In view of the complex networks of oncogenic signaling pathways, global analyses on the entire genome and proteome of cancer patients should hold promises in uncovering diagnostic and therapeutic targets for treatment of HCC. The whole-genome profiling has allowed us to identify gene signatures and mutations that are implicated with diagnostic and prognostic applications in HCC. Few examples are as follows: a 35-Myc regulated gene set for discriminating early HCC tumors from benign nodules (5), a 57-gene set for recurrence prediction (6), a 186-gene set for survival prognostication (3), and microRNA-122 as a key regulator on the mitochondrial metabolic gene network in HCC (7).

On the other hand, proteomics profiling studies have also gained much momentum on HCC (8–10). 2DE followed by mass spectrometry (MS) analysis has been widely adopted as the most classical proteomic approach for many years. By the 2DE profiling analysis and artificial neural network (ANN) and classification regression tree (CART) algorithms (11), we have (1) characterized the epitope specificity of a monoclonal antibody targeted against the Ku70 autoantigen in HCC (12), (2) unraveled the proteomic roadmap during mouse liver embryonic development and the associated oncofetal markers in HCC (13), (3) established a distinctive proteomic pattern that can differentiate cancerous and noncancerous tissues in HCC (14), (4) revealed significant upregulation of heat shock proteins (Hsp 27, Hsp 70, and Grp78) and the associated prognostic significance in HCC (15), (5) identified mortalin/ Grp75/mHsp70 as a predictor to tumor recurrence of which the sensitivity and specificity reaches 90.9 and 71.4%, respectively (16, 17). The latter finding subsequently led us to discover the mortalin–p53 interaction as a promising therapeutic target for HCC therapy (18, 19). Using surface enhanced laser desorption/ ionization (SELDI) technology, we also unveiled a serum proteomic signature that could differentiate HCC and liver cirrhosis patients, with an accuracy of 83% for HCC and 92% for liver cirrhosis. In combination with the conventional serum alpha-fetoprotein (AFP) marker, the proteomic signature could detect HCC with enhanced 95% sensitivity and 98% specificity (20).

Early diagnosis of liver cancer can benefit patients by reducing the risk of tumor recurrence and death (21, 22), and currently ultrasonography represents the primary radiologic HCC surveillance of at-risk populations. However, despite its ability to identify large lesion in livers, ultrasonography is not sensitive enough to detect small HCC especially in obese patients and those with underlying cirrhosis. Indeed, a recent report showed that ultrasonography failed to detect any suspicious nodules when performed 6–12 months before diagnosis in 14 of 39 HCC patients (23). In this context, computed tomography (5) and magnetic resonance imaging (MRI) have emerged as more sensitive and accurate imaging modalities but their uses in HCC surveillances have been limited. CT and MRI are not available in many medical centers, and more importantly, they require administrations of contrasting materials to patients that are potentially harmful (Fig. 1 and 24).

In this context, we have been striving to identify circulating biomarkers of diagnostic and prognostic values. Two biomarkers were identified: vimentin (VIM) and lamin B1 (LMNB1). Circulating vimentin assayed at a cut-off level of 245 ng/ml is able to detect small HCC tumor (≤2 cm) with sensitivity and specificity of 40.91 and 87.50%, respectively. When combined with AFP (≥100 ng/ml), an enhanced detection of 58.77% sensitivity and 98.15% specificity can be achieved (25). For LMNB1, its circulating mRNA level alone can detect early HCC with 76% sensitivity and 82% specificity (26).

In short, proteomic profiling studies on HCC has given us ample opportunities to identify novel markers for early detection and targets for HCC therapeutics using peptide inhibitors, shRNA, and monoclonal antibodies (27–30). Herein, we summarize our methods and technical details using proteomics tools as an example that led us to identify VIM and LMNB1 as circulating biomarkers for early HCC detection.

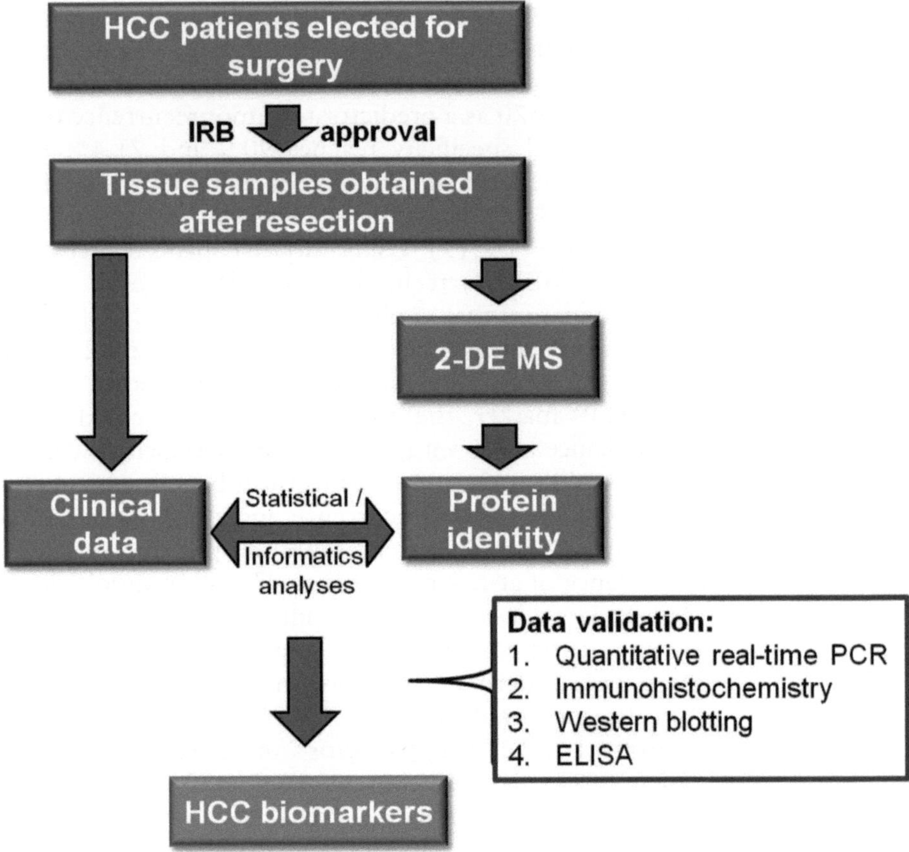

Fig. 1. A schematic diagram shows our typical workflow for the discovery of HCC biomarkers. Tumor and nontumor samples were obtained from surgical patients, and then analyzed for the differential protein expression by 2DE. Proteins-of-interest were identified by the combination of MS and database mining. Candidate markers were then validated on an independent cohort by different approaches such as quantitative real-time PCR, immunohistochemistry, western blotting, and ELISA.

2. Materials

2.1. Human Liver Tissue

1. For both studies on LBMN1 and VIM, we included 40 pairs of tumorous and nontumorous liver tissues and 20 cirrhotic liver (control) samples. We also obtained normal liver tissue samples that were residual donor grafts of subjects with no known liver diseases.
2. We reviewed the histology of all collected tissue samples and confirmed tissue homogeneity (either tumor or nontumor phenotype) using hematoxylin and eosin. Tumors were staged on the American Joint Committee cancer staging system (16). Only tissue samples of which morphologies >90% of their

distinctive nature was further processed. This procedure could ensure the tissue homogeneity that is essential to an informative proteomic study (see Note 1).

3. Methods

3.1. Protein Extraction from Tissue Homogenates and Quantitation

1. We carefully chose the detergent for extracting proteins from tissue homogenate. Although high-pH extraction buffers can effectively solubilize protein complex, they make protein quantitation inaccurate and affect gel electrophoresis (31, 32). Therefore, we employed a commercial extraction kit named ReadyPrep Sequential Extraction Kit (Bio-Rad, Hercules, CA, USA). This kit adopts a sequential extraction approach that allows proteins of differential solubility to be isolated (33).
2. We followed our established protocols for protein extraction in order to reduce variations in sample preparation. After tissue homogenization and centrifugation, we extracted proteins from the obtained pellet using extraction buffer (8 M urea, 4% CHAPS (w/v), 40 mM Tris–HCl pH 8.0, 0.2% carrier ampholytes (w/v), and 2 mM tributyl phosphate).
3. For 2DE analysis, the protein content was determined by the PlusOne 2DE Quant Kit (GE Biosciences, Buckinghamshire, UK). For western blotting and ELISA, Bradford and Lowry assay were performed in 1 ml plastic cuvettes. In general, we did not use 96-well plate for protein measurement because the relative pipetting error is much higher in a 96-well plate than in a single cuvette (see Note 2).
4. We minimized the protein mixtures from degradation by adding protease inhibitor cocktails from Roche (Mannheim, Germany). The phosphoproteome was preserved by using phosphatase inhibitor. In fact, PMSF (*p*henyl*m*ethyl*s*ulfonyl *f*luoride) has long been a popular serine protease inhibitor in biochemical studies (34). However, we did not use PMSF because of its instability (half-life of about 35 min at pH 8.0), and its high cytotoxicity (LD_{50} < 500 mg/kg) (see Notes 3 and 4).

3.2. Two-Dimensional Gel Electrophoresis

1. To rehydrate proteins, we mixed protein mixture of 30 μg with 200 μl of rehydration buffer in ceramic strip holders for 12 h at 30 V. The buffer was from GE Biosciences, and contained 8 M urea, 2% CHAPS (w/v), 0.5% IPG buffer (v/v), 40 mM DTT, and 0.002% (w/v) bromophenol blue.
2. We performed all isoelectric point focusing using the 11-cm strip (pH 4–7) in an IPGphor (GE Bioscience) in the following

conditions: 200 V for 1 h, 500 V for 1 h, 100 V for 1 h, and finally 8,000 V for 2 h.

3. After isoelectric point focusing, we transferred the strip onto 12.5% precast gels of 1 mm think (GE Biosciences). We applied a cooling water-bath system to prevent overheating of the gel.

3.3. Gel Visualization and Spot Picking

1. After gel electrophoresis, we fixed and stained the gel using GE Biosciences PlusOne silver staining kit following the manufacturer's instruction.
2. We carried out several measurements to reduce contamination of gels with keratin and other impurities that potentially interfere with the subsequent mass spectrometry. We disassembled glass plates in a laminar flow hood, and washed all of the gel containers with 70% (v/v) ethanol or acetonitrile. All researchers wore gloves and facemask.
3. We captured gel images with the GS-800 Calibrated Densitometer (Bio-Rad). To minimize the nonexpression related variations in spot intensity, we generally normalized all spot intensities with total valid volume. After intensity normalization, we identified protein spots that were differentially expressed in normal, cirrhotic, and tumor samples with the help of one-way ANOVA test.

3.4. In-Gel Trypsin Digestion and Mass Spectrometry

1. We excised protein spots of interest from gels using 21-gauge needles. After washing the protein spots with 50 mM ammonium bicarbonate, pH 8.0, we destained them using potassium ferricyanate. We then dehydrated the spots with acetonitrile and dried them using a SpeedVac system (Thermo Scientific, Barrington, IL, USA).
2. We digested the protein spots with trypsin of MS grade (Promega, Madison, WI, USA). Protein spots were rehydrated in trypsinization buffer (20 ng/μl trypsin, 20 mM NH_4HCO_3, pH 8.0) on ice for 40 min, and then incubated for overnight at 37°C. We digested all protein spots using the same protocol to avoid variations. This could enhance the reproducibility of proteomic data.
3. After trypsinization, we obtained the resulting peptides by extraction with 25 mM NH_4HCO_3, 50% (v/v) acetonitrile, and 0.1% (v/v) trifluoroacetic acid. Residual unwanted chemicals were then removed by lyophilization. We then dissolved the lyophilized peptides in 0.1% (v/v) trifluoroacetic acid.
4. We then use ZipTip® pipette tips (Millipore, Bedford, MA, USA) to desalt the peptides, which were then re-dissolved in a mixture of 5 mg/ml α-cyano-4-hydroxycinnamic acid matrix, 50% (v/v) acetonitrile, and 0.1% (v/v) trifluoroacetic acid.

5. The desalted peptides were then subjected to MS analysis in ABI 4800 Proteomics Analyzer MALDI-TOF/TOF (Applied Biosystems, Foster City, CA, USA) using a result-dependent acquisition mode. Detailed MS specifications were as described (16). In brief, peptide mass maps were acquired in a positive ion reflector mode with 2,500 laser shots per spectrum. Within the mass range of 1,000–4,000 Da, monoisotopic peak masses were automatically determined with a signal/noise ratio minimum set to 5 and a local noise window width of m/z 200. A maximum of 5 most significant intense tandem mass spectra was collected for MS/MS.
6. In order to retrieve protein identity, we searched the MS and MS/MS spectra against the NCBInr database using the software GPS Explorer, version 3.6 (Applied Biosystems) and MASCOT version 2.1 (Matrix Science, Boston, MA, USA). The following setting was employed: 65 ppm mass accuracy, trypsin cleavage, one missed cleavage allowed, carbamidomethylation set as fixed modification, oxidation of methionines allowed as variable modification, peptide mass tolerance set to 50 ppm, fragment tolerance set to ±0.2 Da, and minimum ion score confidence interval for MS/MS data set to 99%.

3.5. Data Verification

1. We finally identified LMNB1 and VIM as markers for early HCC. To confirm the findings, we recruited an independent cohort including 10 cirrhotic livers, 20 pairs of HCC tumor and nontumor tissues. For the tumor samples, 10 was small HCC (≤2 cm) and 10 was HCC tumors >2 cm.
2. We examined the expression by quantitative real-time PCR, western blotting, and immunohistochemistry following procedures that we routinely used in our laboratory (7, 35, 36) (see Note 5).

3.6. Determination of LMNB1 mRNA in Human Plasma

1. We obtained plasma samples for an independent cohort of 35 HCC patients and 28 nonneoplastic control subjects. To preserve RNA integrity, we made aliquots of the plasma samples, and kept them in −80°C freezer before use.
2. We extracted RNA from the plasma samples using TriZol reagent (Invitrogen, Carlsbad, CA, USA) in a ratio of 1:1 (v/v). First-strand cDNA was synthesized from 100 ng of human plasma RNA using TaqMan Reverse Transcription Reagent (Applied Biosystems). We then performed conventional PCR to determine the presence of LMNB1 mRNA.
3. We optimized the PCR conditions that allowed specific amplification of LMNB1. The condition was as following: initial incubation at 94°C for 5 min; 35 cycles of 94°C for 45 s, 58°C for 45 s, and 72°C for 40 s; and final extension at 72°C

for 10 min. Glyceraldehyde-3-phosphate dehydrogenase was also amplified as an internal control for PCR quality.

3.7. Serum VIM ELISA

1. To test the presence of VIM in patients' sera, we incubated them with coating buffer in duplicates on a 96-well ELISA plate (Corning Incorporated, Corning, NY, USA) overnight at 4°C.
2. Nonspecific binding sites on ELISA plates were blocked by 5% BSA. We probed the patients' sera with VIM antibody (2.5 μg/ml, V9, Santa Cruz Biotechnology, Santa Cruz, CA, USA) for 2 h at 37°C. After incubation, any unbound antibodies were washed away by Tris-buffered saline complemented with 0.5% (v/v) Tween-20 (TBS-T).
3. We detected the bound VIM antibody using HRP-conjugate (Invitrogen). Incubation was allowed for 2 h at 37°C. After washing, we developed the signal by incubating ABTS (Invitrogen) with HRP-conjugate at room temperature for 20 min. Signal at optical density of 415 nm was measured using an ELISA plate reader from Bio-Rad (see Note 6).
4. The VIM content in patients' sera was evaluated using a standard curve, in which recombinant VIM protein was used as the reference, which was produced in the pET-43.1b/His expression system and purified using the HiTrap affinity chromatography (GE Biosciences) (36). Purified vimentin concentration was measured using NanoDrop (Thermo Scientific, Waltham, MA, USA), and then used to generate an ELISA standard curve (10, 70, 140, 281, 526, and 1,125 ng/ml).

4. Notes

1. Laser capture microdissection (LCM) is a method for obtaining a specific population of cells from a heterogeneous sample such as liver tissues. In this method, toluidine blue is used to stain tissue sections, offering an advantage to visualize and microdissect target cells/tissue. The first comprehensive LCM study on HCC tissues was reported by Li et al. In this study, researchers microdissected a specific type of cells from liver tissues to a homogeneity >95%. They then identified and quantitated 261 proteins that have not been able to reveal otherwise (37). LCM has also been utilized to harvest hepatitis B-related tumor tissues, allowing researchers to identify many low-abundance biomarkers (38). In addition, LCM is helpful to procure the different nodules from cirrhotic liver (39).

Table 1
Compatibility of Bradford and Lowry assay to salts and detergents commonly present in tissue homogenates

	Interfering substance	Bradford	Lowry
Salts	Ammonium sulfate	1.0 M	>28 mM
	EDTA	100 mM	1 mM
	Glycine	100 mM	1 mM
	Imidazole	200 mM	25 mM
	Sodium azide	0.5%w/v	0.5%w/v
	Sodium chloride	5 M	1 M
	Tris	2 M	10 mM
Detergents	CHAPS	5%w/v	0.06%w/v
	Deoxycholic acid	0.5%w/v	0.02%w/v
	SDS	0.13%w/v	1%w/v
	Triton X-100	0.13%v/v	0.03%v/v
	Tween-20	0.06%v/v	0.06%v/v

Remark: values presented are obtained from various studies (76–79)

2. We preferred the PlusOne 2DE Quant Kit to Bradford and Lowry because of its higher reproducibility and tolerance to detergents and other interfering substances in tissue homogenates. A study reported that Bradford reagent of different commercial suppliers show substantial variations in responses to proteins (40). In addition, both assays are susceptible to the interference by extreme pH, detergents, salt, lipid, and other impurities in tissue homogenates (Table 1).
3. The major challenge of sample preservation is to preserve the proteome and phosphoproteome at the time of tissue procurement. In a surgical suite, the issue of sample preservation is usually overlooked because patients are the primary concerns of surgical team members. Currently, snap freezing and formalin fixation are the two major approaches for tissue preservation. However, they have been proven not effective enough to protect nucleic acids and proteins from degradation (41). The problem is further complicated by the fact that immediately after the procurement of tissues from patients, cells within are exposed to a variety of stress (e.g. acidosis, electrolyte loss, hypoxia, and, ischemia). These stresses potentially activate many stress-responsive signaling pathways (42). In addition, the residual ex vivo activity of phosphatase can substantially decrease the cellular content of the phosphoproteins (43). The collective outcome is the substantial changes in the compositions of tissue proteomes that are biologically irrelevant. New methods for sample preservation are urgently needed. Svensson

et al. have proposed a simple approach using an instrument called Stabilizer T1. This machine applies heat and pressure to tissues, either fresh or frozen, within a short time period (<1 min) to destroy proteases and phosphatase (44). This approach can preserve not only proteomes of brain (45), heart (46), pancreas (47), lung, and liver (43), but also the peptidome (48).

4. Many of the disease-specific biomarkers are present in blood plasma with ~ng/ml concentration, a level that is difficult to detect (49, 50). In this context, methods have been developed to deplete the high-abundance interfering proteins, and thus to enriching the biomarker content in tissue samples.

   In 2003, a method called immunoaffinity depletion emerged to deplete the high abundance proteins (abundance >95%) from the serum and plasma proteome (51). This technique utilizes immobilized antibodies to capture the high-abundance proteins from serum. These proteins includes albumin, immunoglobulins G and A, transferrin, haptoglobulin, alpha-1-antitrypsin, hemopexin, transthyretin, alpha-2-HS glycoprotein, alpha-1-acid glycoprotein, alpha-2-macroglobulin, and fibrinogen. By using this approach, researchers could detect 350 additional low abundance proteins by 2DE (52). In addition, a study suggested that immunoaffinity column can enrich the amount of potential biomarkers by fivefold (53). Besides, immunodepletion was shown to enhance the detection of biomarkers of which the serum concentrations are as low as 100 pg/ml (54). In addition to immunoaffinity chromatography, other approaches such as N-linked glycopeptide enrichment, (55) cysteinyl-peptide enrichment (56), and magnetic bead separation (57) have also been developed.

   Proteomic technologies have also been developed to enrich integral membrane proteins (IMP), a class of proteins that plays key roles in many important metabolic processes. To date, three methods are available for obtaining plasma membrane from tissue homogenates. They are (a) density gradient centrifugation (58), (b) aqueous polymer two-phase system (59), and (c) cationic colloidal silica technique (60, 61). Despite their usefulness in many proteomic studies, these techniques show some drawbacks. For example, the density gradient centrifugation fails to resolve membranes of endoplasmic reticulum, Golgi's apparatus, and the nucleus (62, 63). Plasma membrane preparation from the different techniques can usually be contaminated by high-abundance nonmembrane proteins such as cytoskeleton (64).

   Innovative techniques have also been developed to purify IMPs from tissue homogenate. One of the proven strategies is streptavidin-resin chromatography that captures biotinylated

Table 2
Protein markers and signatures associated with HCC biology and clinicopathological features

Protein marker	Associated HCC features or potential applications	Number of samples	Discovery platforms	Validation tools	References
Ku70/Ku80 autoantigen	Cell proliferation	Protein extracts of HCC	IP, MALDI-TOF MS	IFC, IHC, WB	(12)
PCNA, HSP7C, HEM6	HCC metastasis	Mouse: fetal (E13.5, E16.5), newborn, and postnatal livers Human: N, 3; NT and T, 59	2DE, MALDI-TOF-MS	qRT-PCR	(13)
CTSB, CYB5A, KRT1, Hp2, HSPB1, PGRMC1	Serum AFP, total protein levels and the Ishak's score	Training: N, 10; NT, 55; T, 55 Validation: NT, 25; T, 25	2DE, MALDI-TOF-MS	IHC, qRT-PCR, WB	(14)
Hsp27 and GRP78	Hsp27: serum AFP GRP78: tumor venous infiltration	N, 12; NT, 67; T, 67	2DE, MALDI-TOF-MS	IHC, WB	(15)
Mortalin	Liver metastasis and recurrence	N, 16; NT, 68; T, 68	2DE, MALDI-TOF-MS	IHC, WB	(16)
Five proteomic peaks (*m/z* at 3,324, 3,994, 4,665, 4,795, and 5,152)	Distinguishing HCC from liver cirrhosis	Serum sample: HCC, 120; LC, 120	SELDI-TOF-MS	CART algorithm	(20)
Vimentin	Small HCC (≤2 cm) detection	Training: NT, 36; T, 40 Validation (serum): CTL: 64; T, 88	2DE, MALDI-TOF-MS	ELISA, qRT-PCR, WB	(25)
LMNB1	Early stage HCC detection	Training: NT, 16; LC, 20; T, 39 Validation (serum): CTL, 12; LC, 16; T, 35	2DE, MALDI-TOF-MS	qRT-PCR	(26)

hPCNA proliferation cell nuclear antigen, *HSP7C* heat shock cognate 71 kDa protein, *HEM6* coproporphyrinogen III oxidase, *CTSB* lysosomal proteinase cathepsin B, *CYB5A* cytochrome b5, *KRT1* keratin I, *Hp2* haptoglobin, *HSPB1* heat shock 27 kDa protein 1, *PGRMC1* progesterone receptor membrane component 1, *Hsp27* heat shock protein 27, *GRP78* glucose-regulated protein 78 kDa, *CTL* nonneoplastic subjects, *LC* cirrhotic liver, *N* normal liver, *NT* adjacent nontumor liver tissue, *T* HCC tumor, *IP* immunoprecipitation, *MALDI-TOF* matrix-assisted laser desorption/ionization-time of flight mass spectrometry, *2DE* two-dimensional polyacrylamide gel electrophoresis, *SELDI-TOF-MS* surface enhanced laser desorption/ionization-time of flight mass spectrometry, *IFC* immunofluorescent cytochemistry, *IHC* immunohistochemistry, *WB* western blotting, *qRT-PCR* quantitative real-time polymerase chain reaction, *ELISA* enzyme-linked immunosorbent assay

Table 3
Strengths and weaknesses of the different proteomic technologies

Proteomic technology	Strengths	Weaknesses
2DE	• Readily provides information on protein isoforms • Relatively low-cost of operation	• Resolve poorly heavily charged proteins, and proteins of too high (>200 kDa) or too low molecular mass (<10 kDa) • Relatively poor reproducibility
MudPIT	• Better coverage on human proteome, more sensitive to low-abundance proteins • Shorter processing time comparing with 2DE	• Provide very little information of posttranslational modifications and isoforms of proteins
SELDI	• Able to identify disease-related proteomic fingerprints • A wide variety of biochips available for different purpose	• Provide very little information of individual proteins and their posttranslations
Protein array	• Multiplexed analysis on proteome and phosphoproteome, providing information of signaling activity	• Applications are always limited by the availability of capturing antibody and bait protein • Relatively poor reproducibility because of the spatial effect of protein bindings

IMPs (65–67). Given that biotin binds lysine residue, this approach has been found not effective in capturing proteins with no or few lysine residues, potentially leading a biased proteomic finding (68). In view of this, lectin affinity chromatography and cell surface capturing (CSC) have been developed to by-pass the problem. Despite the different underlying principles, both methods capture IMPs by interacting with the sugar moieties of the glycosylated IMPs. Using these two approaches, a panel of functionally important IMPs has been identified (55, 69–73).

5. We always perform western blotting and immunohistochemistry to validate our proteomic findings (Tables 1–3). However, it is sometimes quite challenging to find antibodies suitable to the assays. To help researchers obtain antibodies useful for their proteomic researches, online databases providing information on antibody epitopes have been established. One of the examples of the online databases is AntibodypediA (http://www.antibodypediA.org) (74, 75).
6. We found the HRP substrate kits for ELISA and immunohistochemistry from different commercial suppliers show substantial variations in response. For example, in immunohistochemistry

in which the primary antibody is polyclonal, the DAKO substrate kit can usually generate a more intense reactivity. In view of this, it is always important to use substrate kits from a single supplier throughout the study.

References

1. Hao K, Luk JM, Lee NP, Mao M, Zhang C, Ferguson MD, Lamb J, Dai H, Ng IO, Sham PC, Poon RT (2009) Predicting prognosis in hepatocellular carcinoma after curative surgery with common clinicopathologic parameters. BMC Cancer 9:389
2. Spangenberg HC, Thimme R, Blum HE (2009) Targeted therapy for hepatocellular carcinoma. Nat Rev Gastroenterol Hepatol 6: 423–432
3. Hoshida Y, Villanueva A, Kobayashi M, Peix J, Chiang DY, Camargo A, Gupta S, Moore J, Wrobel MJ, Lerner J, Reich M, Chan JA, Glickman JN, Ikeda K, Hashimoto M, Watanabe G, Daidone MG, Roayaie S, Schwartz M, Thung S, Salvesen HB, Gabriel S, Mazzaferro V, Bruix J, Friedman SL, Kumada H, Llovet JM, Golub TR (2008) Gene expression in fixed tissues and outcome in hepatocellular carcinoma. N Engl J Med 359:1995–2004
4. Laurent-Puig P, Legoix P, Bluteau O, Belghiti J, Franco D, Binot F, Monges G, Thomas G, Bioulac-Sage P, Zucman-Rossi J (2001) Genetic alterations associated with hepatocellular carcinomas define distinct pathways of hepatocarcinogenesis. Gastroenterology 120:1763–1773
5. Kaposi-Novak P, Libbrecht L, Woo HG, Lee Y-H, Sears NC, Coulouarn C, Conner EA, Factor VM, Roskams T, Thorgeirsson SS (2009) Central role of c-Myc during malignant conversion in human hepatocarcinogenesis. Cancer Res 69:2775–2782
6. Wang SM, Ooi LLPJ, Hui KM (2007) Identification and validation of a novel gene signature associated with the recurrence of human hepatocellular carcinoma. Clin Cancer Res 13:6275–6283
7. Burchard J, Zhang C, Liu AM, Poon RT, Lee NP, Wong KF, Sham PC, Lam BY, Ferguson MD, Tokiwa G, Smith R, Leeson B, Beard R, Lamb JR, Lim L, Mao M, Dai H, Luk JM (2010) microRNA-122 as a regulator of mitochondrial metabolic gene network in hepatocellular carcinoma. Mol Syst Biol 6:402
8. Sun S, Yi X, Poon RT, Yeung C, Day PJ, Luk JM (2009) A protein-based set of reference markers for liver tissues and hepatocellular carcinoma. BMC Cancer 9:309
9. Sun S, Day PJ, Lee NP, Luk JM (2009) Biomarkers for early detection of liver cancer: focus on clinical evaluation. Protein Pept Lett 16:473–478
10. Sun S, Lee NP, Poon RT, Fan ST, He QY, Lau GK, Luk JM (2007) Oncoproteomics of hepatocellular carcinoma: from cancer markers' discovery to functional pathways. Liver Int 27: 1021–1038
11. Luk JM, Lam BY, Lee NP, Ho DW, Sham PC, Chen L, Peng J, Leng X, Day PJ, Fan ST (2007) Artificial neural networks and decision tree model analysis of liver cancer proteomes. Biochem Biophys Res Commun 361:68–73
12. Luk JM, Su YC, Lam SC, Lee CK, Hu MY, He QY, Lau GK, Wong FW, Fan ST (2005) Proteomic identification of Ku70/Ku80 autoantigen recognized by monoclonal antibody against hepatocellular carcinoma. Proteomics 5:1980–1986
13. Lee NP, Leung KW, Cheung N, Lam BY, Xu MZ, Sham PC, Lau GK, Poon RT, Fan ST, Luk JM (2008) Comparative proteomic analysis of mouse livers from embryo to adult reveals an association with progression of hepatocellular carcinoma. Proteomics 8:2136–2149
14. Lee NP, Chen L, Lin MC, Tsang FH, Yeung C, Poon RT, Peng J, Leng X, Beretta L, Sun S, Day PJ, Luk JM (2009) Proteomic expression signature distinguishes cancerous and nonmalignant tissues in hepatocellular carcinoma. J Proteome Res 8:1293–1303
15. Luk JM, Lam CT, Siu AF, Lam BY, Ng IO, Hu MY, Che CM, Fan ST (2006) Proteomic profiling of hepatocellular carcinoma in Chinese cohort reveals heat-shock proteins (Hsp27, Hsp70, GRP78) up-regulation and their associated prognostic values. Proteomics 6:1049–1057
16. Yi X, Luk JM, Lee NP, Peng J, Leng X, Guan XY, Lau GK, Beretta L, Fan ST (2008) Association of mortalin (HSPA9) with liver cancer metastasis and prediction for early tumor recurrence. Mol Cell Proteomics 7:315–325
17. Lu WJ, Lee NP, Fatima S, Luk JM (2009) Heat shock proteins in cancer: signaling pathways,

tumor markers and molecular targets in liver malignancy. Protein Pept Lett 16:508–516

18. Lu WJ, Lee NP, Kaul SC, Lan F, Poon RT, Wadhwa R, Luk JM (2011) Induction of mutant p53-dependent apoptosis in human hepatocellular carcinoma by targeting stress protein mortalin. Int J Cancer 129: 1806–1814.
19. Lu WJ, Lee NP, Kaul SC, Lan F, Poon RT, Wadhwa R, Luk JM (2011) Mortalin-p53 interaction in cancer cells is stress dependent and constitutes a selective target for cancer therapy. Cell Death Differ 18:1046–1056
20. Chen L, Ho DW, Lee NP, Sun S, Lam B, Wong KF, Yi X, Lau GK, Ng EW, Poon TC, Lai PB, Cai Z, Peng J, Leng X, Poon RT, Luk JM (2010) Enhanced detection of early hepatocellular carcinoma by serum SELDI-TOF proteomicsignaturecombinedwithalpha-fetoprotein marker. Ann Surg Oncol 17:2518–2525
21. Fukuda S, Itamoto T, Nakahara H, Kohashi T, Ohdan H, Hino H, Ochi M, Tashiro H, Asahara T (2005) Clinicopathologic features and prognostic factors of resected solitary small-sized hepatocellular carcinoma. Hepatogastroenterology 52:1163–1167
22. Kikuchi LO, Paranagua-Vezozzo DC, Chagas AL, Mello ES, Alves VA, Farias AQ, Pietrobon R, Carrilho FJ (2009) Nodules less than 20 mm and vascular invasion are predictors of survival in small hepatocellular carcinoma. J Clin Gastroenterol 43:191–195
23. Lok AS, Sterling RK, Everhart JE, Wright EC, Hoefs JC, Di Bisceglie AM, Morgan TR, Kim HY, Lee WM, Bonkovsky HL, Dienstag JL (2010) Des-gamma-carboxy prothrombin and alpha-fetoprotein as biomarkers for the early detection of hepatocellular carcinoma. Gastroenterology 138:493–502
24. Kim MJ (2011) Current limitations and potential breakthroughs for the early diagnosis of hepatocellular carcinoma. Gut Liver 5:15–21
25. Sun S, Poon RT, Lee NP, Yeung C, Chan KL, Ng IO, Day PJ, Luk JM (2010) Proteomics of hepatocellular carcinoma: serum vimentin as a surrogate marker for small tumors (<or =2 cm). J Proteome Res 9:1923–1930
26. Sun S, Xu MZ, Poon RT, Day PJ, Luk JM (2010) Circulating lamin B1 (LMNB1) biomarker detects early stages of liver cancer in patients. J Proteome Res 9:70–78
27. Hu MY, Lam CT, Liu KD, Xu Z, Fatima S, Su YC, Tsang F, Chen J, Pang JZ, Qin LX, Luk JM (2009) Proteomic identification of a monoclonal antibody recognizing caveolin-1 in hepatocellular carcinoma with metastatic potential. Protein Pept Lett 16:479–485
28. Liu LX, Lee NP, Chan VW, Xue W, Zender L, Zhang C, Mao M, Dai H, Wang XL, Xu MZ, Lee TK, Ng IO, Chen Y, Kung HF, Lowe SW, Poon RT, Wang JH, Luk JM (2009) Targeting cadherin-17 inactivates Wnt signaling and inhibits tumor growth in liver carcinoma. Hepatology 50:1453–1463
29. Luk JM, Wong KF (2006) Monoclonal antibodies as targeting and therapeutic agents: prospects for liver transplantation, hepatitis and hepatocellular carcinoma. Clin Exp Pharmacol Physiol 33:482–488
30. Tsang FH, Lee NP, Luk JM (2009) The use of small peptides in the diagnosis and treatment of hepatocellular carcinoma. Protein Pept Lett 16:530–538
31. Speers AE, Wu CC (2007) Proteomics of integral membrane proteins – theory and application. Chem Rev 107:3687–3714
32. Josic D, Clifton JG (2007) Mammalian plasma membrane proteomics. Proteomics 7:3010–3029
33. Rabilloud T, Adessi C, Giraudel A, Lunardi J (1997) Improvement of the solubilization of proteins in two-dimensional electrophoresis with immobilized pH gradients. Electrophoresis 18:307–316
34. James GT (1978) Inactivation of the protease inhibitor phenylmethylsulfonyl fluoride in buffers. Anal Biochem 86:574–579
35. Xu MZ, Chan SW, Liu AM, Wong KF, Fan ST, Chen J, Poon RT, Zender L, Lowe SW, Hong W, Luk JM (2010) AXL receptor kinase is a mediator of YAP-dependent oncogenic functions in hepatocellular carcinoma. Oncogene 30:1229–1240
36. Wong KF, Luk JM, Cheng RH, Klickstein LB, Fan ST (2007) Characterization of two novel LPS-binding sites in leukocyte integrin betaA domain. FASEB J 21:3231–3239
37. Li C, Hong Y, Tan YX, Zhou H, Ai JH, Li SJ, Zhang L, Xia QC, Wu JR, Wang HY, Zeng R (2004) Accurate qualitative and quantitative proteomic analysis of clinical hepatocellular carcinoma using laser capture microdissection coupled with isotope-coded affinity tag and two-dimensional liquid chromatography mass spectrometry. Mol Cell Proteomics 3:399–409
38. Ai J, Tan Y, Ying W, Hong Y, Liu S, Wu M, Qian X, Wang H (2006) Proteome analysis of hepatocellular carcinoma by laser capture microdissection. Proteomics 6:538–546
39. Guedj N, Dargere D, Degos F, Janneau JL, Vidaud D, Belghiti J, Bedossa P, Paradis V (2006) Global proteomic analysis of microdissected cirrhotic nodules reveals significant

biomarkers associated with clonal expansion. Lab Invest 86:951–958

40. Chan JK, Thompson JW, Gill TA (1995) Quantitative determination of protamines by coomassie blue G assay. Anal Biochem 226:191–193
41. Nassiri M, Ramos S, Zohourian H, Vincek V, Morales AR, Nadji M (2008) Preservation of biomolecules in breast cancer tissue by a formalin-free histology system. BMC Clin Pathol 8:1
42. Espina V, Edmiston KH, Heiby M, Pierobon M, Sciro M, Merritt B, Banks S, Deng J, VanMeter AJ, Geho DH, Pastore L, Sennesh J, Petricoin EF III, Liotta LA (2008) A portrait of tissue phosphoprotein stability in the clinical tissue procurement process. Mol Cell Proteomics 7:1998–2018
43. Rountree CB, Van Kirk CA, You H, Ding W, Dang H, Vanguilder HD, Freeman WM (2010) Clinical application for the preservation of phospho-proteins through in-situ tissue stabilization. Proteome Sci 8:61
44. Svensson M, Boren M, Skold K, Falth M, Sjogren B, Andersson M, Svenningsson P, Andren PE (2009) Heat stabilization of the tissue proteome: a new technology for improved proteomics. J Proteome Res 8:974–981
45. Goodwin RJ, Lang AM, Allingham H, Boren M, Pitt AR (2010) Stopping the clock on proteomic degradation by heat treatment at the point of tissue excision. Proteomics 10:1751–1761
46. Robinson AA, Westbrook JA, English JA, Boren M, Dunn MJ (2009) Assessing the use of thermal treatment to preserve the intact proteomes of post-mortem heart and brain tissue. Proteomics 9:4433–4444
47. Scholz B, Skold K, Kultima K, Fernandez C, Waldemarson S, Savitski MM, Svensson M, Boren M, Stella R, Andren PE, Zubarev R, James P (2011) Impact of temperature dependent sampling procedures in proteomics and peptidomics – a characterization of the liver and pancreas post mortem degradome. Mol Cell Proteomics 10:M900229MCP200
48. Rossbach U, Nilsson A, Falth M, Kultima K, Zhou Q, Hallberg M, Gordh T, Andren PE, Nyberg F (2009) A quantitative peptidomic analysis of peptides related to the endogenous opioid and tachykinin systems in nucleus accumbens of rats following naloxone-precipitated morphine withdrawal. J Proteome Res 8:1091–1098
49. Qian WJ, Jacobs JM, Liu T, Camp DG II, Smith RD (2006) Advances and challenges in liquid chromatography–mass spectrometry-based proteomics profiling for clinical applications. Mol Cell Proteomics 5:1727–1744
50. Rifai N, Gillette MA, Carr SA (2006) Protein biomarker discovery and validation: the long and uncertain path to clinical utility. Nat Biotechnol 24:971–983
51. Anderson NL, Anderson NG (2002) The human plasma proteome: history, character, and diagnostic prospects. Mol Cell Proteomics 1:845–867
52. Pieper R, Su Q, Gatlin CL, Huang ST, Anderson NL, Steiner S (2003) Multi-component immunoaffinity subtraction chromatography: an innovative step towards a comprehensive survey of the human plasma proteome. Proteomics 3:422–432
53. Brand J, Haslberger T, Zolg W, Pestlin G, Palme S (2006) Depletion efficiency and recovery of trace markers from a multiparameter immunodepletion column. Proteomics 6:3236–3242
54. Qian WJ, Kaleta DT, Petritis BO, Jiang H, Liu T, Zhang X, Mottaz HM, Varnum SM, Camp DG II, Huang L, Fang X, Zhang WW, Smith RD (2008) Enhanced detection of low abundance human plasma proteins using a tandem IgY12-SuperMix immunoaffinity separation strategy. Mol Cell Proteomics 7:1963–1973
55. Zhang H, Li XJ, Martin DB, Aebersold R (2003) Identification and quantification of N-linked glycoproteins using hydrazide chemistry, stable isotope labeling and mass spectrometry. Nat Biotechnol 21:660–666
56. Liu T, Qian WJ, Strittmatter EF, Camp DG II, Anderson GA, Thrall BD, Smith RD (2004) High-throughput comparative proteome analysis using a quantitative cysteinyl-peptide enrichment technology. Anal Chem 76:5345–5353
57. Baumann S, Ceglarek U, Fiedler GM, Lembcke J, Leichtle A, Thiery J (2005) Standardized approach to proteome profiling of human serum based on magnetic bead separation and matrix-assisted laser desorption/ionization time-of-flight mass spectrometry. Clin Chem 51:973–980
58. Zhang L, Xie J, Wang X, Liu X, Tang X, Cao R, Hu W, Nie S, Fan C, Liang S (2005) Proteomic analysis of mouse liver plasma membrane: use of differential extraction to enrich hydrophobic membrane proteins. Proteomics 5:4510–4524
59. Everberg H, Peterson R, Rak S, Tjerneld F, Emanuelsson C (2006) Aqueous two-phase partitioning for proteomic monitoring of cell surface biomarkers in human peripheral blood mononuclear cells. J Proteome Res 5: 1168–1175
60. Kohnke PL, Mulligan SP, Christopherson RI (2009) Membrane proteomics for leukemia classification and drug target identification. Curr Opin Mol Ther 11:603–610

61. Rahbar AM, Fenselau C (2004) Integration of Jacobson's pellicle method into proteomic strategies for plasma membrane proteins. J Proteome Res 3:1267–1277
62. Zhang LJ, Wang XE, Peng X, Wei YJ, Cao R, Liu Z, Xiong JX, Yin XF, Ping C, Liang S (2006) Proteomic analysis of low-abundant integral plasma membrane proteins based on gels. Cell Mol Life Sci 63:1790–1804
63. Macher BA, Yen TY (2007) Proteins at membrane surfaces-a review of approaches. Mol Biosyst 3:705–713
64. Stasyk T, Huber LA (2004) Zooming in: fractionation strategies in proteomics. Proteomics 4:3704–3716
65. Tang X, Yi W, Munske GR, Adhikari DP, Zakharova NL, Bruce JE (2007) Profiling the membrane proteome of Shewanella oneidensis MR-1 with new affinity labeling probes. J Proteome Res 6:724–734
66. Scheurer SB, Rybak JN, Roesli C, Brunisholz RA, Potthast F, Schlapbach R, Neri D, Elia G (2005) Identification and relative quantification of membrane proteins by surface biotinylation and two-dimensional peptide mapping. Proteomics 5:2718–2728
67. Rybak JN, Ettorre A, Kaissling B, Giavazzi R, Neri D, Elia G (2005) In vivo protein biotinylation for identification of organ-specific antigens accessible from the vasculature. Nat Methods 2:291–298
68. Lu B, McClatchy DB, Kim JY, Yates JR III (2008) Strategies for shotgun identification of integral membrane proteins by tandem mass spectrometry. Proteomics 8:3947–3955
69. Na K, Lee EY, Lee HJ, Kim KY, Lee H, Jeong SK, Jeong AS, Cho SY, Kim SA, Song SY, Kim KS, Cho SW, Kim H, Paik YK (2009) Human plasma carboxylesterase 1, a novel serologic biomarker candidate for hepatocellular carcinoma. Proteomics 9:3989–3999
70. Chaerkady R, Thuluvath PJ, Kim MS, Nalli A, Vivekanandan P, Simmers J, Torbenson M, Pandey A (2008) O Labeling for a quantitative proteomic analysis of glycoproteins in hepatocellular carcinoma. Clin Proteomics 4: 137–155
71. Dai Z, Fan J, Liu Y, Zhou J, Bai D, Tan C, Guo K, Zhang Y, Zhao Y, Yang P (2007) Identification and analysis of alpha1,6-fucosylated proteins in human normal liver tissues by a target glycoproteomic approach. Electrophoresis 28:4382–4391
72. Xu Z, Zhou X, Lu H, Wu N, Zhao H, Zhang L, Zhang W, Liang YL, Wang L, Liu Y, Yang P, Zha X (2007) Comparative glycoproteomics based on lectins affinity capture of N-linked glycoproteins from human Chang liver cells and MHCC97-H cells. Proteomics 7:2358–2370
73. Wollscheid B, Bausch-Fluck D, Henderson C, O'Brien R, Bibel M, Schiess R, Aebersold R, Watts JD (2009) Mass-spectrometric identification and relative quantification of N-linked cell surface glycoproteins. Nat Biotechnol 27:378–386
74. Cottingham K (2008) Antibodypedia seeks to answer the question: "how good is that antibody?". J Proteome Res 7:4213
75. Bjorling E, Uhlen M (2008) Antibodypedia, a portal for sharing antibody and antigen validation data. Mol Cell Proteomics 7:2028–2037
76. Bradford MM (1976) A rapid and sensitive method for the quantitation of microgram quantities of protein utilizing the principle of protein–dye binding. Anal Biochem 72: 248–254
77. Lowry OH, Rosebrough NJ, Farr AL, Randall RJ (1951) Protein measurement with the Folin phenol reagent. J Biol Chem 193:265–275
78. Peterson GL (1979) Review of the Folin phenol protein quantitation method of Lowry, Rosebrough, Farr and Randall. Anal Biochem 100:201–220
79. Stoscheck CM (1990) Quantitation of protein. Methods Enzymol 182:50–68

Chapter 20

Combinatorial Peptide Ligand Libraries to Discover Liver Disease Biomarkers in Plasma Samples

Gian Maria D'Amici, Anna Maria Timperio, Sara Rinalducci, and Lello Zolla

Abstract

High-abundance proteins present in blood plasma make the detection of low-abundance proteins extremely difficult by proteomics technology. Hexapeptide combinatorial ligand libraries can be used to investigate the hidden proteome in depth. Here we describe how liver disease biomarkers can be successfully discovered in blood plasma by two main steps: preparative methods that reduce the dynamic range of protein concentration, and analytic methods that allow resolution of proteins. Thus, blood plasma from hepatitis B virus infected patients were treated with ProteoMiner™ enrichment kit and analyzed by two dimensional gel electrophoresis and mass spectrometry. This approach allowed us to identify plasma gelsolin as possible candidate biomarker for hepatitis B-associated liver cirrhosis.

Key words: Human plasma, Biomarker discovery, Hexapeptide combinatorial ligand library, ProteoMiner

1. Introduction

In the last decade it has been largely documented that the development of an infection can change the host proteome. Host proteome modifications are correlated to multiple factors, both host- and pathogen-related. Proteomic technologies are proving to be particularly useful in identification of biomarkers for disease diagnosis, disease progression, and response to therapy in chronic diseases and cancer (1, 2). The host-specific response to the infection holds the disease progression, thus comparison of proteome between diseased and healthy individuals is a good starting point to discover possible biomarkers of a certain pathology.

Djuro Josic and Douglas C. Hixson (eds.), *Liver Proteomics: Methods and Protocols*, Methods in Molecular Biology, vol. 909,
DOI 10.1007/978-1-61779-959-4_20, © Springer Science+Business Media New York 2012

Blood plasma being in close communication with body tissues would appear ideal for proteomic analysis because it may give information about these tissues and in regards to a possible disease state. Most plasma proteins are biosynthesized in the liver and the majority of these change their structure and abundance in response to liver disease, so plasma analysis is a good method to test the liver health (3). Another advantage of using plasma for biomarker analysis is that plasma samples are readily obtained from patients.

The application of separation methods such high resolution two-dimensional gel electrophoresis (2DE) (4) followed by mass spectrometry protein identification can largely contribute to compare protein signatures in blood plasma from patients and controls (5, 6). However, plasma is surely the most difficult protein-containing sample for proteomic analysis due to its complexity and the extreme dynamic range of abundance of protein species (7). The removal of a small number of high-abundance proteins (HAP), such us albumin (up to 60% of total plasma proteins), immunoglobulins (up to 15%), α-1-antitrypsin, transferrin and haptoglobins in plasma samples before 2DE analysis has been demonstrated to be very effective for attaining more comprehensive proteome coverage. Elimination of HAPs increases the probability to reveal a large number of low-abundance proteins (LAP), including many that are or could be diagnostically significant. Low molecular weight plasma proteins and peptides can be a rich source of disease biomarkers, however it is well known that they can complex HAPs, so quantitative removal of these latter species may prevent disease biomarker discovery (8).

A novel LAP enrichment technology, which has outstripped old prefractionation methods (9), has been recently marketed under the trade name of ProteoMiner™ (10). This innovative sample preparation tool is based on a large number of variegate bead-based hexapeptide combinatorial ligand libraries that simultaneously dilute HAPs and concentrate LAPs. When plasma samples are treated with ProtoMiner, the HAPs quickly saturate their high affinity ligands and the excess protein is washed away. At the same time, LAPs are immobilized on their specific ligands focusing themselves. This reduces the dynamic range of protein concentration while maintaining representative of all proteins within the original sample. ProteoMiner dramatically increases the resolution and spot count in 2D gels providing a tool to investigate the deep proteome (11).

The aim of this chapter is to provide working protocols to obtain high-quality 2D gel electrophoresis of blood plasma samples. The protocols are based on the use of the commercially available ProteoMiner™ protein enrichment kit, applied to compare plasma samples from patients with liver cirrhosis associated to hepatitis B virus (HBV) infection and chronic HBV-infected patients who were asymptomatic. Treated samples were then analyzed by

2D IEF-SDS-PAGE followed by MS/MS to identify possible candidate biomarkers for HBV cirrhosis liver disease (12).

2. Material

Protocols here described are suitable for human blood plasma (see Note 1).

2.1. Plasma Isolation and Treatment with ProteoMiner™ Kit

ProteoMiner™ enrichment kit may be purchased from Bio-Rad (Hercules, CA, USA).

1. Spin column: each column contains 500 μL bead slurry (20% beads, 20% aqueous EtOH, 0.05% w/v sodium azide).
2. Collection tubes: 2 ml capless or capped centrifuge tubes.
3. Wash buffer: 150 mM NaCl, 10 mM NaH_2PO_4, pH 7.4. Make 50 mL of it.
4. Elution reagent: lyophilized urea CHAPS [8 M urea, 2% 3-[(3-cholamidopropyl)dimethylammonio]-1-propanesulfonate (CHAPS)].
5. Rehydration reagent: 5% v/v acetic acid.
6. Collection tube: capped 2 mL centrifuge tube.
7. Rehydrated elution reagent: prepare elution reagent by adding 610 μL of rehydration reagent to one vial of lyophilized elution (see Note 2).

2.2. First Dimension: Isoelectric Focusing

1. Best-quality reagents and very high-quality water (double distilled water, ddH_2O; 18.2 MΩ) should be used for all buffers.
2. First dimension (IEF) gel apparatus from Bio-Rad or GE Healthcare (Buckinghamshire, UK).
3. Precipitating solution: 80% (v/v) acetone.
4. IPG strip: immobiline DryStrip pH 4–7, 18 cm (GE Healthcare).
5. Sample preparation buffer/IPG strip rehydration buffer (25 mL): 8 M urea, 4% CHAPS, 0.5% v/v carrier ampholyte pH 4–7 (Bio-Lyte; Bio-Rad), 40 mM Tris–HCl base (pH 8.8). Dissolve 10.51 g urea in 12 mL of ddH_2O. Add 1 g CHAPS, make the volume up to 25 mL. Add 312 μL of 40% (v/v) carrier ampholyte pH 4–7 (Bio-Lyte; Bio-Rad) (see Note 3).
6. 2 M Iodoacetamide (IAA): dissolve 56 mg IAA in 150 μL of sample preparation buffers.
7. 2 M Dithioerythritol (DTE): dissolve 47.27 mg DTE in 150 μL of sample preparation buffers.

8. 1.9 M Tris–HCl, pH 8.8: dissolve 21.7 g Tris–HCl in approx. 80 mL ddH_2O, pH to 8.8 using concentrated hydrochloric acidio and make up to 100 mL with ddH_2O.
9. 4.5 M Tributylphosphine (TBP, Sigma Aldrich, Poole, UK) should be added during protein extraction.

2.3. Second Dimension: SDS-PAGE

Second dimension (SDS-PAGE) gel apparatus may be purchased from Bio-Rad.

1. 1.5 M Tris–HCl, pH 8.8: dissolve 171.21 g Tris–HCl in 800 mL of distilled H_2O (dH_2O), pH to 8.8 using concentrated hydrochloric acid and make up to 1 L with dH_2O.
2. 0.5 M Tris–HCl pH 6.8: dissolve 57.07 g Tris–HCl in 800 mL of dH_2O, pH to 8.8 using concentrated hydrochloric acid and make up to 1 L with dH_2O.
3. 10× SDS-PAGE running buffer (5 L): consists of 1.87 M glycine, 0.25 M Tris-base, 1% SDS. Dissolve 700 g glycine, 152 g Tris-base, and 50 g SDS in dH_2O and make up to 5 L with dH_2O.
4. Agarose overlay (100 mL): mix 1 g agarose in 100 mL running buffer. Add Bromophenol blue to 0.03% (30 mg/100 mL).
5. 10% (w/v) SDS (100 mL): dissolve 10 g SDS in 90 mL dH_2O and make up to 100 mL.
6. 10% (w/v) Ammonium persulfate (APS): dissolve 1 g APS in 10 mL dH_2O, make up the volume required, use fresh, and discard excess.
7. *N,N,N,N'*-Tetramethyl-ethylenediamine (TEMED).
8. IPG strip equilibration buffer 1 (500 mL): 50 mM Tris–HCl (pH 6.8), 6 M urea, 30% (v/v) glycerol, 3% (w/v) SDS. Mix 13.3 mL of 1.9 M Tris–HCl, pH 8.8, 180 g urea, 300 mL of 50% (v/v) glycerol, 15 g SDS and make up to 500 mL with dH_2O, pH to 6.8 using concentrated hydrochloric acid and make up to 1 L with dH_2O (see Note 4).

3. Methods

3.1. Plasma Isolation and Treatment with ProteoMiner™ Kit

1. Collect 10 mL of venous blood in EDTA tubes. Centrifuge for 10 min at 1,000×*g* (4°C). Collect the supernatant (plasma) and repeat centrifugation for 10 min at 10,000×*g* (4°C). Recover the supernatant (blood plasma) and discard the pellet, aliquot (1-mL volumes) and store at 80°C.
2. Remove the top cap and then snap off the bottom cap from the spin column (see Note 5). Place the spin column in a capless

collection tube and centrifuge for 2 min at 1.000 × *g* (RT) to remove the storage solution. After centrifugation discard the solution collected in the capless collection tube.

3. Replace the bottom cap and add 1 mL dH_2O, then replace the top cap. Rotate the spin column end to end over a period of 5 min. Remove top end bottom caps, place the spin column in the capless collection tube and centrifuge for 2 min at 1,000 × *g* (RT) to remove H_2O. Discard collected material. Repeat centrifugation to remove any remaining materials.
4. Repeat step 3 using 1 mL of wash buffer in place of dH_2O. Replace bottom cap on spin column and it is now ready for sample binding.
5. Add 1 mL of sample to spin column. Replace the top cap and rotate the spin column end to end over a period of 2 h at room temperature (see Note 6). Remove top end bottom caps, place the spin column in the capless collection tube and centrifuge for 2 min at 1,000 × *g* at room temperature. Discard collected material and repeat centrifugation to remove any remaining materials (see Note 7).
6. Replace the bottom cap and add 1 mL of wash buffer to spin column. Replace the top cap and rotate the spin column end to end over a period of 5 min at room temperature. Remove caps, place the spin column in the capless collection tube and centrifuge for 2 min at 1.000 × *g* at room temperature. Discard collected material and repeat centrifugation to remove any remaining materials. Discard collected material.
7. Repeat step 6 two more times.
8. Replace the bottom cap and add 1 mL of dH_2O. Replace the top cap and rotate the spin column end to end over a period of 1 min at room temperature. Remove caps, place column in a capless collection tube and centrifuge at 1,000 × *g* for 2 min to remove dH_2O. Discard collected material and repeat centrifugation to remove any remaining materials.
9. Attach the bottom cap to the column, add 100 μL of rehydrated elution reagent to the column and replace the top cap. Vortex for 5 s. Incubate column at room temperature for 15 min by vortexing every 3 min. Remove caps, place the column in a clean collection tube and centrifuge at 1,000 × *g* for 2 min (see Note 8).
10. Repeat step 9 two more times (see Note 9), store the eluate at −20°C or proceed with 2D IEF-SDS-PAGE analysis.

3.2. First Dimension: Isoelectric Focusing

1. After treatment with the ProteoMiner kit, the protein enriched solution also contains lipids, salts, and other contaminants. These can be readily removed by precipitation of the proteins before being redissolved for subsequent analysis (see Note 10).

2. Proteins should be precipitated from a desired volume containing 400 μg of proteins. Add 80% (v/v) of cold acetone and incubate overnight at –20°C. Pellet the precipitated proteins using a microcentrifuge at 15,000 × *g* for 15 min. Wash the pellet three times with 100 μL ice-cold acetone. Air-dry the pellet, which can then be redissolved in the appropriate solution for subsequent analysis.
3. Resuspend the pellet in 250 μL of sample preparation buffer, add 21.8 μL of Tris–HCl (1.9 M) and agitate at room temperature for 15 min.
4. Add 0.56 μL of 4.5 M TBP and agitate at room temperature for 1 h.
5. Add 5 μL of 2 M IAA and agitate at room temperature for 1 h.
6. Add 5.5 μL of 2 M DTE and agitate at room temperature for 1 h.
7. Precipitate the protein solution in adding 1,000 μL of cold acetone (final 80%v/v) and incubate for 90 min at –20°C. Pellet the precipitated proteins using a microcentrifuge at 10,000 × *g* for 10 min. Wash the pellet three times with 100 μL ice-cold acetone. Air-dry the pellet, which can then be redissolved in the appropriate solution for subsequent analysis.
8. Resuspend the pellet in 315 μL of sample preparation buffer and agitate at room temperature for 15 min.
9. Rehydrate overnight 17 cm pH 4–7 immobiline DryStrips in reswelling tray (Ge-HAlcare) using 315 μL of protein solution (see Note 11).
10. Perform protein focalization using an appropriate IEF apparatus. Separation of proteins should be undertaken for 80,000 of kVh for each strip at 20°C.

3.3. Second Dimension: SDS-PAGE

Several manufacturers produce excellent gel running apparatus; these include the Protean II from Bio-Rad with 165 × 155 mm gels. Assembly of the glass plates and gel casting is fully explained in the manufacturer's instructions. Glass plates should be washed, rinsed with dH_2O, and then dried before use. They can be kept clean until use in a plastic rack in 30% (v/v) nitric acid. They will just need rinsing (distilled water then 95% ethanol) to remove the acid and air-dry.

1. Prepare a 1.5 mm, 11.5% gel by mixing 15 mL acrylamide/bis solution, with 10 mL of 1.5 M Tris–HCl pH 8.8, 14.6 mL of dH_2O, 400 μL SDS solution, 300 μL 10% (w/v) ammonium persulfate solution and 26.6 μL TEMED. Pour the gel solution, leaving space for a stacking gel, and overlay with isobutanol. The gel should polymerize in about 30 min. When polymerization is

complete, pour off the isobutanol and raise the top of the gel twice with water.

2. Prepare the staking gel by mixing 1.65 mL acrylamide/bis solution with 2.5 mL of 0.5 M Tris–HCl pH 6.8, 5.75 mL of water dH_2O, 100 μL SDS solution, 100 μL 10% (w/v) ammonium persulfate solution and 7 μL TEMED. Pour the gel solution, leaving space to load the first dimension gel strip, and overlay with isobutanol. The stacking gel should polymerize in about 30 min.
3. Put the IPG strip into a Petri plate containing 15 mL of fresh IPG strip equilibration buffer. Incubate the strip for 30 min with gentle agitation (use a rocking platform).
4. Place the strip on the stacking gel (see Note 12). No space should be between the IPG strip and the stacking gel. Overlay the strip with about 1 mL of hot (50–100°C) agarose overlay. After 5 min the sealing buffer should be a gel.
5. Prepare the running buffer by diluting 200 ml of running solution (10×) with 1.8 L of MQ water. Complete the assembly of the gel unit and connect to the power supply. If the cooling system is available for the gel unit then set the temperature at 13°C. The gel runs at 11 mA overnight then during the day at 300 V, until the dye front reaches the bottom of the gel.
6. The protein spots can be visualized using Coomassie R-250, colloidal Coomassie and also by silver staining (13).

4. Notes

1. This protocol has been optimized for plasma samples with protein concentration >50 mg/mL. When working with human plasma it is important to follow biohazardous material handling guidelines.
2. Following rehydration, each vial will contain enough elution reagent for processing two samples; if preparing an uneven number of columns, remaining material may be stored at -20°C for up to 1 week.
3. The solution (with omission of carrier ampholyte) may be aliquoted (1-mL volumes) and stored at –80°C.
4. The solution may be aliquoted (10-mL volumes) and stored at -20°C.
5. First remove the top cap followed by the bottom cap and do not discard them, they will be reused throughout the protocol.
6. With plasma samples clumping may occur after 1 h of agitation.

7. Take caution to ensure that the bottom cap is tightly attached.
8. This elution contains your eluted proteins; do not discard it.
9. Elution may be pooled or analyzed individually.
10. Prior to any 2D IEF-SDS-PAGE analysis, quantify the amount of protein in your sample. We recommended the Quick Start Bradford protein assay (Bio-Rad).
11. Place the coffin onto the reswelling tray. Pipet the sample along the whole length of the coffin. Care should be taken not to introduce any air bubbles. Remove the protective backing film from the IPG strip and lay the strip, gel side down, onto the sample fluid in the coffin, to enable it to soak up the sample. At this point, care should be taken not to displace the fluid over the top of the strip and that the full length of the strip is in contact with the sample fluid. Overlay the strip and sample fluid with drystrip cover fluid to prevent dehydration. Run the strips following the manufacturer's recommendations for the type of IPG strip used. After isoelectric focusing is completed, the IPG strips can be run by SDS-PAGE immediately. Alternatively, IPG strips may be stored at 80°C for extended periods with little loss of resolution. Prior to freezing, excess immersion oil should be removed from the strips.
12. You can use a gel spacer of 0.75 mm to facilitate IPG strip loading.

References

1. Conrads TP, Hood BL, Petricoin EF III, Liotta LA, Veenstra TD (2005) Cancer proteomics: many technologies, one goal. Expert Rev Proteomics 5:693–703
2. Petricoin E, Wulfkuhle J, Espina V, Liotta LA (2004) Clinical proteomics: revolutionizing disease detection and patient tailoring therapy. J Proteome Res 3:209–217
3. Zolla L (2008) Proteomics studies revel important information on small molecule therapeutics. Drug Disc Today 13:1042–1051
4. Anderson L, Anderson NG (1977) High resolution two-dimensional electrophoresis of human plasma proteins. Proc Natl Acad Sci USA 74:5421–5425
5. Patton WF (2002) Detection technologies in proteome analysis. J Chromatogr B Analyt Technol Biomed Life Sci 771:3–31
6. Zhou G, Li H, DeCamp D, Chen S, Shu H, Gong Y, Flaig M, Gillespie JW, Hu N, Taylor PR, Emmert-Buck MR, Liotta LA, Petricoin EF III, Zhao Y (2002) 2D differential in-gel electrophoresis for the identification of esophageal scans cell cancer-specific protein markers. Mol Cell Proteomics 1:117–124
7. Righetti PG, Castagna A, Antonioli P, Boschetti E (2005) Prefractionation techniques in proteome analysis: the mining tools of the third millennium. Electrophoresis 26: 297–319
8. Geho DH, Liotta LA, Petricoin EF, Zhao W, Araujo RP (2006) The amplified peptidome: the new treasure chest of candidate biomarkers. Curr Opin Chem Biol 10:50–55
9. Adkins JN, Varnum SM, Auberry KJ, Moore RJ, Angell NH, Smith RD, Springer DL, Pounds JG (2002) Toward a human blood serum proteome: analysis by multidimensional separation coupled with mass spectrometry. Mol Cell Proteomics 1:947–955
10. Boschetti E, Righetti PG (2008) The ProteoMiner in the proteomic arena: a non-depleting tool for discovering low-abundance species. J Proteomics 71:255–264

11. Sennels L, Salek M, Lomas L, Boschetti E, Righetti PG, Rappsilber J (2007) Proteomic analysis of human blood serum using peptide library beads. J Proteome Res 6:4055–4062
12. Marrocco C, Rinalducci S, Mohamadkhani A, D'Amici GM, Zolla L (2010) Plasma gelsolin protein: a candidate biomarker for hepatitis B-associated liver cirrhosis identified by proteomic approach. Blood Transfus 8:105–112
13. Blum H, Beier H, Gross HJ (1987) Improved silver staining of plant proteins, RNA and DNA in polyacrylamide gels. Electrophoresis 8:93–99

Chapter 21

Isolation of Urinary Exosomes from Animal Models to Unravel Noninvasive Disease Biomarkers

Javier Conde-Vancells and Juan M. Falcon-Perez

Abstract

In the last years, disease biomarker discovery has highly evolved thanks to the application of high-throughput technologies such as proteomics. However, due to the elevated complexity and abundance of some of the proteins in the samples the analysis of subcellular compartments has been revealed to be fundamental in order to identify underrepresented clinically relevant proteins. In this sense, extracellular microvesicles including exosomes that are present in different body fluids constitute a suitable and convenient subcellular compartment to identify disease biomarkers. On the other hand, animal models offer numerous advantages over human samples in order to accelerate the identification of candidate biomarkers. In this chapter we provide a detailed methodology to purify and analyze urinary exosomes that can be applied to the study of different diseases that have good animal models.

Key words: Microvesicles, Exosomes, Biomarkers, Urine, Mouse, Rat, Animal model

1. Introduction

Today, there is considerable interest in the discovery of new biomarkers preferably measured in a noninvasive manner to assist in the early diagnosis of diseases, assessment of prognosis, and monitoring of a particular therapeutic regimen, as well as to guide new drug development to therapeutic targets.

Cells are continuously subjected to a variety of stimulus at each stage of their life. These stimuli usually lead to the release to the extracellular milieu of biomolecules packaged enveloped in lipid vesicles termed microvesicles (MVs). Owing to the plasticity of cellular membranes, lateral organization of lipids and proteins may vary according to a specific stimulus resulting in an inclusive or exclusive sorting of biomolecules. Thus, the same cell may release

Djuro Josic and Douglas C. Hixson (eds.), *Liver Proteomics: Methods and Protocols*, Methods in Molecular Biology, vol. 909,
DOI 10.1007/978-1-61779-959-4_21, © Springer Science+Business Media New York 2012

MVs of different composition depending on the stimulus that generates them, including deleterious processes that may develop into a disease. The extracellular microenvironment, including body fluids, contains a mixed population of MVs with distinct structural and biochemical properties. Of particular interest are exosomes, which are nanometer-sized vesicles (30–100 nm) of endocytic origin that participate in cell-to-cell communication processes (1). Thus, exosomes may function as a vehicle to exchange genetic material between cells or as a shuttle for infectious agents to escape immune surveillance (2, 3). These vesicles are formed by inward budding of the limiting membrane of multivesicular bodies (MVBs) and are released to the extracellular milieu upon fusion of MVBs with the plasma membrane. When negative-stained and analyzed by transmission electron microscopy (TEM), exosomes present a characteristic "cup-shaped" morphology (4). The protein composition of exosomes derived from various cell types and body fluids has been extensively analyzed revealing the presence of common as well as cell type-specific proteins (5). The conserved set of proteins includes components of the endosomal sorting complex required for transport (ESCRT) (AIP1/Alix, Tsg101), cytoskeleton-related proteins, cell adhesion molecules (integrins, ICAM-1, MFG-E8), proteins involved in membrane transport and fusion (annexins, rab proteins), signal transducing proteins (14-3-3, heterotrimeric G proteins), heat-shock proteins (HSPs), metabolic enzymes, and tetraspanin molecules (CD81, CD63, and CD9). This common set of proteins found in exosomes suggests useful and preserved functions of these vesicles. Moreover, specific proteins that are characteristic of the cell from which they are derived have been also shown to be present in exosomes (6), and may be useful to identify the origin of a particular exosome population.

Cellular processes generate several by-products including MVs that require elimination from the bloodstream through urine excretion. Thus, analysis of urine may provide putative new biomarkers of diverse human diseases. However, the complete protein composition of urine remains largely uncharacterized since it is a very complex mixture containing thousands of proteins. Furthermore, some urinary proteins are abundant, hampering the identification of low-component molecules that might be clinically relevant of a particular disease. In this regard, analysis of urinary MVs—specially exosomes—represents a means of reducing the complexity of the urinary proteome ten or more orders of magnitude, the end result being a very large enrichment for low-abundant proteins.

In the last 5 years exosomes isolated from human urine samples have been extensively analyzed by protein mass spectrometry resulting in the identification of a few disease-associated proteins (7, 8). Among them, several proteins associated with the EGF receptor pathway that were found in urinary exosomes were proposed as potential biomarkers of bladder cancer while activating transcription factor 3 (ATF3) was proposed as a novel renal tubular cell injury biomarker to diagnose acute kidney injury (AKI) in human patients (9). Notwithstanding, several recent reports have noted that the quantity and quality of exosomes present in human urine samples are highly variable (10), a primary cause to slowing and even avoiding the identification of disease biomarkers using human samples. To overcome this variability issue, studies in animal models of diseases are of great advantage because of the large degree of control over experimental conditions and sampling. Clinical samples are largely heterogeneous, and hence, animal models have a smaller degree of heterogeneity than clinical samples, a characteristic that facilitates the identification of candidate biomarkers. Furthermore, rodent animal models are of particular interest since most human genes have functional rodent counterparts and the genome is organized in a similar manner. In addition, rodents are easy to breed and maintain in a laboratory, and allow investigation of disease stages that would be inaccessible or nonviable in humans. Finally, there are already many genetically and environmentally generated mouse and rat models available for a myriad of diseases, and many that are under development that will be very useful in the future (11).

In this chapter, we detail methodologies to purify MVs (mainly exosomes) from rodent urine samples. These methodologies were successfully applied to identify putative protein biomarkers of liver disease using animal models of acute (d-galactosamine (GalN)) and chronic liver injury (glycine *N*-methyltransferase (*GNMT*) knockout mouse model) (12). GalN is a potent hepatotoxin that induces liver damage in rats very similar to that of acute human viral hepatitis (13). On the other hand, genetic defects in key enzymes of the metabolism of *S*-adenosyl methionine (SAMe) such us GNMT or methionine adenosyltransferase 1A (MAT1A) are considered *bona fide* models for chronic liver disease because they recapitulate many of the clinical manifestations found in human liver patients such as steatosis, steatohepatitis, fibrosis, and hepatocellular carcinoma (14, 15).

2. Materials

2.1. Urine Collection

1. Rats or mice with a genetic background and age of interest.
2. Metabolic cages (any metabolic cage adapted to collect urine and measure water intake is suitable).
3. 50 ml Falcon tubes.

2.2. Isolation of Exosome-Like Vesicles from Urine Samples

1. Phosphate-buffered saline (PBS).
2. Appropriate ultracentrifuge (e.g., Optima™ L-100XP Biosafe Centrifuge System, Beckman Coulter).
3. 50 ml Falcon tubes.
4. Vacuum filter system. Other devices provided with a 0.22 μm pore size membrane may be used instead.
5. Polycarbonate centrifuge bottles, 38 mm × 102 mm, 70 ml, with cap assemblies (Beckman Coulter).
6. Type 45, fixed angle, titanium, rotor 6 × 94 ml, 45,000 rpm, 235,000 × *g* (Beckman Coulter).

2.3. Fractionation of Urine-Derived Exosome-Like Vesicles on Continuous Sucrose Gradients

1. Sucrose solution: 20 mM 4 (2-hydroxyethyl)-1-piperazineethanesulfonic acid (HEPES), 2.5 M protease-free sucrose. Adjust pH to 7.4 with HCl. Store at 4°C up to 2 years.
2. HEPES solution: 20 mM HEPES. Adjust pH to 7.4. Store at 4°C up to 2 years.
3. Ultra-Clear™ tubes, thinwall, 17 ml, 16 mm × 102 mm (Beckman Coulter).
4. Centrifuge tubes, polycarbonate thick wall, 13 mm × 56 mm, 3.2 ml (Beckman Coulter).
5. SW32.1 Titanium Rotor, swinging bucket, 6 × 17 ml, 32,000 rpm, 187,000 × *g* (Beckman Coulter).
6. TLA-110 Titanium Rotor, fixed angle, 8 × 5.1 ml, 110,000 rpm, 657,000 × *g* (Beckman Coulter).
7. Auto-densi flow density gradient fractionator (Labconco).
8. SG-15 gradient maker, 15 ml total volume (GE Healthcare).
9. Magnetic stirrer.
10. Appropriate ultracentrifuges [e.g., Optima™ L-100XP Biosafe Centrifuge System (Beckman Coulter) and Optima™ TLX ultracentrifuge (Beckman Coulter)].
11. Refractometer (PCE instruments).

2.4. Purification of Exosomes from Urine Samples

1. PBS.
2. Tris/sucrose/D_2O solution: 15 g protease-free sucrose, 1.2 g Tris-HCl base, 25 ml deuterium oxide (D_2O). Adjust pH to 7.4 with HCl. Adjust volume to 50 ml with D_2O. Store at 4°C up to 2 months.
3. Centrifuge tubes, thickwall, polyallomer, 32 ml, 25 mm × 89 mm (Beckman Coulter).
4. SW32 Titanium Rotor, swinging bucket, 32,000 rpm, 6 × 38.5 ml, 175,000 × *g*.
5. Appropriate ultracentrifuge [e.g., Optima™ L-100XP Biosafe Centrifuge System (Beckman Coulter)].

6. Syringes.
7. 18-G needles.
8. Centrifuge tubes, thickwall, polycarbonate, 32 ml, 25 mm × 89 mm (Beckman Coulter).
9. Type 45 Titanium rotor, fixed angle, 6 × 94 ml, 45,000 rpm, 235,000 × *g* (Beckman Coulter).

2.5. Sample Preparation for Proteomic Quantitation

2.5.1. Preparation of Samples for In-Gel Digestion

1. Bradford protein assay or any other assay suitable to measure protein concentration.
2. Loading buffer: We use commercially available loading buffer (Invitrogen), but any loading buffer suitable for running sodium dodecyl sulfate (SDS)-PAGE may be used instead.
3. Thermoblock.
4. Appropriate centrifuge for Eppendorf tubes.
5. Appropriate cuvette for running SDS-PAGE.
6. Optional: NuPAGE® 4–12% Bis-Tris Gel, 1.0 mm × 12 well (Invitrogen).
7. Running buffer: We use commercially available running buffer (Invitrogen), but any running buffer suitable for running SDS-PAGE may be used instead.
8. Protein molecular weight marker.
9. Fixing solution: 50% methanol, 7% acetic acid.
10. Coomassie Blue staining solution: We use commercially available Coomassie Blue staining solution (Invitrogen), but any staining solution based on Coomassie Blue may be used instead.
11. Sterile scalpel.

2.5.2. Preparation of Samples for In-Solution Digestion

1. Bradford protein assay or any other assay suitable to measure protein concentration.
2. Lysis buffer: 2% SDS, 40 mM Tris-HCl base (it is highly recommended to prepare this buffer with highly pure water suitable for LC-MS applications).
3. Exchange buffer: 10 mM Tris–HCl, pH 7.0. (It is highly recommended to prepare this buffer with highly pure water suitable for LC-MS applications.)
4. Water of LC-MS grade.
5. Thermoblock.
6. Appropriate centrifuge for Eppendorf tubes.
7. Amicon® Ultra-0.5 ml centrifugal filter unit with Ultracel®-10K membrane (Millipore).

3. Methods

3.1. Urine Collection

Urine from rats or mice can be collected for prolonged times by means of metabolic cages. Metabolic cages allow monitoring of the animals under standardized conditions, collection of urine and feces, and measurement of food and water intake. Collection of urine from rats or mice using metabolic cages is detailed below.

1. Maintain animals in an environmentally controlled room at 22°C on a 12-h light/dark cycle and provided with standard diet (Rodent Maintenance Diet, Harlan Teklad Global Diet 2014) and water ad libitum.
2. Before drug administration for comparative studies collect urine for analysis of the basal state of the animal. For urine collection, animals are kept in metabolic cages with just water for 12 h (see Note 1).
3. Shortly after drug administration, animals are again kept in metabolic cages with just water for 12 h. Constantly examine the general state of the animal upon drug administration.
4. After collection, label and immediately freeze urine at –80°C in 50 ml Falcon tubes. Do not fill up the tube entirely or it may break off upon freezing.

3.2. Isolation of Exosome-Like Vesicles from Urine Samples

As has been mentioned before, body fluids contain a mixed population of MVs with distinct structural and biochemical properties. Frequently, different nomenclatures have been used in the literature to name MVs including exosomes, microparticles, shedding microvesicles, ectosomes, argosomes, dexosomes, etc. However, most studies have mainly focused on two different types of MVs: exosomes and microparticles (MPs). Exosomes are nanometer-sized vesicles (30–100 nm) of endocytic origin that form by inward budding of the limiting membrane of MVBs. MPs are larger in size with a diameter that ranges from 100 to 1,000 nm. MPs originate from outward budding and excision of membrane vesicles from the plasma membrane of a wide variety of cell types (16).

The procedure commonly used to isolate exosomes mainly involves differential ultracentrifugation. Hence, it is not exclusive for this type of vesicles. In this procedure samples are filtrated through a 0.22 μm pore size membrane prior to vesicle isolation by differential ultracentrifugation. In this step, most MPs are usually retained on the filter membrane because of their larger size. Thus, preparations resulting from this procedure are highly enriched in exosomes, and hence, named as "exosome-like vesicles." The standard procedure used to isolate "exosome-like vesicles" is described below and a schematic representation is provided in Fig. 1.

1. Thaw urine samples at 4°C (see Note 2). From this step on, keep the samples always on ice. Try to maintain the rotors to be

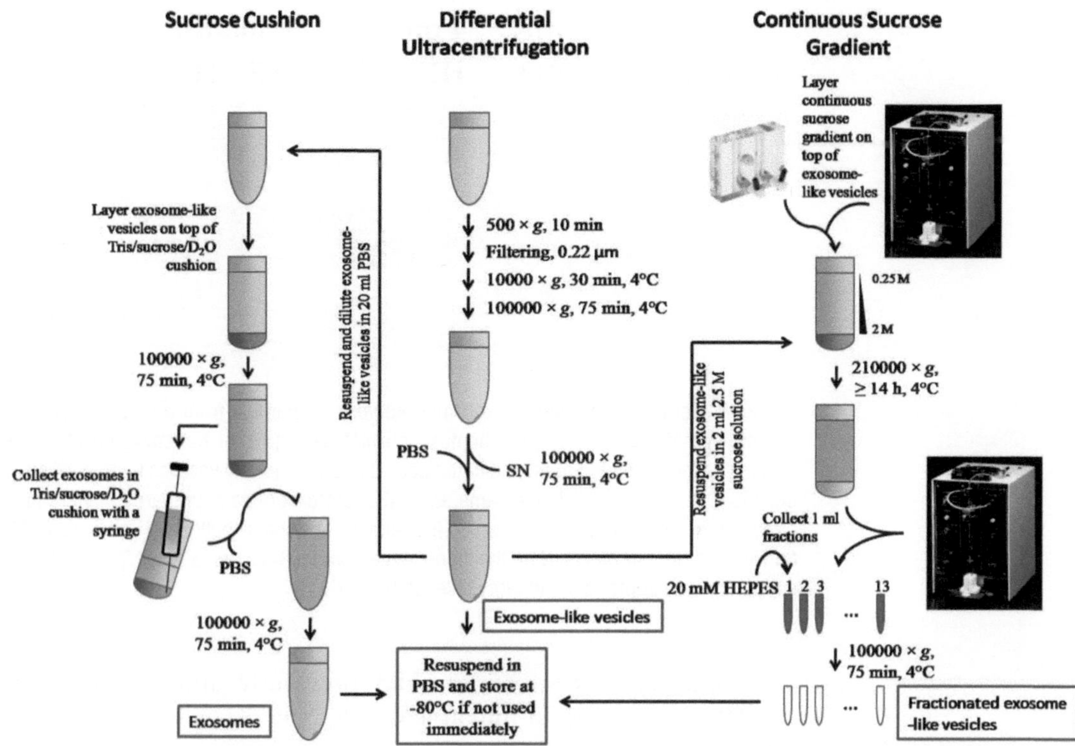

Fig. 1. Schematic representation of the different methodologies used to purify exosomes and exosome-like vesicles from urine samples. For isolation of exosome-like vesicles samples are subjected to filtration through 0.22 µm pore size membranes and differential ultracentrifugation. Last step involves washing of exosome-like vesicles in PBS and again ultracentrifugation to pellet vesicles. if highly pure exosomes preparations are desired, isolated exosome-like vesicles can be further purified by ultracentrifugation onto a 30% sucrose cushion are collected, diluted with PBS, and ultracentrifuged to pellet exosomes. However, if fractionation of exosome-like vesicles according to their density is required, a continuous sucrose gradient ranging from 0.25 to 2 M sucrose is layered onto exosome-like vesicles dissolved in 2.5 M sucrose solution by means of a gradient maker and a fractionator. After ultracentrifugation, 1 ml fractions are collected using the fractionators, diluted with HEPES solution, and ultracentrifuged to pellet vesicles.

used for centrifugation steps at 4°C. Otherwise, just make sure that the rotors have been cooled down before use by putting them into the centrifuge under vacuum conditions at 4°C.

2. Transfer samples to 50 ml Falcon tubes if required and centrifuge at 1,500 × *g* for 30 min at 4°C.
3. Filter supernatant through 0.22 µm pore filter membranes (see Note 3). This step may take a few minutes.
4. Transfer the cleared urine samples to polycarbonate bottles and balance them using cold PBS if necessary.
5. Mark one side of each bottle with a waterproof marker and place the bottle in the rotor with the mark facing up, so you know where the pellet will be located after centrifugation.
6. Centrifuge at 10,000 × *g* for 30 min at 4°C (see Note 4).
7. Slowly pour the supernatant into clean polycarbonate bottles trying not to disturb the pellet.

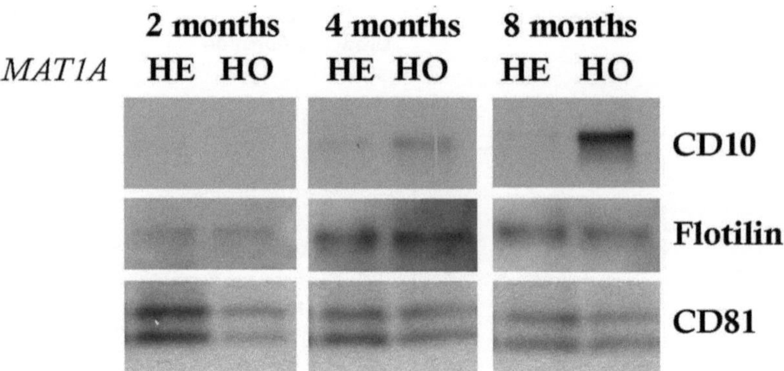

Fig. 2. Western-blot analysis of exosome-like vesicles purified from a mouse model for chronic liver disease. Urine from homozygote *MAT1A*$^{-/-}$ (HO) or heterozygote *MAT1A*$^{+/-}$ (HE) mice were collected at the indicated ages and exosome-like vesicles were purified as described in this chapter. Later, same amount of proteins of each preparation were analyzed by Western blotting with antibodies against exosomal hallmarks such as Cd81, flotillin, and Cd10 proteins. Note the increased amount of Cd10 in samples from MAT1A$^{-/-}$ supporting this approach to identify disease biomarkers.

8. Mark the bottles as you did previously and centrifuge at 100,000 × *g* for 75 min at 4°C (see Note 4).
9. Slowly pour off the bottle of supernatant trying not to disturb the pellet.
10. Resuspend the pellet in 500 μl cold PBS. Pool the resuspended pellets if required and fill each bottle to be centrifuged with at least 30 ml of cold PBS.
11. Balance bottles and centrifuge at 100,000 × *g* for 75 min at 4°C (see Note 4).
12. Slowly pour off the bottle of supernatant trying not to disturb the pellet.
13. Resuspend the pellet in 50–100 μl cold PBS depending on its size and further experiments to be performed.
14. Aliquot into Eppendorf tubes and store at −80°C if not used immediately.

An example of the application of this methodology to study a mouse model of chronic liver disease is provided in Fig. 2.

3.3. Fractionation of Exosome-Like Vesicles on Continuous Sucrose Gradients

MVBs in the cell have two possible fates: fusion with lysosomes for degradation or fusion with the plasma membrane and consequent release of exosomes to the extracellular milieu. Thus, different types of MVBs must coexist in the cell because of the distinct cellular fates of this organelle. However, whether MVBs contain a specific set of vesicles destined either for degradation or secretion or a mixed population of vesicles is still unclear. Evidences for the coexistence of distinct types of MVBs in the same cell have already been reported.

Consequently, releasing of different exosomes (in terms of protein content) by a determined cell type is expected (1). Furthermore, previous studies in humans have highlighted the existence in urine of exosomes secreted by distinct cellular types of the renal epithelia that can be discriminated by their protein composition.

One characteristic hallmark of exosomes is their floatation on sucrose gradients. A methodology to separate exosomes contained in the urine by means of their floatation on continuous sucrose gradients is described. The different fractions obtained are vesicles of distinct composition that can be further analyzed by other experimental techniques such as Western blot. The density at which exosomes float has been determined to range from 1.13 to 1.21 g/ml. We strongly recommend to become familiar with the procedure before applying it to your preparations. A schematic representation of the following procedure used to fractionate urinary exosome on a continuous sucrose gradient is provided in Fig. 1.

1. Put the gradient maker on a magnetic stirrer. Couple the outer tubing of the gradient maker to the outer tubing of the density gradient fractionator. Load some water in both the distal and the proximal compartments. Turn on the density gradient fractionator and let the water flow to remove all bubbles that otherwise would interfere in the formation of the continuous gradient. Repeat this step two or three times.
2. Resuspend vesicles in 2 ml (2.5 M) sucrose solution (see Note 5). Mix thoroughly and place the suspension at the bottom of the centrifuge tube. Keep on ice.
3. For each gradient, prepare 6 ml of 2 and 0.25 M sucrose solutions by diluting the sucrose solution with HEPES solution (see Note 6).
4. Make sure that there is no water left in the gradient. Close both valves and load 6 ml of 2 M sucrose solution in the proximal compartment (located closer to the outer tubing) and 6 ml of 0.25 M sucrose solution in the distal compartment (located farthest from the outer tubing). Open the valves placed between the proximal and distal compartments. Turn on the magnetic stirrer at a low speed.
5. Turn on the density gradient fractionator and set it to deposit a gradient. Place the centrifuge tube with the vesicle suspension in the holder and adjust it so that the tube is completely fixed. Lower the probe till it is in contact with the vesicle suspension and set the probe to go up while depositing the gradient (this way the probe is always on top of the layer). Depositing speed should be low for formation of a uniform continuous sucrose gradient.
6. Open the valve between the proximal compartment and the outer tubing, and layer a continuous gradient on top of vesicle suspension. When the gradient is completed, clean the tubing

of the density gradient fractionator by letting some water to flow through two or three times. If more than one gradient needs to be made always do this step in between (see Note 7).

7. Centrifuge gradients at 32,000 rpm at 4°C. Tubes containing gradients should be balanced with each other or with tubes containing the same volumes of the solutions used. After ≥16 h, allow the rotor to come to a stop without using the brake since this may disturb the continuous sucrose gradient.
8. Turn off the centrifuge (see Note 8).
9. To collect fractions set the density gradient fractionator to withdraw liquid. Again place the centrifuge tube with your gradient in the holder and adjust it so that the tube is fixed. Lower the probe till it is in contact with the liquid and this time set the probe to go down while removing the layers (this way the probe is always on top of the layer).
10. Collect 1 ml fractions in Eppendorf tubes (mark the walls of the tubes to know when to shift tubes). Keep on ice.
11. Use 10 μl from each fraction to measure the refractive index. Use the table given at the ultracentrifugation catalog from Beckman to convert the measured refractive index to density and sucrose concentration (http://www.bio.huji.ac.il/upload/Centrifuges%20Rotors_Manual.pdf).
12. Transfer 1 ml fractions to polycarbonate thick wall tubes, add 2 ml of HEPES solution, and pipet thoroughly to make sure that the sucrose is uniformly mixed with HEPES.
13. Mark one side of each tube with a waterproof marker and place the tube in the rotor with the mark facing up so that you know where the pellet will be located after centrifugation.
14. Centrifuge at 100,000 × *g* for 75 min at 4°C (see Note 4).
15. Slowly decant the supernatant taking care not to disturb the pellet. Make sure that all the liquid is removed from the tube by placing a pipette on the opposite side where the pellet is located and collecting the remaining liquid (see Note 9).
16. Resuspend each pellet no more than 25 μl of cold PBS depending on the amount of exosome-like vesicles previously used.

3.4. Purification of Exosomes from Urine Samples

The procedure commonly used to isolate exosomes involves differential ultracentrifugation, and hence, is not exclusive of this type of vesicles. Although samples are filtrated through 0.22 μm pore size filter membranes to remove large cell debris and MPs, protein aggregates (smaller than 0.22 μm in size) may also flow through the filter membrane along with exosomes. Since these protein aggregates are large enough to sediment at the centrifugal speeds used to isolate exosomes, they will contaminate the exosome preparations. Thus in order to obtain highly pure exosome preparations it is necessary to perform an extra purification step involving sedimentation onto a 30% sucrose cushion. Exosomes exclusively float at the density of this sucrose cushion, and hence, other contaminants such as protein aggregates are eliminated. The procedure for the purification of exosomes in a 30% sucrose cushion is detailed below and a schematic representation is provided in Fig. 1.

1. Dilute exosome-like vesicles obtained by the differential ultracentrifugation procedure described above in 20 ml cold PBS. If pellets were kept frozen at –80°C let them slowly defrost on ice.
2. Load 4 ml of the Tris/sucrose/D_2O solution (sucrose cushion) at the bottom of a polyallomer tube. Try to avoid touching the walls of the tube.
3. Slowly layer the diluted vesicles on top of the sucrose cushion. Try not to disturb the cushion so that an interface is clearly visible (see Note 10).
4. Centrifuge tubes in a swinging bucket rotor at 100,000 × *g* for 75 min at 4°C. In this centrifuge step, avoid using brake to stop the rotor since this may cause disturbance of the continuous sucrose gradient.
5. After centrifugation, remove some of the upper phase consisting mostly of PBS. Collect 3.5 ml from the bottom of the tube using a syringe fitted with a 18-G needle. Place the needle on the side of the tube to collect the exosomes since other contaminants may accumulate at the bottom.
6. Transfer the collected volume to a polycarbonate tube, add 20 ml of cold PBS, and mix thoroughly.
7. Mark one side of each tube with a waterproof marker and place the tube in the rotor with the mark facing up so that you know where the pellet will be located after centrifugation.
8. Centrifuge at 100,000 × *g* for 75 min at 4°C (see Note 4).
9. Slowly decant the supernatant taking care not to disturb the pellet and resuspend exosomes in 50–100 μl of cold PBS.

3.5. Sample Preparation for Proteomic Quantitation

The advent of protein mass spectrometry has enabled a new approach to biomarker discovery. This approach consists of the identification of putative protein biomarkers by large-scale profiling of proteins in complex samples such as urine. Many biomarker discovery studies have followed a qualitative approach in which proteins are either present or absent. However, subtle changes in the level of particular proteins may be more informative of the stage of the disease, and therefore, very valuable as putative biomarkers. Thus, quantitation of protein abundance is an important prerequisite in biomarker discovery. The quantitation of protein in cells, tissues, and organisms is one of the major challenges in the post-genomic era. Two-dimensional gel electrophoresis (2-DE) may provide accurate quantitation based on either spot density comparison or using fluorescence labeling methods that allow comparison of two samples in the same gel (17). However, distinct quantitation approaches have become available during the last years increasing the popularity of LC-MS/MS tandem mass spectrometry coupled to a liquid chromatographic system as the method of choice for biomarker discovery. Two different methodologies exist for protein quantitation in LC-MS/MS: label-free methods and stable isotopic labeling methods. Following label-free methodologies, samples are analyzed in separate LC-MS/MS runs and proteins are quantified according to the intensity of the peaks generated in the mass spectrometer for each protein peptide. Stable isotopic labeling methods use stable isotopes to allow the combination of the different samples and their analysis in a single LC-MS/MS run. Stable isotopes are used to label protein peptides before combining the samples being compared. Proteins from each sample are distinguished because the masses of their isotopically labeled peptides are shifted in the spectrum generated, thus allowing their quantitation according to the intensity of the peak measured in the mass spectrometer.

Depending on the approach being used for quantitative proteomics, it may be desirable to prepare your samples for either in-gel or in-solution digestion. Preparation of sample for either procedure is detailed below. We recommend asking your proteomic facility for advice regarding the quantitative approach and the type of digestion that best suits your study. Both procedures are a very cost-effective way to remove several contaminating compounds such as salts and detergents that may further interfere in the LC-MS/MS analysis. In either procedure it is highly recommended to perform the protocol in a clean room and protect yourself with lab coat and gloves since samples are easily contaminated with keratin and cloth fibers, which again may further interfere in the LC-MS/MS analysis.

3.5.1. Preparation of Samples for In-Gel Digestion

SDS-polyacrylamide gel electrophoresis.

1. Measure protein concentration in your preparation/s by means of Bradford protein assay or other suitable assay (see Note 11).
2. Transfer the desired amount of protein (at least 10 μg) from your preparation/s to a new Eppendorf tube/s and add the appropriate volume of loading buffer.
3. Heat samples in a thermoblock for 5 min at 65°C, followed by a 10-min incubation at 95°C.
4. Let samples to cool down and centrifuge at 15,000 × *g* for 10 min at room temperature (see Note 12).
5. Prepare the cuvette and gel for electrophoresis (see Note 13). Introduce your gel into the cuvette and fill it with running buffer (see Note 14). Wash the wells by pipetting running buffer up and down in each well.
6. After centrifugation, immediately load your samples into the wells of the gel along with a molecular weight protein marker.

7. Run the SDS-PAGE at the desired voltage. (Precast gels handle high voltages for fast run times. However, it is always recommended to run your gels at a low voltage for a better resolution and separation of your protein bands.)
8. After running, rinse the gel with distilled water and incubate with fixing solution.
9. Let the proteins fix in the gel for 30 min and rinse the gel with distilled water for at least 1 h.
10. Stain the gel overnight with Coomassie Blue to visualize protein bands (see Note 15).
11. After 16 h, cut as many bands from the gel as desired with a sterile scalpel, and place each band in a sterile Eppendorf tube filled with 250 μl of highly pure water (see Note 16).
12. Subject each gel piece to in-gel digestion following standard protocols.

3.5.2. Preparation of Samples for In-Solution Digestion

1. Measure protein concentration in your preparation/s by means of Bradford protein assay or another suitable assay (see Note 11).
2. Transfer the desired amount of protein (at least 10 μg) from your preparation/s to a new Eppendorf tube/s and adjust the concentration to 1% SDS, 20 mM Tris–HCl. (Adjust the volume using lysis buffer.)
3. Heat samples in a thermoblock at 65°C for 20 min.
4. After heating, let the samples cool down and centrifuge them at 15,000 × *g* for 10 min at room temperature.
5. Transfer supernatant to a centrifugal filter device to exchange buffer (see Note 17). We use a 10K cutoff device so that sample recovery is very high and spinning times are not very long.
6. Fill up the device to 500 μl with exchange buffer.
7. Insert the filter device into the provided microcentrifuge tube and place the capped filter device into the centrifuge.
8. Centrifuge capped filter device at 15,000 × *g* for the required time at room temperature (see Note 17).
9. Discard the flow through from the microcentrifuge tube and repeat steps 6–8 at least three more times (see Note 18).
10. Once the buffer is exchanged, recover your sample insert immediately by inserting the filter device upside down into a clean microcentrifuge tube. Place the tube in the centrifuge so that open caps are aligned towards the center of the rotor.
11. Centrifuge at 1,000 × *g* for 2 min at room temperature.
12. Transfer sample to a sterile Eppendorf tube and subject to in-solution digestion following standard protocols.

Table 1 shows a list of proteins found in urinary exosomes that have been associated with different diseases.

Table 1
List of proteins associated with different diseases that were detected in urinary

Accession number	Protein name	Cancer	Cardiovascular disease	Connective tissue disorders	Dermatological diseases and conditions	Developmental disorder	Endocrine system disorders	Gastrointestinal disease	Genetic disorder
ACY1A_RAT	Aminoacylase 1	X						X	X
PPBT_RAT	Alkaline phosphatase, liver/bone/kidney	X	X				X		X
AMPN_RAT	Alanyl (membrane) aminopeptidase	X						X	
ANXA2_RAT	Annexin A2	X	X					X	X
ANXA5_RAT	Annexin A5	X	X	X			X		X
DPP4_RNT	Dipeptidyl-peptidase 4	X	X	X	X	X	X	X	X
ENOA_RAT	Enolase 1, (alpha)	X		X					X
EZRI_RAT	Ezrin	X						X	
FABPH_RAT	Fatty acid-binding protein 3, muscle and heart	X	X			X		X	X
FABP4_RAT	Fatty acid-binding protein 4, adipocyte	X	X	X			X		X
G3P_RAT	Glyceraldehyde-3-phosphate dehydrogenase	X							X
GGT1_RAT	Gamma-glutamyltransferase 1	X		X		X			X
GNAS2_RAT	GNAS complex locus	X	X	X		X	X	X	X
GPC5C_RAT	G protein-coupled receptor, family C, group 5, member C	X							X
GSTO1_RAT	Glutathione S-transferase omega 1	X	X					X	X
GSTP1_RAT	Glutathione S-transferase pi 1	X	X		X			X	X
MEP1A_RAT	Merprin A, alpha	X	X					X	X
MEP1B_RAT	Merprin A, beta	X							X
MIF_RAT	Macrophage migration inhibitory factor	X	X	X	X	X	X	X	
NEP_RAT	Membrane metallo-endopeptidase	X		X				X	X

exosomes purified by using a sucrose cushion from normal rat urine samples

Hematological disease	Hepatic system disease	Hypersensitivity response	Immunological disease	Infection mechanism	Infectious disease	Inflammatory disease	Inflammatory response	Metabolic disease	Neurological disease	Nutritional disease	Ophthalmic disease	Organismal injury and abnormalities	Psychological disorders	Renal and urological disease	Reproducitve system disease	Respiratory disease	Skeletal and muscular disorders
	X							X									
	X							X									
				X	X	X	X					X				X	
X			X	X	X		X		X			X			X		X
X			X	X		X	X	X	X						X		X
X		X	X	X	X	X	X	X	X	X			X	X	X	X	
X			X			X									X	X	X
															X		
																	X
X	X	X															
	X																
X																	
	X																
X	X																
X																	

(continued)

Table 1
(continued)

Accession number	Protein name	Cancer	Cardiovascular disease	Connective tissue disorders	Dermatological diseases and conditions	Developmental disorder	Endocrine system disorders	Gastrointestinal disease	Genetic disorder
RASN_RAT	Neuroblastoma RAS viral (v-ras) oncogene homolog	X				X		X	X
PARK7_RAT	Parkinson disease (autosomal recessive, early onset) 7	X							X
PEBP1_RAT	Phosphatidylethanolamine-binding protein 1	X							X
PROF1_RAT	Profilin 1	X	X					X	X
KPYM_RAT	Pyruvate kinase, muscle	X						X	X
PRDX1_RAT	Peroxiredoxin 1	X		X					X
PRDX6_RAT	Peroxiredoxin 6	X		X					X
RB11A_RAT	RAB11A, member RAS oncogene family								X
S22A5_RAT	Solute carrier family 22 (organic cation/carnitine transporter), member 5	X					X	X	X
S47A1_RAT	Solute carrier family 47, member 1	X							
SODC_RAT	Superoxide dismutase 1, soluble	X	X			X	X	X	X
TMM27_RAT	Transmembrane protein 27	X							X
TPIS_RAT	Triosephosphate isomerase 1	X						X	X
THIO_RAT	Thioredoxin	X	X	X	X	X	X	X	X
1433E_RAT	Tyrosine 3-monooxygenase/tryptophan 5-monooxygenase activation protein, epsilon polypeptide	X	X			X		X	X
1433Z_RAT	Tyrosine 3-monooxygenase/tryptophan 5-monooxygenase activation protein, zeta polypeptide	X							X

	X		X						X	X					X	Hematological disease
		X	X		X	X	X			X		X				Hepatic system disease
		X														Hypersensitivity response
		X	X		X											Immunological disease
		X						X								Infection mechanism
																Infectious disease
		X	X		X		X									Inflammatory disease
		X			X											Inflammatory response
		X	X		X		X									Metabolic disease
X	X	X	X		X		X	X								Neurological disease
					X											Nutritional disease
		X			X											Ophthalmic disease
		X			X											Organismal injury and abnormalities
																Psychological disorders
																Renal and urological disease
X	X		X		X											Reproducitve system disease
	X	X	X	X												Respiratory disease
X		X			X			X								Skeletal and muscular disorders

4. Notes

1. It is recommended to adapt animals to the cages over several days for a few hours per day to reduce stress and related alteration in behavior. Between 10–20 ml and 1.5–3 ml of urine can be collected from a single rat and mouse, respectively.
2. It is recommended to defrost samples by leaving them overnight in the refrigerator.
3. Vacuum filter systems as those supplied by Corning® can be used to speed up the filtration process. Otherwise any device provided with a 0.22 μm pore size membrane is suitable for this step.
4. We recommend not using the brake to stop the centrifuge because the MV pellet may dissolve into the supernatant. This is particularly important when the amount of vesicles contained in the sample is very low.
5. If exosome-like vesicles have been previously frozen you might observe some visible aggregates, so some pipetting may be required before to disperse them in the sucrose solution.
6. Always prepare more solution than you need.
7. If more than two gradients are going to be made the same day, be aware of the time required for each gradient to be processed the next day.
8. Alternatively, you may program the centrifuge to stop at least 1.5 h before you arrive to the laboratory if you do not want to wait for the rotor to stop.
9. It is very important that no liquid is left in the tube to assure that all the pellets from each experiment are in the same volume of cold PBS.
10. We recommend using a 5 ml pipette for depositing the diluted vesicles on top of the cushion. The pipette is placed just above the cushion and with the help of a pipetboy the diluted vesicles are slowly added.
11. If detergents are going to be added to your preparation before measurement, make sure that the assay used to measure protein concentration is compatible with the detergent/s and its concentration.
12. This step is very important to remove highly abundant lipids in exosome preparations that may interfere with SDS-PAGE.
13. We recommend using precast gels since these give the best resolution and the most consistent results. These gels require special cuvettes. Before use remove the comb from the wells and white strip from the bottom of the gel. Wash the gel extensively with distilled water.

14. If precast gels are going to be used, check whether you have the appropriate running buffer to run your gel.
15. Other staining solutions suitable for polyacrylamide gels such as SYPRO stains may be used instead.
16. Centrifugal filter devices are commonly used to concentrate samples. However, they also provide a fast and clean system to exchange buffer with high recovery rates.
17. A table indicating typical concentrate volume/concentration factor versus centrifugation time comes along the data sheet provided with the devices.
18. It is important to make sure that SDS is completely removed from your sample since it may later interfere in the LC-MS analysis.

Acknowledgements

This work was supported by grants from Foundation the Institute of Health Carlos III (06/0621 and 09/00526 to JMFP), the program Ramon y Cajal (JMFP). Centro de Investigación Biomédica en Red en el Área temática de Enfermedades Hepáticas y Digestivas (CIBERehd) is funded by the Institute of Health Carlos III.

References

1. Simons M, Raposo G (2009) Exosomes – vesicular carriers for intercellular communication. Curr Opin Cell Biol 21(4):575–581 [Pubmed: 19442504]
2. Valadi H, Ekstrom K, Bossios A, Sjostrand M, Lee JJ, Lotvall JO (2007) Exosome-mediated transfer of mRNAs and microRNAs is a novel mechanism of genetic exchange between cells. Nat Cell Biol 9(6):654–659 [Pubmed: 17486113]
3. Nguyen DG, Booth A, Gould SJ, Hildreth JE (2003) Evidence that HIV budding in primary macrophages occurs through the exosome release pathway. J Biol Chem 278(52):52347–52354 [Pubmed: 14561735]
4. Andre F, Escudier B, Angevin E, Tursz T, Zitvogel L (2004) Exosomes for cancer immunotherapy. Ann Oncol 15(Suppl 4):iv141–iv144 [Pubmed: 15477298]
5. Mathivanan S, Ji H, Simpson RJ (2010) Exosomes: extracellular organelles important in intercellular communication. J Proteomics 73(10):1907–1920 [Pubmed: 20601276]
6. Simpson RJ, Lim JW, Moritz RL, Mathivanan S (2009) Exosomes: proteomic insights and diagnostic potential. Expert Rev Proteomics 6(3):267–283 [Pubmed: 19489699]
7. Pisitkun T, Shen RF, Knepper MA (2004) Identification and proteomic profiling of exosomes in human urine. Proc Natl Acad Sci USA 101(36):13368–13373 [Pubmed: 15326289]
8. Gonzales PA, Pisitkun T, Hoffert JD et al (2009) Large-scale proteomics and phosphoproteomics of urinary exosomes. J Am Soc Nephrol 20(2):363–379 [Pubmed: 19056867]
9. Zhou H, Cheruvanky A, Hu X et al (2008) Urinary exosomal transcription factors, a new class of biomarkers for renal disease. Kidney Int 74(5):613–621 [Pubmed: 18509321]
10. Mitchell PJ, Welton J, Staffurth J et al (2009) Can urinary exosomes act as treatment response markers in prostate cancer? J Transl Med 7:4 [Pubmed: 19138409]
11. Schughart K (2010) SYSGENET: a meeting report from a new European network for systems genetics. Mamm Genome 21(7–8):331–336 [Pubmed: 20623354]
12. Conde-Vancells J, Rodriguez-Suarez E, Gonzalez E et al (2010) Candidate biomarkers

in exosome-like vesicles purified from rat and mouse urine samples. Proteomics Clin Appl 4(4):416–425 [Pubmed: 20535238]

13. Keppler D, Lesch R, Reutter W, Decker K (1968) Experimental hepatitis induced by d-galactosamine. Exp Mol Pathol 9(2):279–290 [Pubmed: 4952077]
14. Lu SC, Alvarez L, Huang ZZ et al (2001) Methionine adenosyltransferase 1A knockout mice are predisposed to liver injury and exhibit increased expression of genes involved in proliferation. Proc Natl Acad Sci USA 98(10):5560–5565 [Pubmed: 11320206]
15. Martinez-Chantar ML, Vazquez-Chantada M, Ariz U et al (2008) Loss of the glycine *N*-methyltransferase gene leads to steatosis and hepatocellular carcinoma in mice. Hepatology 47(4):1191–1199 [Pubmed: 18318442]
16. Cocucci E, Racchetti G, Meldolesi J (2009) Shedding microvesicles: artefacts no more. Trends Cell Biol 19(2):43–51 [Pubmed: 19144520]
17. Pisitkun T, Johnstone R, Knepper MA (2006) Discovery of urinary biomarkers. Mol Cell Proteomics 5(10):1760–1771 [Pubmed: 16837576]

Index

A

Djuro Josic and Douglas C. Hixson (eds.), *Liver Proteomics: Methods and Protocols*, Methods in Molecular Biology, vol. 909, DOI 10.1007/978-1-61779-959-4, © Springer Science+Business Media New York 2012

MIX
Papier aus verantwortungsvollen Quellen
Paper from responsible sources
FSC® C105338

If you have any concerns about our products, you can contact us on
ProductSafety@springernature.com

In case Publisher is established outside the EU, the EU authorized representative is:
Springer Nature Customer Service Center GmbH
Europaplatz 3, 69115 Heidelberg, Germany

Printed by Libri Plureos GmbH
in Hamburg, Germany